A comprehensive volume for zoologists and their advanced students, this valuable book explores the nature and history of ocean life and our studies of it. Its focus is on the ecology of organisms and communities, with individual chapters devoted to a comparative study of the plankton, the benthos, and the fauna between tidemarks. Emphasis is given to the adaptations and biology of these organisms as well as the physical and biological factors which govern their lives. A separate chapter traces the history of marine biological investigation and its problems, achievements, and methods. In its original German edition this book has already become a standard text.

BIOLOGY SERIES

General Editor: R. Phillips Dales
Reader in Zoology in the University of London at Bedford College

U.S. Editor: Arthur W. Martin
Professor of Physiology, Department of Zoology, University of Washington

---

Practical Invertebrate Zoology
F. E. G. Cox, R. P. Dales, J. Green, J. E. Morton,
D. Nichols, D. Wakelin

Animal Mechanics
R. McNeill Alexander

The Biology of Estuarine Animals
J. Green

Structure and Habit in Vertebrate Evolution
G. S. Carter

IN PREPARATION:

Principles of Histochemistry
W. G. Bruce Casselman

The Investigation of Natural Pigments
G. Y. Kennedy

Molecular Biology and the Origin of Species
Clyde Manwell and C. M. Ann Baker

Description and Classification of Vegetation
David W. Shimwell

Developmental Genetics and Animal Patterns
K. C. Sondhi

MARINE BIOLOGY

An introduction to its problems and results

by

HERMANN FRIEDRICH

Director of the Overseas Museum, Bremen

*Translated from the German
by Gwynne Vevers*

UNIVERSITY OF WASHINGTON PRESS

SEATTLE

Library of Congress Catalog Card Number 71-93028

*Published in the United States by
the University of Washington Press, 1969*

First published in Great Britain 1969

Copyright © 1965 by Gebruder Borntraeger, Berlin

*Copyright © in this translation 1969 by
Sidgwick and Jackson Limited*

Originally published in Germany by Gebruder
Borntraeger, Berlin, under the title "Meeresbiologie"

Printed in Great Britain

Dedicated to my Wife

CONTENTS

FOREWORD

The biology of the sea has been investigated on a systematic basis for the last 150 years; before that time various individual observations had been made over a longer period. An incredible amount of information has been discovered, described and illustrated. This store of knowledge is being continuously enriched by the use of a variety of technical methods, and there has been a marked intensification of marine research during the last two decades, the results of which can be found in the literature.

In many cases these investigations have been primarily concerned with such important practical interests as navigation, coast protection, the production of raw materials and food, and the preservation of the coast for recreation in the face of pollution by oil, industrial effluents and the dangerous waste material from atomic power stations. These interests and those of a military nature require a thorough knowledge of physical, chemical and biological conditions in the sea. In view of the steady increase in our knowledge it is difficult for anyone not directly concerned with research to obtain a general view of the aims, problems and results of marine biology.

This is the aim of the present work which is addressed primarily to students and to biologists who are not in direct contact with marine biological problems; it is hoped, however, that even specialists may find something of value in this book. The contents should also be understandable to those trained in other scientific disciplines. For those who wish to study the subject in more detail there is a good bibliography, and attention may also be drawn to the extensive summary of marine ecological literature produced by Wieser (1960). Owing to limitations of space no attempt has been made to explain problems of nomenclature and classification.

In view of the wide range of available information it is almost impossible for any one person to produce a uniform account of all the different aspects of marine biological research, but an effort has been made to cover the subject as comprehensively as possible. I shall always be grateful for constructive criticism and for information on any results that extend our knowledge. The preparatory work for this book was mainly completed about the end of 1963.

I have availed myself of the advice and help of friends and colleagues and am grateful for the stimulation received from discussion and written sources. I would like to thank the publishers for complying with my wishes, particularly with regard to the preparation of the originals for the illustrations. I am much indebted to my wife for her help, particularly in the task of proof-reading.

Bremen,   HERMANN FRIEDRICH
Autumn 1964

INTRODUCTION

The study of marine organisms has made important contributions to our understanding of many general biological problems. Several phyla, classes and orders are only represented in the sea; knowledge of the larvae of many marine forms and their ontogeny and comparative anatomy have provided an indispensable background for the development of present-day ideas on systematics and phylogeny. A great amount of purely physiological work has been carried out on marine organisms and this has led to many fundamental advances in the field of general physiology. And there is no doubt that marine organisms will continue to be used in the future for the investigation of general biological problems.

The biological investigation of marine organisms is not however synonymous with what we mean by marine biology, which is concerned with biological phenomena as integral parts of all the processes going on in the sea. These include the dependence of living organisms upon the abiotic (physical and chemical) conditions, the interdependence of various biological processes and also their formative influence upon the abiotic environment in the sea. In the widest sense marine biology is therefore the ecology of marine organisms. In principle it requires on the one hand a knowledge of plant and animal systematics, since well-defined basic units—the species and groups of species—are quite essential and on the other hand the results of physiological and ethological studies which are indispensable for an understanding of biological processes in the sea. It is also obvious that marine biology must take account of the physical and chemical conditions as they are at any given time and also of their short- and long-term fluctuations.

Research may be undertaken in several different ways: the ecology of the species being investigated can be worked out by observations on the organisms in their own locality and habitat with simultaneous analysis of the exogenous conditions prevailing in the area. Comparison may also be made of the behaviour of the organisms in different areas with differing conditions. Where comparative observations reveal correlations between behaviour and changes in the external conditions, one can infer a causal connection. If comparative observations are concentrated on a single exogenous factor, such as salinity, temperature or the oxygen content of the medium, one can ascertain the ecology of the organism in relation to that factor.

Such methods are considerably limited by the fact that the behaviour of the organisms at any given time will be determined not by a single exogenous factor, but by the influence of several factors working simultaneously. The sum of the relationships between an organism and its environment is known as its ecological niche. From the total spectrum of the conditions prevailing at any given time, the ecological niche of a single species represents the range of values for each in-

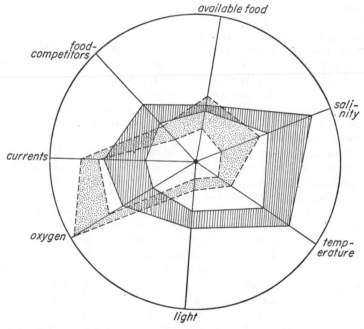

Figure 1. Diagram showing the ecological niche of a single species at different localities or of two species at the same locality; explanation in the text (Original)

dividual factor within which the species is ecologically viable. The value for one factor may however shift when another is changed; in other words an ecological niche is not an absolute measure but may differ from area to area as conditions change.

Figure 1 is intended to illustrate this point. The radial lines represent the possible variation in the exogenous factors at two different localities.

For locality A the ecological niche is represented by the striped polygon, for locality B by the dotted polygon. From this it follows that the ecological status ascertained by observation will apply for one factor at one locality for the individual population at this locality, but not necessarily for a second locality with different conditions. This is perhaps an example of the general principle formulated by Kinne (1956:) 'Species that are stable (euryplastic) to abiotic factors are often labile (stenoplastic) to biotic factors and vice versa.' Investigation of the total ecology of a species requires both observation at a number of localities and also an analysis of the biotic and abiotic factors.

It follows that the physiological potentialities of a species taken as a whole are greater and more diverse than would be inferred from ecological observations made in a single locality. This point has already been confirmed by numerous experimental investigations.

The advantage of experiment is that under well-defined and quantifiable conditions one or more of the factors can be altered. One can thus assess the physiological tolerance towards the altered factor and at the same time the extent to which the altered factor is dependent upon the other factors that are kept constant.

Numerous physiological investigations on the effect of changes in the salinity, temperature and oxygen content of the water have already yielded important results in this field and have provided opportunities for the interpretation of ecological observations.

There are however relatively narrow limits to the use of experimental and physiological methods. They can only be employed with a few individuals from selected populations and one has to generalize on the behaviour of the total population. Only single quantifiable factors can be analysed; the numerous mutual relationships between the different species which inhabit a locality can only be worked out very incompletely by experimental methods. Finally one must reckon that in a natural locality with a system of relationships subject to several factors there will be combinations of factors which cannot be reproduced experimentally.

A further limit to the value of the experimental method is that not only the ecological status but also the physiological tolerance of different populations may vary. It has long been known that meristic characters, that is the size and number of individual parts of the body, are determined by the conditions under which the organisms live during a critically sensitive stage of development. It is therefore possible for ecological conditions in early life to affect anatomical structures in such a way that they may no longer be completely appropriate for subsequent ecological conditions. The physiological potentialities of an organism have also been shown to be affected in this way. From this it must be concluded that tolerance ranges established in physiological experiments do not necessarily correspond with the potentiality of the species, but may already have been restricted by the conditions prevailing in an earlier stage. Therefore, in order to analyse the observed ecological status of a given species throughout its total range it is necessary to ascertain its species-specific potentiality; this could be done by breeding the species concerned under different conditions and then testing experimentally the physiological tolerance of the different groups. By these means one would learn not only the extent to which ecological conditions may have influenced the organism's physiology but also whether there are any hereditary differences between populations or races from different localities. It seems to me that the process of conditioning to ecological conditions may be involved more frequently than genetic differentiation, particularly in those species which produce pelagic larvae, allowing frequent opportunities for exchange between separated populations.

Excluding this conditioning process which takes place in the juvenile stage and lasts throughout life, the short-term physiological reaction will often be determined by the way in which the animals have become adapted to the conditions under which they lived before the start of the experiment. Such adaptational phenomena have frequently been recognized and investigated, particularly as regards the dependence of living processes upon temperature (summary in Precht et al.; see Figs. 27, 28, 29); corresponding data also exist for salinity and exposure to light, and it would be worth having similar information in the case of all the other abiotic environmental factors.

The far-reaching importance of such a process of conditioning has been clarified, for instance, by the much-discussed work on the homing ability of salmon in their native waters (particularly by Hasler 1954 ff.). The sexually mature fish migrating from the sea to the coast react to specific olfactory substances in the waters in which they lived when young. In this case the conditioning must be retained during the years in which they have been living in the sea.

In the long run, therefore, a combination of these three research methods are necessary (cf. Redfield 1958). For marine organisms this has been done only in a very few cases, and it can scarcely be expected that more extensive material will be available within a reasonable space of time. From the knowledge gained in the few cases that have been worked out it seems that great care must be taken in interpreting ecological findings, but on the other hand such results do provide an opportunity for deduction. Thus on this basis one could for example provide a preliminary explanation of the differing behaviour of juvenile and adult animals in a population with different generations in a single year or in the populations from different geographical regions, but this would of course require experimental verification.

In principle what has been said applies to ecological investigations in general. The need to complement ecological observation by physiological experiment and breeding is understandably easier to fulfil with terrestrial and freshwater (including brackish-water) organisms, where it has frequently been done, than it is for marine plants and animals. As a special discipline within the science of biology, marine biology is comparable to limnology; it must, however, proceed along different lines, partly on account of the different kinds of organisms involved but also because the different nature of the conditions and the spatial extent of the marine environment demand different working methods and special points of view.

CHAPTER I

THE HISTORY OF MARINE BIOLOGICAL RESEARCH

A. The basic problems

The biological investigation of the sea involves several lines of development which more or less cut across each other. It seems to me that a brief summary of the development of the subject, giving only the important points, is important for a general understanding, but no attempt has been made to approach completeness.

With the end of the Middle Ages and the beginning of modern times there appeared reports of animals and plants collected more or less at random, some of which were accurate while others were mixed with fantasy (cf. Gesner 1516–1565, Rondelet 1507–1566, Marsigli 1658–1730). During the 18th century there were systematic investigations of a few areas of the sea, as for instance those of O. F. Müller (1730–1784) who in *Zoologica Danica* described the bottom-living animals of Danish and Norwegian waters. The occurrence of fishes and whales far from the coasts was known from the reports of sailors, fishermen and whale-hunters. Nevertheless, during his circumnavigation of the world on H.M.S. *Beagle* (1831–1835), Charles Darwin still regarded large areas of the oceans as deserts. By the turn of the 18th and 19th centuries there was already a considerable knowledge of marine forms and their anatomy. About this time too it was also generally known that the sea floor in the vicinity of the coast was populated with animals and plants and that the composition of such populations could vary considerably from place to place.

During his search for the North-west Passage, Sir John Ross took bottom samples in Baffin Bay (1817–1818) at depths down to almost 2,000 metres, and discovered also that living animals also occurred in areas distant from the coasts. Using a dredge in the Antarctic in 1839–1843 his nephew Sir James Clark Ross also found living animals at great depths. Nevertheless, from his own dredge hauls in the Aegean Sea, Edward Forbes in 1843 postulated that no animal life would be found below a depth of 550 metres and that the depths of the oceans would be azoic. Twenty years later, in 1862, G. C. Wallich reported on catches of animals at greater depths, which he had obtained during the survey voyages of the *Bulldog* prior to the laying of a submarine cable across the North Atlantic. In 1860 a cable between Sardinia and North Africa had been raised from the sea bed for repair; it was covered with a number of different sessile animals, which must therefore have been living at depths of about 2,000 metres. During the 'sixties M. Sars and G. O. Sars caught animals off the Lofotens in depths of over 550 metres, among them an echinoderm which was closely related to forms previously known only as fossils.

On the one hand these results contradicted Forbes's theory that the depths of the sea were azoic, and on the other hand they aroused particular interest in view of the publication in 1859 of Darwin's theory of natural selection. In the circumstances it is therefore quite understandable that the initiative of Wyville-Thomson fell on fruitful soil: after his preliminary expeditions on the *Lightning* (1868) and *Porcupine* (1869–1870) came the great pioneer expedition by H.M.S. *Challenger* (1872–1876).

Johannes Muller had already used a fine-mesh net at Heligoland for the capture of small organisms floating and swimming freely in the water. Muller made known a completely new world of living organisms although V. Thompson had used a similar method in investigations on the Irish coast as long ago as 1828 (see Hardy 1953). Muller gave these organisms the collective name: *Auftrieb*. The investigation of this *Auftrieb* in regions of the ocean far from land was in fact one of the tasks of the Challenger Expedition.

The rich collections of the Challenger Expedition brought to light a very great number of new forms and species. This led to a period of detailed systematic and morphological research on marine organisms. This was supplemented by work on the development and life history of many species. This field of zoological research reached its peak in the decades between 1870 and 1900, during which almost all the civilized countries sent out large expeditions for the investigation of marine organisms. It was regarded as a national duty to work on important scientific problems even if no practical use was apparent (cf. p. 97). At the same time there was considerable discussion on the theory of evolution: many species were known which were of particular interest for a natural system of living organisms; the larval stages of many animals provided clues to evolutionary relationships and it was expected that, particularly in the deep sea, living representatives would be found of animal groups hitherto known only as fossils.

Systematic and morphological investigations on marine organisms have continued up to the present time and this line of research will always retain its importance because a knowledge of the organisms

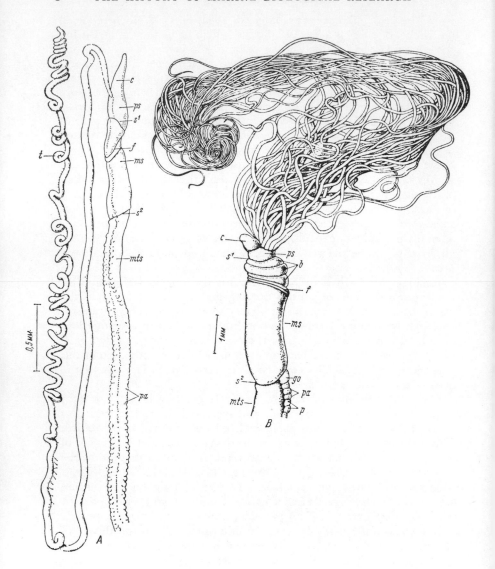

Figure 2. (a) Front part of the body of *Siboglinum caulleryi*, viewed from the left side and somewhat ventrally; (b) Front end of a male of *Polybrachia barbata* from the right (after Ivanov 1963)
b = secondary mesosomal rings, *c* = cephalic lobe, *f* = frenulum (bridle), *go* = genital papilla, *ms* = mesosoma, *mts* = metasoma, *p* = cuticular plaques, *pa* = papillae, *ps* = protosoma, *s¹* = groove between protosoma and mesosoma, *s³* = groove between mesosoma and metasoma, *t* = tentacle

themselves forms an indispensable basis for work on many other biological problems. This is confirmed especially by the continuing discoveries of new species and types of organization (Ax 1960). Here we may recall the discovery of the coelacanth *Latimeria* and the less sensational finding of the Pogonophora (cf. Ivanov 1963 and Fig. 2), of the Mystacocarida (Fig. 3) and Cephalocarida among the crustaceans and of *Neopilina* and *Berthelinia* among the molluscs. As our knowledge of the individual forms has progressed several new problems have arisen which have led to changes in the principal fields of research.

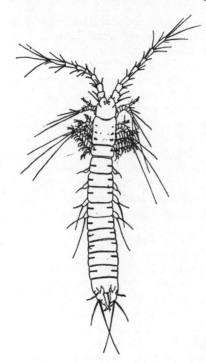

Figure 3. Dorsal view of the mystacocarid *Derocheilocaris remanei katesae* (after Noordt 1954 from Ax 1960)

The collections made by the *Challenger* had made it likely that there was a world of free-floating animals in the open sea and even at great depths. Whereas Alexander Agassiz believed that organisms of this type did not exist below 3,200 metres, Wyville-Thomson believed that they were restricted to the water layers close to the bottom. According to these views the immense volume of water in between would be almost unpopulated. However, this concept was disproved by Chun who used closing nets on the Valdivia Expedition 1898 and showed that all the water layers of the oceans had populations of living organisms. This of course raised the problem of the horizontal and vertical distribution of these organisms. In his *Zoogeography of the sea* Sven Ekman (1953) has summarized the main results of marine biogeographical investigations. Owing to the extent of the seas and the limitations of catching methods it is obvious that some interpretations were provisional. Investigations since then have yielded much supplementary information and made it necessary to modify certain conclusions. To obtain an accurate picture of the distribution of species one ideally requires a very large number of individual records taken at closely spaced stations arranged in a network; this is a field which still offers wide opportunities for research.

The early work in this field was concerned mainly with the individual species and was mainly qualitative. It was Victor Hensen, the Kiel physiologist, who brought a quantitative approach to the subject during the eighties of the last century. He started from the fact that in the western part of the Baltic Sea the flatfishes spawn at certain more or less well-defined places, but that their free-floating eggs were found fairly uniformly distributed outside the spawning areas. From this finding he sought to generalize on the whole world of floating organisms. He introduced the word plankton—those that drift—as a collective term for those floating or drifting organisms which possess little or no ability to move about on their own. After a preliminary expedition on the *Holsatia* he led the Plankton Expedition (1889) on the *National*, working in the North Atlantic Ocean and believed that he had found a relatively uniform distribution of plankton species within their range.

The first very severe critic of this view was Haeckel, who from 1890 onwards carried out numerous investigations on the quantitative distribution of plankton. The development of new methods was of great importance for this work (see p. 28). In general the results of this and later work have shown that, in contrast to the view of Hensen, we can now say that the distribution of the individual species and of the plankton as a whole is by no means homogeneous. However, we still do not have a completely satisfactory picture, partly because the basic methods are still inadequate and partly because some oceanic regions have been more intensively investigated than others. Nevertheless it is possible, for example for the Atlantic Ocean, to produce a chart showing the relative densities of the plankton populations (Fig. 4, cf. also Fig. 163).

The non-homogeneous distribution of plankton and the investigations designed to analyse the reasons for this have both resulted in fundamental changes in the design and execution of expeditions. In the older investigations samples were taken along the route of a voyage and the results obtained were applied to wide areas of the oceans. It is obvious, however, that this method takes no account of

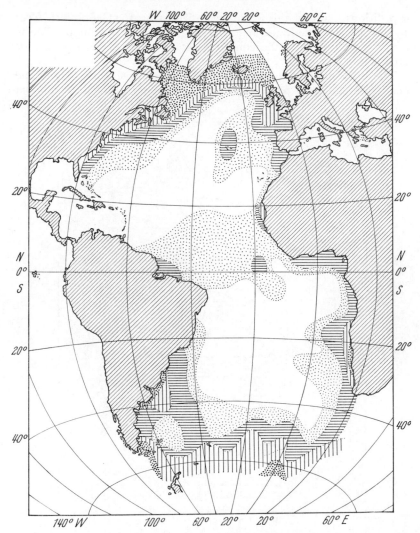

Figure 4. Population density in the Atlantic Ocean, expressed in percentages, relative to the area enclosed by a 5° field at 30°N 40°W. White = 1–15%, fine stipple = 15–30%, horizontal hatching = 30–60%, vertical hatching = 60–200%, coarse stipple = 200% (after Friedrich 1950)

seasonal differences between the various stations. The German Atlantic Expedition on the *Meteor* (1925–1927) was the first to cover a large area with a closely spaced network of stations, and the results to a certain extent eliminated accidental factors in the distribution. Other expeditions followed this example until, during the Geophysical Year 1957–1958, the simultaneous use of several ships made it possible to record simultaneously the conditions at different places over a wide area (Fig. 5).

The first attempt to explain the quantitative distribution and density of planktonic populations was made by recording the conditions

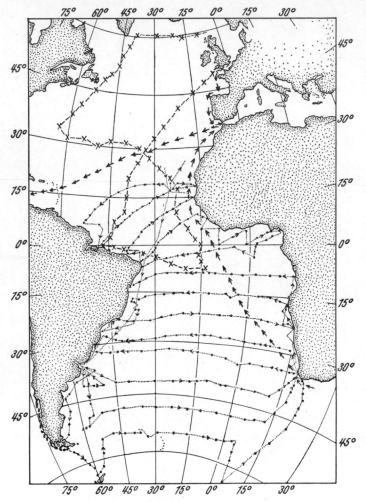

Figure 5. The routes covered by various expeditions: crosses = Plankton Expedition; arrows = Dana Expedition with a few profiles; dots = German Atlantic *Meteor* Expedition with numerous profiles (compiled from Bohnecke & Schumacher 1950, Friedrich 1950)

found at a given time. This soon led to the concept that, of the essential factors, the controlling one is that present in the smallest amount. According to Brandt (1899) the development of the phytoplankton is dependent upon the chemical factors present and is limited by the factor which first sinks below the minimum level required for growth and reproduction. Nathanson (1906), on the other hand, emphasized the importance of currents for the development of such populations. This view added a dynamic aspect to the study of the quantitative data, but the working principles involved still remained vague because there was no synthesis of these two viewpoints. It was only gradually that the views of Brandt and Nathanson were combined to give a unified picture, from which it followed that the processes of water exchange lead to a renewal of the substances essential for life in the productive zones of the sea. The conditions observed at a given time were regarded as being dependent upon the biological, physical and chemical factors at that time. This important fusion of marine biology with research in the field of physical and chemical oceanography found increasing and fruitful use in numerous investigations made at the Plymouth Laboratory (e.g. by Atkins 1923 et seq., Cooper 1933 et seq., Armstrong 1955–1960). The importance of this concept was also shown conclusively by the extensive results obtained by the Meteor Expedition: charts of the population density of the phytoplankton correspond closely with those showing the distribution of phosphate in sea water, and these are closely related to the movements of the water masses (Fig. 6).

Brandt's views also raised the problem of the physiology of plant nutrition and thus opened a chapter which today embraces a very large part of marine biological research, namely the extent of the primary production of organic matter by plant assimilation. It is on this production that animal life at all depths of the oceans ultimately depends, and this of course also applies to life on land and in fresh waters. Estimates of the production of utilizable animals in the sea or future measures for increasing such production depend upon an exact knowledge of the primary production and of the associated processes which determine its course and extent. Extensive research on the amount of phytoplankton and on its dependence upon seasonal conditions have been supplemented by several experimental investigations, some carried out at sea, others in the laboratory. The use of modern methods such as the addition of radio-active isotopes (^{14}C) has already yielded important results.

Hensen's investigations were aimed at determining the number of organisms in a given volume of water, but it was Chun in particular who developed a qualitative viewpoint, relating the specific form and structure of the different planktonic organisms to their way of life and

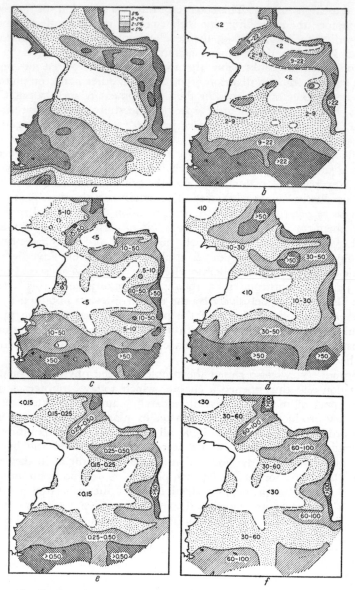

Figure 6. South Atlantic Ocean. Distribution of the following characteristics: (a) colour of sea water (from Schott); (b) phosphate in mg/m³ in the top 50 m; (c) planktonic organisms in thousands/1 in the top 50 m (from Hentschel & Wattenberg 1930); (d) zooplankton metazoans, number /4 l in the top 50 m (from Hentschel 1933); (e) gross production in summer in gC/m²/day; (f) annual net production in gC/m² (from Steemann Nielsen & Jensen 1957)

to their occurrence in certain parts of the total living environment. In this way there came a realization that the form of an organism was of great importance and it was in this context that Ostwald put forward his formula for the speed of sinking of planktonic organisms. The recognition of types of 'life-form' were also applied with success to other marine habitats, e.g. by Remane for the fauna of the interstitial fauna living in the spaces in sand. Some of these discussions have scarcely progressed beyond the formulation stage. In relation to problems of physiology, however, this method of approach ought to contribute significantly to our understanding of the relationship between organism and environment.

In addition to research on marine plankton a great amount of work has been done on the animal life associated with the sea floor, apart from the purely systematic and geographical viewpoints. Forbes, whose postulation of azoic depths has already been mentioned, had predecessors such as Audouin and Milne-Edwards (1832) who outlined a scheme showing the zonation of organisms on the coasts and in areas of shallow sea. Almost contemporaneously with Forbes, Oersted (1844) published his findings on regional settlement in the Öresund. Lorenz (1863) recognized animal communities in the Gulf of Quarnero (Adriatic), the composition of which appeared to be dependent on the substrate. He introduced the terms supra- and sublittoral. This work and the problems raised by it were considerably stimulated by the investigations of Alexander von Humboldt on the distribution and classification of the terrestrial fauna and flora.

From 1865 onwards Möbius working from Kiel investigated the marine fauna of the western Baltic Sea and the North Sea. In his work on oysters he contributed new knowledge on animal communities and introduced the concept of biocoenosis which has today found general use in ecology. The distributional problems raised certainly emphasized quantitative and qualitative aspects and soon took their place alongside the more popular systematic investigations. During the period 1911–1918 a great impetus was given by the work of Petersen in Denmark, who invented a bottom sampler which made it possible to take quantitative samples of the bottom fauna for an evaluation of productivity. From a large series of numerical analyses he was able to recognize and characterize several different living communities. A large number of similar investigations have been carried out since then and one would expect that this field of study will continue to occupy an important place in marine biology.

Petersen's work also acted as a stimulant in other ways, both directly and indirectly. He pointed out the great importance of dead plant material, or detritus, for the nutrition of bottom-living animals. Detritus may be present in large quantities particularly in shallow coastal areas, but it has also been found in the deep sea and doubtless plays a significant role in producing cloudiness in open water. Detritus is important in several different ways: it may serve directly as food for many animals, it acts as an adsorbent for organic matter in solution and as a substrate for micro-organisms, and it also reduces the transmittance of light.

Following on Putter's theory that aquatic animals are to a great extent nourished by dissolved organic matter, a certain amount of research was carried out on this subject, while Krogh, in particular, investigated the amount of organic matter present in dissolved form, and its chemical nature and physiological importance. The latter subject is still the subject of intensive research, which has already yielded important results (see pp. 399 f.).

There is one further branch of marine biological research which was originally inspired by Brandt but which has only been followed up intensively in recent years. Brandt raised the question of the importance of nitrifying and de-nitrifying bacteria in the nitrogen cycle in the sea. He was convinced that the paucity of inorganic compounds in the warmer seas in comparison with the condition in colder areas was due to the increased activity of de-nitrifying bacteria, leading to the transformation of nitrate, nitrite and ammonia into gaseous nitrogen. This theory has remained unproved but it soon led to work on marine bacteriology, a field which has only been intensively investigated during the last 30–40 years, particularly by Waksman and Zobell among others. A further intensification of marine bacteriological research would, however, be very desirable, because theoretical considerations suggest that bacteria have very important functions in relation to the economy of matter and food in the sea. Other marine heterotrophic organisms (yeasts, Phycomycetes, Ascomycetes, Fungi imperfecti, flagellates) will probably also have similar importance, which may be even greater quantitatively; the general occurrence of these organisms in the sea has really only been confirmed in recent years (Höhnk 1952–1959, Kriss 1958–1959 among others). Apart from their frequency and distribution there are several physiological problems requiring investigation, which might lead to a deeper understanding of the dynamics of biological processes in the oceans.

It will be apparent, therefore, that marine biological research should develop along several lines. First, there must be an understanding of the systematics of the organisms involved, and secondly a quantitative evaluation of populations leading to an interpretation of the functional relations between the various components of the major marine communities.

B. The great oceanic expeditions

Observations in the open sea have raised scientific problems in the same way as the results obtained in coastal areas. This has led to an increasing need for oceanic expeditions which have been undertaken by several nations, often at considerable expense. The work of the scientists and sailors on these expeditions has contributed significantly to our understanding of the biology of the sea. It therefore seems worth while to provide a list of the more important of these undertakings, giving the title of the expedition (where there was one), the names of the ships and leaders, the area of investigation and the dates. I cannot unfortunately give a detailed list of all the publications resulting from these expeditions, although this would be extremely desirable, because it seems that much of the specialist material described in these works has not yet been fully evaluated.

Fig. 5 gives only a few of the routes taken during investigations in the Atlantic in order to show the extensive areas where work has been done.

In addition to these large expeditions, oceanic investigations have taken place in all parts of the world, but these cannot all be mentioned here owing to lack of space. We may, however, mention the names of American research vessels such as *Albatross III, Atlantis, Caryn, Horizon, Baird, Crawford, Brown Bear, Chain*. From 1932 onwards the Russians under the leadership of Deryugin have undertaken extensive investigations in the Japan Sea, the Sea of Okhotsk, the Bering Sea and in neighbouring parts of the Pacific Ocean. This work has been continued under the direction of Zenkevich in large areas of the Atlantic, Indian and Pacific Oceans and in the Antarctic, much attention having been paid to the fauna living at great depths. The French research ship *Calypso* is known particularly for its diving work.

At the present time a broadly based international undertaking is starting a detailed investigation of the Indian Ocean, in which a number of new research vessels are being used.

TABLE I

List of important oceanic expeditions

Year	Country	Names of ships	Title of expedition	Names of leaders	Area investigated
1772–1775	Britain	Resolution		J. R. Forster James Cook*	Particularly Pacific coral-reefs
1800–1804	France			de St-Vincent Peron Lesueur Baudin*	Circumnavigation
1815–1818	Russia	Rurik		Eysenhardt Eschschotz A. v. Chamisso Kotzebue*	Marshall and Hawaiian Is., circumnavigation
1817–1818 1817–1820	Britain France	Uranie Physicienne		Sir John Ross* Quoy Gaimard Freycinet*	Baffin Bay Circumnavigation
1826–1829	France	Astrolabe (earlier La Coquille)		Quoy Gaimard Lesson Dumond d'Urville*	Circumnavigation
1826–1829	Russia	Senjavin		Mertens Ruprecht Graf Lütke*	Circumnavigation
1831–1836	Britain	Beagle		C. Darwin Fitzroy*	Circumnavigation
1838–1842	U.S.A.	Porpoise		Dana Pickering Wilkes*	East Pacific
1842–1843	Britain	Erebus Terror		J. Hooker Sir James C. Ross*	South Atlantic
1860	Britain	Bulldog		G. C. Wallick M'Clintock*	N. Atlantic
1867–1869	U.S.A.	Corvin Bibb		Graf Pourtalès L. Agassiz Backe	E. coast of N. America
1868	Britain	Lightning		W. Thomson W. B. Carpenter May*	Faeroes, Shetlands
1869–1870	Britain	Porcupine		W. Thomson W. B. Carpenter Calver*	E. Atlantic, Mediterranean

continued

Year	Country	Names of ships	Title of expedition	Names of leaders	Area investigated
1871–1872	U.S.A.	Hassler		L. Agassiz Graf Pourtalès Steindachner	E. & W. coasts of N. America
1871–1872	Germany	Pommerania		V. Hensen K. Möbius Reinecke Hoffmann	Baltic, North Sea
1872–1876	Britain	Challenger		W. Thomson J. Murray Nares*	Circumnavigation
1874–1876	Germany	Gazelle		Th. Studer Börgen v. Schleinitz*	Circumnavigation
1874–1875	U.S.A.	Tuscarora		Belknap*	Pacific, W. coast of N. America
1876–1877	Norway	Vöringen	Norwegian N. Atlantic Expedition	G. O. Sars H. Mohn Wille*	Iceland, Jan Mayen, Spitzbergen
1877–1880	U.S.A.	Blake		A. Agassiz Sigsbee* (1874–1879) Bartlett* (1879–1880)	Gulf of Mexico, Caribbean, E. coast of N. America
1878	Sweden	Vega		Nordenskiöld	N. coast of Asia, Bering Sea
1880–1881	France	Travailleur		A. & H. Milne-Edwards, Jeffreys, Norman E. Richard*	Bay of Biscay, Mediterranean
1881	Italy	Washington		Managhi*	Mediterranean
1882	France	Talisman		A. Milne-Edwards Parfait*	East and Middle Atlantic
1882–1885	Italy	Vettor Pisani		Lt Chiercha Palumbo*	Circumnavigation
1884–1897	India	Investigator		A. Alcock	Indian Ocean
1885–1914	Monaco	L'Hirondelle L'Hirondelle II Princesse Alice Princesse Alice II		Prince Albert I of Monaco	26 expeditions, Mediterranean and N. Atlantic

continued

Year	Country	Names of ships	Title of expedition	Names of leaders	Area investigated
1889	Germany	National	Plankton Expedition	V. Hensen K. Brandt O. Krümmel Heekt*	N. and Middle Atlantic
1891–1905	U.S.A.	Albatross		A. Agassiz Townsend	Pacific & tropical Indian Ocean; Maldives
1895–1896	Denmark	Ingolf	Danish Ingolf Expedition		Iceland and Greenland
1897	Austria-Hungary	Pola	Pola Expedition	C. Grobben E. v. Marenzeller W. Mörth*	Red Sea
1898–1899	Germany	Valdivia	German Deep-sea Expedition	C. Chun	Atlantic and Indian Oceans
1898–1899	Belgium	Belgica	Belgian Antarctic Expedition	de Gerlache	Antarctic
1899–1900	Netherlands	Siboga	Siboga Expedition	M. Weber M. Nierstrasz J. Verslüys M. Tydeman*	Indonesian Archipelago
1901–1903	Germany	Gauss	German South Polar Expedition	E. v. Drygalski Vanhöffen H. Rüser	S. Atlantic Indian Ocean Antarctic
1902–1903	Sweden	Antarctica	Swedish South Polar	Larsen*	Antarctic
1902–1904	Scotland	Scotia		W. S. Bruce*	Antarctic
1908–1910	Denmark	Thor	Danish Thor Expeditions 1908–1910	J. Schmidt Ostenfeld Paulsen	Mediterranean, N. Atlantic
1910	Norway	Michael Sars	Michael Sars N. Atlantic Deep-Sea Expedition	Sir John Murray J. Hjort H. H. Gran Helland-Hansen E. Koefoid Th. Iversen*	North and Mid-Atlantic
1910	Britain	Terra Nova	British Antarctic Expedition		

continued

Year	Country	Names of ships	Title of expedition	Names of leaders	Area investigated
1911–1912	Germany	Deutschland	Voyage out to German Antarctic Expedition	Drygalski Lohmann	N. & S. Atlantic
1911–1936	France	Pourquoi-pas			N. Atlantic
1920–1936	Denmark	Dana I and II		J. Schmidt	Atlantic Pacific
1925–1927	Germany	Meteor	German Atlantic Expedition	Boehnecke Hentschel Merz Wattenberg Wüst Spiess*	Mid and South Atlantic, North Atlantic
From 1925	Britain	Discovery I			Antarctic seas, Atlantic
From 1930	Britain	Discovery II			Indo-Pacific
1929 ff.	U.S.A.	Carnegie			Atlantic, Pacific
1931 ff.	U.S.A.	Atlantis			North Atlantic
1929–1930	Netherlands	Willibrord Snellius			Indonesian seas
From 1933	France	President Th. Tessier			North Atlantic
1933–1934	England/ Egypt	Mahabiss	John Murray Expedition	Seymour-Sewell	Red Sea, Indian Ocean
1945–1946	Denmark	Atlantide		Bruun Knudsen Wolff Fraser Kraemmer*	Tropical W. Africa
1947–1948	Sweden	Albatross		Pettersson Kullenberg	Circumnavigation
1950–1952	Denmark	Galathea		Bruun Steemann Nielsen	Circumnavigation
1955 ff.	Germany	Anton Dohrn			N. Atlantic
1957 ff.	U.S.-S.R.	Vitiaz Michail Lomonossov Ob Ametiste			Atlantic, Indian and Pacific Oceans, Antarctic

1957–1960 International synoptic investigations in the International Geophysical Year, Atlantic Polar Front Programme, Overflow Programme.

B—M.B.

C. Research stations, laboratories and organizations

The development of marine biology and its present-day work is as closely tied to the institutes and organizations devoted to marine biological problems as it is to the expeditions. Heuss (1940) has described in masterly fashion the aims and the sometimes uphill struggle of Anton Dohrn in establishing the Zoological Station at Naples. His example was soon followed by many other countries, so that today there are large numbers of this kind of research establishment. These have indeed borne the brunt of marine research to an increasing extent. Some of them are independent establishments, others work as outstations of universities, museums or other institutions. In many cases they have working places for scientists of their own country and from elsewhere.

Various organizations have been founded to deal with practical problems and these have functioned on a national or international basis in initiating and executing important projects in marine science and in supplying facilities for liaison.

Only a few of these institutions can be mentioned, and the selection is not intended to imply any degree of merit.

The first marine biological laboratories were established in France. In 1863 the physician Hameau founded the Société scientifique d'Arcachon in the town of that name at the mouth of the river Gironde, and shortly afterwards a biological station was opened there. It was not until 1901 that this laboratory was officially affiliated to the University of Bordeaux, although there had been close collaboration between the two from the start.

In 1872 H. de Lacaze-Duthiers founded a station in Roscoff and later one at Banyuls so that very early on France had stations in the English Channel, on the Atlantic coast and in the Mediterranean. After 1882 these were supplemented by the Dinard Laboratory (situated first on the island of St-Wast la Hougue, later in Saint Servan) on the Atlantic and later by the Marine Station of Endoume on the Mediterranean; since 1884 there has also been the station at Villefranche-sur-Mer on the Mediterranean coast of France, which was originally founded by Korotneff as the Station Zoologique Russe.

However, the most lasting impact has been produced by the Naples Zoological Station, founded by Anton Dohrn, its first building being opened in 1874. Dohrn's ideas bore fruit very rapidly because right from the start he made the facilities of his station available to visiting scientists from other countries. In fact this station was for a long time the only meeting place for many marine biologists, who were working in a great number of general and special fields (Heuss 1940).

In Britain the Marine Biological Association of the United Kingdom established its station at Plymouth in 1888 and today this is one of the largest and most successful of all the marine laboratories. This was followed by the Scottish Marine Biological Association's station at Millport in the Firth of Clyde, and by other stations such as those at Lowestoft, Cullercoats and Port Erin.

At an early stage the Scandinavian countries were also engaged on systematic investigations in marine stations. Sweden has had a station in the Gullmar Fjord since 1884; in Norway there have been laboratories for a long time at Drobak, Trondheim, Tromso and Bergen, of which the last-named has during recent years acquired increased space and working facilities at Espegrend. The wealth of material from the highly successful Danish expeditions has been worked on in the Zoological Museum in Copenhagen; the laboratory at Skalling on the coast of Jutland is concerned especially with the problems of coastal areas subjected to the influence of tides, and a new marine biological station has very recently been established at Elsinore on the Öresund. In Finland various investigations have been carried out in the Baltic Sea from the station at Tvaerminne.

In Germany the Heligoland Biological Station was founded in 1892; this was destroyed during the Second World War but has now been rebuilt. The Prussian Commission for Marine Research (see below) founded the Marine Institute of Kiel in 1935 and a similar establishment at Bremerhaven in 1947.

The Zoological Station of Den Helder, which has worked on a great number of marine biological problems, was established by the Zoological Society of the Netherlands.

The extensive marine biological investigations carried out by Albert I of Monaco culminated in the erection of the Oceanographical Museum and Aquarium at Monaco in 1910 and the foundation of the Oceanographic Institute in Paris a year later.

To complete the picture for Europe one should mention that there are marine stations and laboratories in Spain, Portugal, Italy and Jugoslavia. Marine biological investigations are in fact being carried out from the coasts of Europe in all the neighbouring areas of sea, and several institutes sited inland also take part in this work.

In North America there has been a steady increase in the number of stations since the days of Louis and Alexander Agassiz, and today some of these are the largest establishments of their kind in existence. In Canada marine biological research in the Atlantic Ocean is carried out from St Andrews in New Brunswick and recently also from Halifax. On the Atlantic coast of the United States there are research institutes at Woods Hole, Milford (Connecticut) and in Florida, and there are also stations in Bermuda and the Bahamas (Bimini); from 1904 to

1940 the Carnegie Institution had a station at Dry Tortugas and the University of Texas now maintains an Institute of Marine Sciences at Aransas in the Gulf of Mexico. Similarly there is a string of stations on the Pacific coast of North America. In Canada the Fisheries Research Board has a laboratory at Nanaimo on Vancouver Island. The United States has marine stations at Friday Harbour in Puget Sound, at Pacific Grove (the Hopkins Marine Station), and at La Jolla (the Scripps Institution of Oceanography); there are stations at Corona de Mar near Los Angeles and Dillon Beach, San Francisco, while Los Angeles also has the Allen Hancock Foundation.

In the northern part of the central Pacific marine research is carried out from Honolulu in the Hawaiian Islands, thus providing a link between the North American stations and the numerous Japanese and Soviet stations from which oceanographic, marine biological and fisheries research is carried out. Of the establishments in Japan we may mention the Seto Marine Biological Station, Akkeshi Marine Biological Station, the Tokyo University of Fisheries, and the Faculties of Fisheries in the Universities of Kagoshima and of Mie.

Only an approximate idea can be given of the establishments working in Soviet Russia. Zenkevitch (1955) has reported that there are about 100 institutes, biological stations and observatories working in the field of oceanography, most of which are concerned with marine biological problems. Zenkevitch names the most important as the Institutes for Oceanology, for Marine Hydrophysics and for Zoology and the Biological Stations at Sevastopol and Murmansk, all of which belong to the Academy of Sciences of the U.S.S.R.; the U.S.S.R. and Pacific Scientific Institutes for Fisheries and Oceanography; the Government Institute for Oceanography and the Arctic Institute. One has the impression that these central institutes work mainly on the material collected by the expeditions, while the smaller stations working independently are less highly developed. The results of the research work carried out over a period of some thirty years is recorded in a large number of publications.

There are also several research stations in the southern parts of the world's oceans, some of which have been established quite recently. Thus the University of Cairo has a laboratory in the Red Sea; in India fisheries and marine biological research is carried out mainly from Madras. At Noumea in New Caledonia there is the Institut Français d'Océanie, in New Zealand the Oceanographic Institute in Wellington and in Australia the Division of Fisheries and Oceanography of the Commonwealth Scientific and Industrial Research Organization at Cronulla in New South Wales. The Palao Tropical Biological Station was founded in 1935 but was unfortunately destroyed during the war. In Africa a marine biological station was established in Mozambique in 1951, while South Africa has had an active programme of research in marine biology and fisheries for many years.

Thus there is a world-wide network of large and small research laboratories of which I have only given a selection, without attempting to list them all. This network has developed gradually over the last 100 years or so and will doubtless be augmented by new laboratories in the future.

The responsibilities and tasks of these establishments are extraordinarily varied. Some stations were founded by scientific societies as a result of the initiative of certain individuals. In many cases the state has taken over responsibility at a later stage, or it may have supported such establishments from the beginning, either as a cultural activity of the country concerned or directly as the responsibility of a government department if economic problems were involved.

As the commercial fisheries increased and extended into areas of open sea there came a need for more detailed knowledge of the habits of the fish that were being caught. These problems were first tackled by zoologists, but there soon developed the special discipline of fisheries biology which must now be regarded as a constituent part of marine biology. International competition in the field of fisheries and the common problems involved have brought a need for co-ordinated investigations and the exchange of scientific information. Thus not only national institutions but also international organizations have been established. Since 1902 there has been international co-operation through the work of the International Council for the Exploration of the Sea (ICES) which has its permanent headquarters in Copenhagen; it is primarily concerned with the north-east Atlantic Ocean and the adjacent seas. With this as a model, a parallel organization has been formed for the Mediterranean Sea. The International Commission for North Atlantic Fisheries (ICNAF) and the North Pacific Commission (NORPAC) fulfil similar functions in their own areas. The International Council of Scientific Unions (ICSU) has a Special Committee on Oceanic Research (SCOR), which is supported by national organizations. UNESCO has an International Advisory Committee on Marine Sciences (IACOMS), and UNO takes an active part in problems of marine and fisheries biology through the Food and Agricultural Organization (FAO). This multiplicity of organizations and the great amount of work which they stimulate and carry out are of great value to the development of marine biology, and this applies particularly to those organizations which are concerned with basic practical problems. An important task of these organizations is also to develop and obtain agreement on standardized methods and definitions; this becomes increasingly necessary as research is extended and becomes more diversified (cf. publications such as *Rapports et Proces-verbaux des Reunions*, Vol. 107, 1938; Vol. 144, 1957).

Symposia devoted to a limited field and congresses of a more comprehensive nature have also proved extremely fruitful for the exchange of information and for stimulating and apportioning research problems. In addition, international undertakings such as the Geophysical Year 1957–58, the Overflow Programme 1960 on the Iceland-Faeroes Ridge and the current investigation of the Indian Ocean, in which several ships have taken part, have been of exceptional importance. The simultaneous employment of several ships and the standardized evaluation of the data obtained yield results on the status and dynamics of hydrographic and biological phenomena to an extent that could not otherwise be achieved.

D. Apparatus and methods

We must now give a short outline of the methods that can be used; a more detailed treatment of this subject is beyond the scope of this introduction, but more comprehensive summaries are available, particularly in Barnes (1959 a and b).

The size of the nets used depends partly on the boats or ships employed, but is also of course dependent upon the strength of the material used. The bag of netting of a plankton net towed in open water acts as a filter and is subject to a considerable pressure of water; this is dependent upon the speed of towing, the size of the mesh and the relation of the net opening to the area of the filtering surface. In general, plankton nets are conical and approximately three times as long as the diameter of the opening. At the distal end of the net there is a jar, fastened by a screw or bayonet joint, in which the catch collects (Fig. 7). Different fractions of the plankton can be caught by the use of netting, whether of silk or synthetic gauze, of different mesh sizes. A ring trawl having a mouth with a diameter over 1 metre and a correspondingly long net of hempen material is used for the capture of larger animals. For collecting at great depths where the populations are sparse the Discovery Expeditions used nets with openings having a diameter of 4·5 metres (Marr 1938).

A quantitative evaluation of the catches is only possible to a limited extent. Assuming that the total column of water fished actually passes through the net and is filtered, it is possible to determine from vertical hauls the quantity of organisms present beneath a given area of the surface. The distribution of the population at different depths can be determined by taking hauls in stages, the numbers caught in the shallower horizons being substracted from those taken at greater depths. The results will be more exact if the net is fitted with a closing mechanism so that it can be closed after it has passed through the depth under investigation; it is also possible to make simultaneous catches at different depths by fixing a series of nets on the same cable and this too prevents falsification of the results when the nets are being hauled up.

Bé (1962) constructed a quantitative multiple plankton sampler with opening and closing nets for taking serial samples of the zooplankton at different depths; this apparatus employed a pressure-sensitive release mechanism which allowed horizontal or oblique towing at preselected depths. The use of a sampler of this type should save time, safeguard the different catches from contamination and provide accurate data on the positions at which the catches have been made.

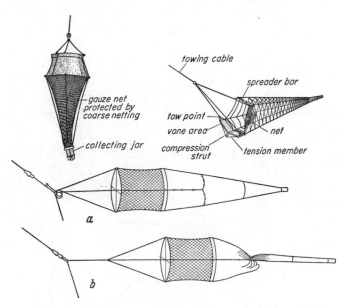

Figure 7. (a) and (b) closing net for horizontal hauls, open and closed (after Marr 1938); (c) plankton net for vertical hauls (after Hensen; (d) Isaacs-Kidd midwater trawl for horizontal hauls (after Hedgpeth 1957)

The great variation in the construction of closing and non-closing plankton nets will be apparent from the list prepared by Linger 1962 (cited in Bé 1962) which gives no fewer than fifty-seven different types. Furthermore, the enormous number of new types right up to the present time bears witness to the uncertain nature of the existing methods for obtaining plankton material.

Flow-meters, which record the amount of water filtered have also been used in the quantitative evaluation of plankton catches.

The conventional conical type of plankton net has the disadvantage that it can only be towed at low speeds, and furthermore it is difficult to keep it on a horizontal track at a given depth. At greater speeds the net is endangered, the catch is more easily damaged and the efficiency reduced because a 'plug' of water builds up in front of the net opening.

For horizontal hauls at higher towing speeds use is made of a solid casing in which the net (frequently made of non-corroding metal gauze) is fixed. The opening in the casing for the entry of the water is relatively small, and in addition there may be stabilizers and an otter board so that the apparatus can be kept in the correct position at the desired depth. A closing mechanism will prevent material from other depths being caught while the net is being hauled up, and a built-in flow-meter will give data on which to base the quantitative evaluation of the catch (Fig. 8) (Gehringer 1952, Kinzer 1962).

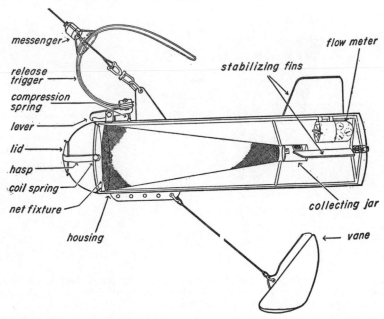

Figure 8. Plankton sampler 'Hai' in towing position after release of the closing mechanism; semi diagrammatic (after Kinzer 1962)

The Continuous Plankton Recorder of Hardy (1935) embodies an original principle. It has a moving filtration surface on which the catch is continuously fixed (Fig. 9). This prevents the mesh from becoming clogged and the distribution of the organisms along the line being sampled can be read off the band of netting. However, the condition of the catch on the rolled band is not always satisfactory, the apparatus does not work for the large planktonic animals and it does not catch the smallest organisms.

Quantitative estimations of plankton caught in nets may not be sufficiently accurate, partly because it is impossible to measure with sufficient accuracy the actual amount of water filtered and partly because the smallest planktonic organisms mostly escape through the meshes of the net or are destroyed when caught because they are so delicate. Accordingly, methods have been developed which attempt to estimate the plankton content of a given sample of water; at first, attempts were made to determine the individual numbers of each species present by means of time-consuming counts, but more recently chemical methods have been developed in which the amount of organic matter present in a given volume of water is determined (for details see Chapter VI, 1).

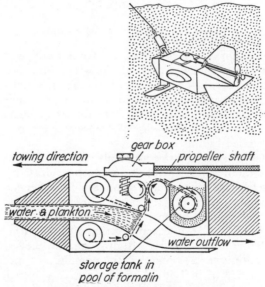

Figure 9. Hardy's plankton recorder

Catching benthonic animals was at first carried out exclusively with dredges and trawls similar to those used in commercial fisheries. Dredges consist of a strong triangular or quadrangular metal frame with smooth or more or less coarsely toothed edges turned outwards. A short bag of coarse, medium-mesh netting receives the bottom material scraped up by the frame and the items of interest are collected by sieving or by hand-picking.

With apparatus of this kind it is possible to obtain organisms living on the surface of the sediment or in its upper layers. On soft bottoms usually only the macrofauna is obtained; on sand and gravel bottoms some of the microfauna will be obtained, but it is not possible to obtain information on vertical distribution within the sediment.

The animals living on and immediately above the bottom, particularly in deep water, can be collected by a special trawl of the type designed by Agassiz and named after him (Fig. 10). This can also be used with success in shallow water areas with growths of eel-grass and algae. The mesh size can be chosen according to the particular organisms to be caught.

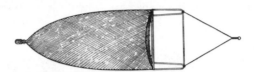

Figure 10.   Sketch of an Agassiz trawl

Fishes are caught with larger nets, similar to those used in commercial fisheries and of course a mass of biological data can be collected from the analysis of the non-fish part of commercial hauls. An interesting method is that used by Maul in Madeira who collected several species, particularly pelagial forms, from the stomachs of fish caught by the local fishermen. Knowledge of deep-water forms could be considerably extended by this method.

The catches of trawl and dredge hauls vary considerably according to the size of the net, the size of the mesh, the duration and speed of the haul, the type of sediment and so on. Such catches cannot be more than fragmentary samples of the fauna present in the narrow stretch traversed by the haul. On hard or uneven bottoms the catches are usually small and to a large extent restricted to the surface fauna, because the infauna can escape by withdrawing into the sediment. Consequently quantitative assessments of the population density of an area and the frequency of individual species are very questionable. The catches made by dredges that dig deep into the sediment can be compared quantitatively with those collected by nets trawled along the surface of the sediment but quantitative comparison is quite impossible because it is in practice not possible to assess the size of the area sampled.

On account of these uncertainties, Riedl (1960) constructed a dredge running on sledge runners with a kinemeter fixed in front. By means of an antenna touching the bottom the kinemeter recorded the unevennesses of the bottom and at the same time registered the distance travelled on a metering device or 'bottom-walker'; in addition the net opening could be automatically closed by previous selection of the distance to be travelled.

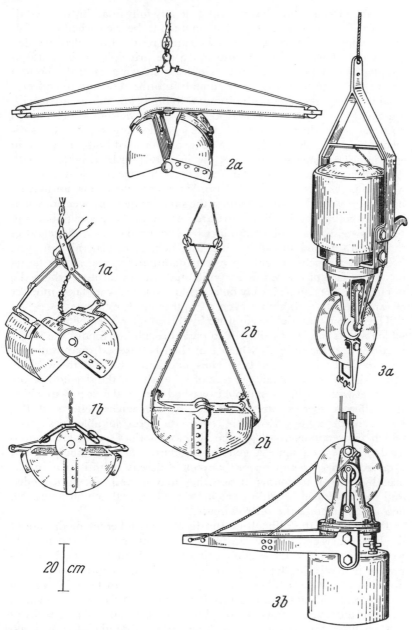

Figure 11. Three types of bottom sampler: (a) open, (b) closed: 1, Petersen grab; 2, van Veen grab; 3, Knudsen grab (after Hedgpeth 1957, somewhat modified)

As far as I know there has not yet been any fully quantitative investigation with this apparatus, and it remains to be seen whether it fulfils its aims. The known defects in catch results from dredges, led Petersen & Jensen (1911) to introduce the principle of the bottom-sampler. This type of apparatus samples a known area of the bottom and at the same time digs deep enough into the substrate to collect a considerable proportion of the infauna living below this area. As with a grab-dredge this apparatus is lowered to the bottom with the jaws open and when the cable is hauled the jaws close and dig into the sediment (Fig. 11). Clearly this may not work on hard bottoms where in certain circumstances it may not dig in at all; on hard bottoms with stones, the jaws may jam or fail to close so that the sediment collected will escape as it is being hauled up. With strong currents, an uneven bottom or when the ship is drifting the sampler may tip over on reaching the bottom and fail to operate. Attempts have been made to compensate for these disadvantages by varying the size and weight of the apparatus and modifying its basic design; in general, however, it is still extremely difficult to assess quantitatively the motile fauna on the surface of the sediment (epifauna) and in an area where the populations are patchy it may be necessary to take a very large number of samples before obtaining even an approximate estimate.

The bottom material brought up by the sampler is washed in a sieve which allows the finer sediment to pass through but retains the larger particles and the animals which can then be assessed quantitatively. This method is however at best semi-quantitative because the working of the sieve depends upon the size of its mesh, on the strength of the water jet with which the sample is washed and on the degree to which the different species can withstand the mechanical effects of this treatment. According to Reish (1959) the most suitable mesh size, which will retain over 90% of the biomass (both of species and individuals), is 0·85–1·4 mm. A general standardization has not yet been attained and so comparisons between different quantitative results are always open to doubt; at best they may be valid for hard-shelled and tough-skinned organisms such as molluscs and echinoderms. No account is taken of the microfauna.

To overcome the difficulties and limitations inherent in the use of dredges, trawls and bottom-samplers various attempts have been made to use photography, the earliest being that of Boutan in the Mediterranean in 1893. Developments in diving techniques and the design of apparatus nowadays allow free-swimming divers to take still photographs and cinematograph films down to depths of 60–80 metres. Modern automatic cameras can also be used to take photographs of the sea floor and this has been done down to depths of over 5,000 metres. Pictures of this kind give a general idea of the appearance of the sea floor and provide information on the species present and on the density of the populations; they also show animal tracks, the significance of which can in most cases only be guessed. In some cameras the camera lights and the shutter can be activated by the animal itself when it takes a bait (Laughton 1957).

McIntyre (1956) has given an interesting comparison between the numbers of animals caught with an Agassiz trawl and a van Veen bottom-sampler and the numbers recorded photographically.

TABLE 2

Locality	Trawl	Sampler	Camera	Relative numbers
A	1,093	19,418	39,542	1 : 17·8 : 36·1
B	1,992	41,919	32,702	1 : 21 : 16·4
C	215	—	15,577	1 : ? : 72·4
D	524	79,686	61,330	1 : 152 : 117
E	290	10,455	4,916	1 : 36 : 16·9

These relative numbers give no real indication that any one of these methods is unquestionably superior. It is probable that an approximate picture of the true facts should be based on a combination of all three methods, because in spite of the small absolute numbers caught by the trawl it actually gives the best results for the Decapoda Natantia, the Brachyura and the lamellibranchs.

As in the plankton, the microfauna and microflora of the benthos are of considerable importance; here too their members often provide the first links in the food chains and they also yield much of interest in the fields of systematics, autecology and biogeography. It is quite easy to collect the animals from samples of marine vegetation by leaving the material in a dish of water. Unless they are sessile, the animals present, which are usually small, will soon collect at the surface as the oxygen becomes deficient. Material can also be obtained from bottom samples by this method, although some forms will remain on the surface of the sediment and have to be picked out by pipette (cf. Riedl 1955).

There is no difficulty in obtaining samples from easily accessible parts of the eulittoral, and provided the number of samples is sufficient one can obtain a very exact picture of the qualitative composition of the microfauna. Quantitative investigations on the number of individuals per unit area are much more difficult and few satisfactory results have so far been obtained.

Several different methods have been developed for obtaining samples from the sea floor at great depths and some of these can be used for quantitative assessment. In general, bottom-samplers dig too deep into soft sediments, for the microfauna prefers to live in the uppermost few centimetres and so it then becomes very difficult to separate out the microfaunal organisms. The unreliability of bottom-samplers on sandy bottoms is particularly noticeable when it comes to collecting the microfauna. One method of obtaining samples of the microfauna is to drag a bar along the bottom, thus stirring up the upper layers of sedi-

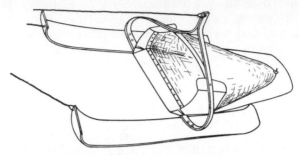

Figure 12. Sledge dredge (epibenthic dredge) (after Purasjoki 1953)

ment which are then collected in a fine-mesh gauze net which is protected by runners or metal plates (a type of sledge dredge). Or the apparatus may be constructed with a cutting edge mounted on the net frame between the runners, which directly planes off the uppermost layers of sediment (sledge plane) (Fig. 12).

Roughly quantitative results can be obtained with a bottom suction

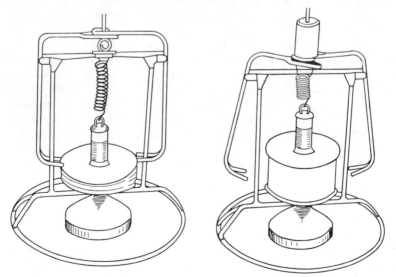

Figure 13.   Bottom suction apparatus (after Purasjoki 1953)

device. In this apparatus a wide-mouthed funnel is lowered to the bottom; fixed over the narrow end of the funnel is a piece of apparatus which when triggered by a falling weight sucks up the surface sediment immediately below the funnel (Fig. 13). With this piece of equipment it is at least possible to obtain samples that are comparable with each other (cf. Purasjoki 1953, who also gives further references). Jones (1961) used a bottom core-sampler (Fig. 14) in which a glass tube

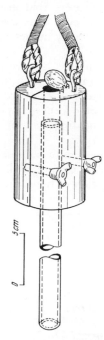

Figure 14. Core sampler (after Jones 1961)

with an external diameter of 22 mm projects about 30 cm from a cylindrical, bored bipartite lead casing.

At present there is no way of collecting the sessile fauna of rocky bottoms, in particular the microfauna, by indirect methods.

For the collection of the heterotrophic flora (fungi, yeasts) from the bottom in deep water (down to 4,000 m in the North Atlantic) Höhnk (1955) used the core-sampler of Pratje. In this apparatus the sample is collected in a tube of plexiglass, which is then hauled up and cut into pieces so that each can be sealed to prevent the ingress of secondary infections. Subsequent examination in the laboratory allows the identification of the organisms even in the deeper sections of the core.

Photography not only supplies a supplementary quantitative method but also has the advantage that it gives a true picture of the habitat and of the behaviour patterns of individual animals. Thus, for example, the fish *Benthosaurus* was regarded as bathypelagic and the elongated rays of the pectoral fins were interpreted as taste organs until photography showed that it uses these as stilts to walk on the bottom (Fig. 128, p. 252).

In the same way as photography, underwater television is able to give vivid impressions of life on the sea floor, and certainly offers still further possibilities for obtaining knowledge on the relationship of the organisms, their distribution and their habits in areas that are otherwise inaccessible. It should however be remembered that the intense artificial illumination is unnatural and may strongly influence the behaviour of the animals.

Photography and television always have the disadvantage, however, that the specific identity of the objects observed can only be given in exceptional cases; this naturally further limits their value as evidence. On the other hand it would be possible to combine some form of capture apparatus with a television camera, thus supplementing the blind sampling by grab and dredge (cf. Riedl 1963, Fig. 15).

The combined results obtained from all kinds of nets and also from photography still only give an indirect and incomplete picture of the population; in particular we still lack clear presentation of the distri-

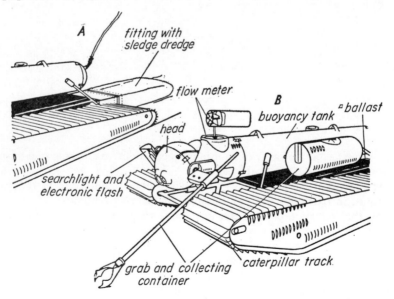

Figure 15. Projected television tank with collecting apparatus: (a) with sledge dredge; (b) with remote-controlled grab and collecting container flushed by sea water (after Riedl 1963)

bution and of the behaviour of the animals under natural conditions. So it is understandable that efforts have been made to fill these gaps by direct observation and by collecting during diving operations. (On the development of diving for scientific purposes see Wasmund 1938 and Riedl 1963.) According to Rouch (1959), Milne-Edwards was one of the first zoologists who went down with the help of a diving apparatus to observe and collect on the sea bottom. In his extensive biocoenotic investigations in the Gullmar Fjord, Gislén also used diving apparatus. Beebe (1934) reached depths of almost 1,000 metres with his bathysphere lowered from a ship, which enabled him to observe the distribution of the light, the density of the plankton and certain aspects of animal behaviour.

In every case the diver was relatively immobile because he had to remain connected with auxiliary apparatus at the surface. Efforts to attain as much mobility as possible under water have led to the development of special diving equipment such as goggles, compressed-air apparatus and so on. The investigations and often surprising observations made with the aid of this equipment have been partly brought to a wider public by the films of Hass and Cousteau; this type of equipment is widely used today, but even with very experienced divers it only allows dives down to depths of 60–80 metres.

It was only after the last war that A. Piccard could put into practice his idea, conceived in 1938, of going down into great depths in an unattached diving apparatus. The 'bathyscaphe' consists of a pressure-proof observation chamber suspended from a float filled with a fluid of low specific gravity. Since then numerous dives have been made in the two bathyscaphes *F.N.R.S. 2** and *Trieste* down to depths of 11,500 metres (see e.g. Pérès 1958, 1959). They have enabled direct observations and measurements to be made and also properly directed photography, which are not possible with free or unattended cameras.

The use of the bathyscaphe is only worth while at great depths, from about 500 m downwards; the free diver is restricted to shallow depths, and so Cousteau has constructed a 'diving saucer' for the intermediate depths. This has already been used for taking faunistic and biocoenotic photographs in the Mediterranean and Red Seas (see Pérès 1960).

Observations made from closed diving apparatus (bathysphere, bathyscaphe, diving saucer) naturally preclude an intimate knowledge of the expected fauna. It is quite understandable that these methods have their limitations: the numerous minute differences between related species often make it impossible to identify the organisms observed, and furthermore the need to illuminate a small area artificially must have its effect on the behaviour of the animals.

* *F.N.R.S.1* was Piccard's stratosphere balloon, the construction of which was also made possible by the Fonds National de la Recherche Scientifique Belge.

Direct observation supplemented by photography and the determination of captured material will certainly yield much valuable information on biological processes in the deep sea.

Following the introduction of the echo-sounder for navigation and marine research there have been numerous observations on sound-reflecting layers in open water (deep scattering layers = D.S.L.; see pp. 95, 302). In many cases the occurrence of these reflecting layers has been traced back to accumulations of organisms, because in the fisheries fish shoals and their distribution have been located with great success. We are still not at all clear about the organisms which give this kind of deep echo or about the hydrographical conditions under which such accumulations take place. Diving operations at the depths recording such deep echoes failed to confirm significant accumulations of organisms (Pérès 1959).

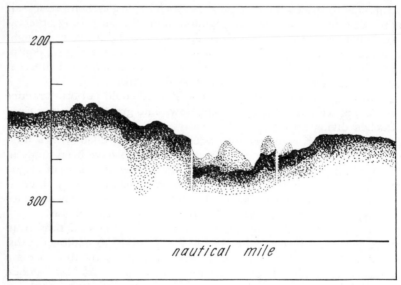

Figure 16. Echo-sounder trace from the Gulf of Mexico with indications of the reef of *Lophelia prolifera* (after Moore & Bullis 1960)

On the other hand, Moore & Bullis (1960) found two remarkable traces on an echogram of the bottom in the Gulf of Mexico, the significance of which was not immediately apparent. By the use of a dredge these proved to be an extensive reef of the coral *Lophelia prolifera* at depths of 230–280 metres.

The echo-sounder is therefore an instrument which can supplement the other methods of observation and capture.

CHAPTER II

MORPHOLOGY OF THE OCEANS

A. Geomorphological divisions and depths of the ocean

Approximately 71% of the surface of the earth is covered by the sea. The remaining 29% is divided up into numerous more or less extensive areas of land together with archipelagos or small islands, whereas the seas extend from pole to pole and are connected with each other east to west right round the world. The different sea-areas of the world are therefore in communication with each other at least at their shallower parts. This continuity contrasts with the discontinuity of the land masses and the even greater discontinuity of fresh waters, so we should expect considerable differences in the biology.

In the northern hemisphere the oceans are mainly separated from each other by the continents which border the Atlantic and Pacific Oceans, while in the southern hemisphere the continents form a ring round the Indian Ocean. In the south the boundaries between the three oceans are defined by convention: the meridian 20°E is taken as the boundary between the Atlantic and the Indian Oceans, while the meridian 147°E separates the Indian and the Pacific Oceans. The warm-water regions of the three oceans are therefore separated from each other by the continents, whereas in the south there is a continuous connection between them. The Atlantic and Pacific Oceans also meet at the Bering Straits, the North Polar Sea being regarded as part of the Atlantic Ocean. In volume and surface area the Pacific Ocean is larger than the other two taken together.

The oceans are considered to be ancient formations in the morphology of the earth. The adjacent seas, on the other hand, which are to varying extents separated from the oceans by island chains or by submarine sill-like ridges rising from the sea floor, have been subjected to considerable changes in the course of geological history. They may be classified according to their relationship to the neighbouring continents:

Marginal seas, adjacent to the continents:

Bering Sea	North Sea
Sea of Okhotsk	Gulf of St Lawrence
East China Sea	Irish Sea
Andaman Sea	

Intercontinental, Mediterranean seas:
Arctic Mediterranean Sea
Australasiatic Mediterranean Sea
American Mediterranean Sea (Gulf of Mexico and Carribbean Sea)
European Mediterranean Sea (with Black Sea)

Intracontinental seas:

Hudson Bay	Baltic Sea
Red Sea	Persian Gulf

Owing to their more or less extreme separation from the oceans the adjacent seas have certain special characteristics, which are closely correlated with the reduced exchange of water masses; they show considerable differences between each other, because their water masses react more markedly than the open oceans to the local climatic conditions. Figure 17 demonstrates this by a comparison of the Baltic Sea

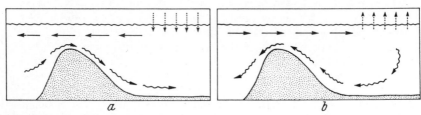

Figure 17. Diagram to show the circulation in adjacent seas with excess (a) inflow of fresh water (e.g. Baltic Sea), (b) evaporation (e.g. Mediterranean)

and the Mediterranean; in the Baltic there is an excess of precipitation and fresh water inflow relative to evaporation and this leads to an overspill of low salinity surface water into the North Sea, while high salinity water streams in along the bottom. In the Mediterranean, on the other hand, evaporation is high, so that, particularly in the east, there is a high salinity water which sinks to the bottom and flows out over the ridge at the Straits of Gibraltar, while Atlantic water flows in at the surface.

From this one must expect that biological comparisons between such adjacent seas will show more marked differences than comparisons between the corresponding oceanic regions. In addition to this the geological history of the adjacent seas from the Pleistocene onwards, e.g. of the Baltic and Mediterranean Seas, is scarcely comparable.

Apart from the major divisions the small-scale differentiation of the coasts is of great importance in many biological problems. Bays, lagoons, estuaries, steep and flat coasts with beaches of sand, boulders or mud are regions which are all very different biologically and which exhibit diverse biological phenomena due to differences in their detailed structure. These are dependent partly on the structure itself and partly on the processes by which the structure has been formed and modified (cf. A. Guilcher 1958).

The form of the sea bottom is also of direct and indirect importance for biological processes in the sea, and we may differentiate here between major and minor morphological types. The major bottom types include the Continental Shelf, the Continental Slope, the deep-sea basins, ridges and sills, and the deep-sea troughs. The Shelf area is the girdle of shallow sea which surrounds the continents and slopes away gently to varying distances. In depths of about 200 metres there is a marked increase in the gradient and the Shelf passes into the Continental Slope, from the foot of which extend the deep-sea basins. The deep-sea basins are bounded by ridges and sills with a very rough relief and they form an enormous interconnecting system which can be traced through all three oceans; most of the oceanic islands lie on these ridges. The deep-sea troughs are elongated depressions cut into the deep-sea basins; they all have a depth exceeding 6,000 metres and are characterized by the fact that they lie adjacent to continents or to sills and ridges; these great depths are therefore spatially isolated from each other. This isolation as well as the relatively restricted extent of the troughs and their great depth account for the peculiarities in their living populations.

The minor types of sea floor are, for example, the isolated rises with steep slopes, which sometimes emerge at the surface of the sea in the form of sea-mounts, banks and shoals. In recent times submarine canyons have become the subject of special interest; these cut into the Continental Slope as furrows leading down into deep water. In certain cases they may be continuations of mainland valleys but usually they are erosion valleys. Submarine erosion occurs when sediment on the Continental Slope is gathered together by a descending current and passed at speed into greater depths together with water which has become denser owing to the presence of this sediment in suspension.

The varied nature of the sea floor as briefly outlined here is an important factor in the study of marine biology. The deep-sea basins are indeed very extensive, but they comprise only a part of the oceans possessing diverse hydrographic characteristics; in many cases the ridges and sills separating the troughs function as barriers to distribution that are difficult to overcome, so that the concept of a cosmopolitan deep-sea fauna requires revision. The deep-sea troughs like the sea-mounts and banks are geomorphological formations isolated from one another and this may favour divergent evolution among their inhabitants. Steep slopes offer opportunities for settlement that are different from those in troughs full of sediment. Currents carrying suspended matter will deposit in the deep-sea basins a sediment which may be quite different in character—sometimes even sandy—from the usual deep-sea sediments (e.g. Locher 1953); in addition, organisms carried along with the suspended matter must eventually reach deep water, where they will either perish, or if they survive will contribute to the living population of deep-sea areas, provided they are pre-adapted. In warm-water areas sea-mounts and banks that rise up into well-lit surface layers may become the basis for the development of coral reefs and atolls, which in their turn offer a chance for numerous other organisms to settle.

For these reasons marine biology must follow very closely any advances in the investigation of sea-bottom morphology. The results of research on sediments also deserve special attention, because those differing in structure or chemistry will provide different environments for settlement.

Therefore, although the water in the great ocean basins is practically continuous, the sea floor is surprisingly varied in form.

B. Main ecological divisions

The whole marine environment, therefore, has a vast horizontal and vertical extent, combined with an almost continuous watery medium. Biological investigations soon showed differences in the populations of different areas whether close to or far from the coast, at the surface or in deep water, and these led to many attempts at a bio-ecological classification (e.g. Audouin & Milne-Edwards 1832, Forbes 1843, Oersted 1844, Lorenz 1863); in the same way large-scale divisions based on physical characteristics have also been proposed. Any classification of the marine environment into different zones or regions is only of importance for the study of marine biology if the development and distribution of the ecological factors and their gradations are of biological significance.

In view of the great variation in the ecological factors, which will be discussed in more detail in the next chapter, any large-scale classification can only serve as an approximation. This is because, for instance, zones that can be well defined in one area may be ill-defined or even lacking in another, and boundaries can never be laid down as lines or surfaces but only as more or less extensive transition zones.

Amongst others Hedgpeth (1957) has attempted an ecological classification of the marine environment. In recent years the results of the Galathea Expedition (Bruun 1955) and of the Russian investigations in the north-west Pacific (Zenkevich 1954) have supplied important contributions on this subject.

The boundary surfaces between water and other media can be clearly defined from the ecological point of view, so that we can differentiate:

the pelagial=open water; the animals present are pelagic = pelagos
the benthal= the sea floor with the benthonic organisms, which form the benthos
the pleustal= the boundary layer water/air with pleustonic organisms= the pleuston.

An ecologically based vertical classification of the offshore pelagial is relatively easy to produce. The primary production of organic matter by plant assimilation is essential to all biological phenomena and since this depends on light one can define an upper layer in the pelagial. I define the euphotic zone as the layer in which the average light intensity is great enough for assimilation to exceed dissimilation. The upper limit of the euphotic zone is the surface of the water and its lower limit will lie at the depth of the compensation point (the compensation point is the depth at which assimilation and dissimilation are equal, so that the quotient assimilation : dissimilation $=1$). The penetration of light fluctuates, so that this lower limit will vary between the coast and the open sea and according to latitude (see Chapter II). In areas with illumination that is largely constant the lower limit of the euphotic zone could theoretically be flat and level, but in practice it is more often in the form of a rather indistinct transition zone. In the euphotic zone other ecological factors such as temperature, salinity and water movement are subject, at least locally, to relatively large periodic and non-periodic fluctuations.

The greater depths of the sea are completely beyond the influence of sunlight and are therefore accurately defined as aphotic. Here the other factors show very great uniformity. One should probably also define a lowermost zone influenced by its proximity to the bottom, but in my opinion the necessary basis for this is lacking and the data so far available do not yet appear to be sufficient.

Between the euphotic and aphotic zones one can recognize a transitional zone of varying breadth, in which daylight is still present but of such a strength that dissimilation exceeds assimilation. This zone can be designated as mesophotic and it seems to be characterized not only by the poor light intensity. It contains relatively large amounts of dead plant matter derived from the euphotic zone (see Chapter VI, 1). During the day this zone will be inhabited by those animals which undertake vertical migrations into the euphotic zone as night approaches, some even reaching the surface (see Chapter VI, 2).

Thus it would seem to me that a tripartite classification into euphotic, mesophotic and aphotic zones is quite natural. These are approximately equivalent to the frequently used terms epipelagial, mesopelagial and bathypelagial.

In the vertical classification of the pelagial in the tropics and subtropics Bruun takes into consideration the distribution of temperature, light and hydrostatic pressure and arrives at the following divisions:

I: Thermosphere (temperature above *c.* 10°C)
    1. Epipelagic zone = zone of photosynthesis
    2. Mesopelagic zone = aphotic zone, its upper limit the influence of daylight, lower limit the discontinuity layer (see Chapter III, B 4).

II: Psychrosphere (temperature below *c.* 10°C).
    1. Bathypelagic zone = temperature 4–10°C
    2. Abyssopelagic zone = temperature below 4°C
    3. Hadopelagic zone = zone of the deep-sea troughs with hydrostatic pressures of over 600 atmospheres.

The division into thermosphere and psychrosphere is not applicable in the cool temperature and cold regions, where even the surface water belongs to the psychrosphere, so that in discussing regional differences one can only speak of a thermospheric or a psychrospheric epi- and mesopelagial.

We owe to Haeckel the horizontal division of the pelagial into a neritic region near to the coast and an oceanic region at a distance from the coast. The neritic can be defined as the area above the Continental Shelf, so that the separation between the two regions contains both a horizontal and a vertical component. From the viewpoint of their contents the separation of these two regions is related to the extent to which the biology of the pelagial is influenced by the coastline and the sea floor. Here the ecological factors themselves scarcely provide a consistently usable yardstick, so that one has to rely on special biological phenomena. For instance, many benthonic animals spend part of their development in the surface layers and this has a considerable influence on the composition of the fauna and flora of the pelagial.

This influence not only concerns the species-specific composition of the fauna. The movement of organisms from the benthal into the pelagial also involves a considerable transfer of matter from the bottom into the water layers above. With increasing depth and distance from the coast the number of species and individuals of pelagic stages of benthonic animals decreases, and there is an increase in the percentage of purely pelagic forms. From this it follows that the neritic region could be characterized as that part of the pelagial in which the fauna at certain times of the year will regularly be determined by the larvae of benthonic animals.

This demarcation is of course somewhat indistinct, and it is complicated by the fact that the reproductive cycles are determined by the seasons; furthermore, in cold-water regions, for instance, a large percentage of the benthonic animals do not have pelagic plarvae (see Chapter IV, 6). The idea of using the presence of larvae to define the pelagic fauna is therefore not a very precise one, added to which the larvae may of course be moved about by onshore and offshore currents. Nevertheless, I believe that this concept still provides the best demarcation between the neritic and the oceanic regions.

In considering the biology of the pelagial it is of particular importance to realize that the water filling a given area is in no way uniform, but consists of a more or less large number of 'water masses'. As defined by Krey 1952 d, p. 1, these may be viewed as 'individual units more or less sharply delimited from each other and having a long life'. Water masses may differ from each other in origin and extent and in the sum total of their ecological factors, including direction and speed of currents (cf. Kalle 1937). Correspondingly their biological composition also differs and they can frequently be identified by specific numbers of their populations (indicator species, pages 74, 142, 356). Where different bodies of water come into contact mixing will naturally take place, and the extent to which this occurs will depend upon hydrodynamic conditions; that is, on the nature of the currents.

Water masses with given characteristics may contain smaller 'water pockets' with different compositions which arise by intercalation; these have a shorter life and gradually pass by mixing into the main water masses.

Continual observations and measurements at fixed stations reveal the kaleidoscopic passage of different water masses and water pockets. Samples taken at the surface show a picture that is more like a mosaic, the pattern of which is determined partly by the conditions discussed and partly by the closeness of the sampling stations.

The benthal extends from the upper limit of the sea down to the great depths and it obviously requires further subdivision. The marginal zone is generally known as the littoral. Its upper limit can be defined as the line where the regular and dominant influence of sea water ceases; on an exposed coast it therefore lies in the upper part of the spray zone. This is therefore a boundary zone which will show very different characters depending on whether the coast is flat or steep, and on the presence or absence of tides. It is difficult to define the lower limit of the littoral in biological terms. In my opinion it should conform with the divisions of the pelagial, because particularly in the marginal zone the sea bottom and the waters above it form parts of a living environment which should be looked at as a whole. I therefore regard as littoral those areas of the bottom whose fauna at certain times determines the composition of the pelagic fauna by the presence of its larvae. Naturally this means that the lower limit of the littoral is somewhat indistinct, but it cannot be defined more accurately by other criteria, unless one were merely to define it in metres. By and large this lower limit of the littoral corresponds very approximately with the edge of the Continental Shelf.

More detailed ecological divisions of the littoral are discussed in later chapters.

Those parts of the benthal which lie beyond the littoral are difficult to differentiate on the basis of our present knowledge. According to the results of recent research the deep-sea troughs represent a special zone with a specific fauna, which Bruun has designated as the hadal (derived from Hades). The area between the littoral and hadal is commonly divided into an upper part, the bathyal, and a lower part, the abyssal. The approximate boundary between the two is taken as about 2,000 m. Hedgpeth gives as an ecological characteristic of the abyssal that the temperature never rises above 4°C. In my opinion the bathyal would correspond approximately with the mesophotic zone of the pelagial. I believe, however, that we need far more information before we can give a classification of the deep benthal based on firm biological and ecological evidence; Pérès (1957), for example, gives the following

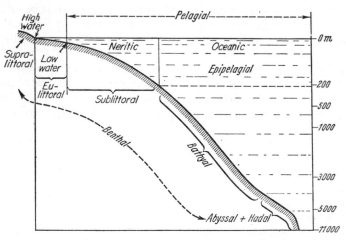

Figure 18. Diagrammatic section to show the main ecological divisions of the sea (based on Hedgpeth 1957)

TABLE 3

Some proposals for the vertical classification of the benthal

Stephenson 1949	Giordani-Soik 1950 a	Yonge	Pérès 1957	Ercegovic 1957	Pérès & Piccard 1958	
Supralittoral zone		Splash zone	Étage supralittoral	Étage holophotique	Étage supralittoral	Supralittoral
Supralittoral fringe	Zone intercotidale supérieure	Upper shore				Transition zone
Midlittoral zone	Zone intercotidale moyenne	Middle shore	Étage mesolittoral	Étage talantophotique	Étage mediolittoral	Eulittoral
Infralittoral fringe	Zone intercotidale inférieure	Lower shore	Étage infralittoral	Étage mégaphotique / Étage métriophotique	Étage infralittoral	Transition zone
Infralittoral zone			Étage circalittoral	Étage oligophotique		Sublittoral
			Étage bathylittoral	Étage méiophotique	Étage circalittoral	Elittoral
			Étage épibathyal	Étage amydrophotique	Étage bathyal	Archibenthal
			Étage mésobathyal	Étage aphotique		Upper / Lower } Abyssal
			Étage infrabathyal		Étage infrabathyal	
			Étage hadal		Étage hadal	Hadal

divisions of the deep benthal for the Mediterranean in which the temperature shows a considerably smaller drop with depth:

epibathyal ⎫
mesobathyal ⎬corresponding to the bathyal
infrabathyal   (corresponding to the abyssal)
hadal

Table 3 gives a comparison between some of the divisions which have been proposed, and Fig. 18 shows once more the main divisions used here.

The idea of defining the boundary layer between air and water as a living environment is justified by the occurrence of characteristic animals and plants. In addition there is here a certain loss of material to the land on drifting material and from the activities of sea birds. The air-breathing marine vertebrates are more or less tied to this boundary area and it is also used for flight by flying-fish. The pleustal does, of course, overlap with the pelagial in so far as the latter extends to the water surface and pleustonic organisms dive or move down into the pelagial which provides them with most of their food.

CHAPTER III

THE ECOLOGICAL FACTORS
A. The basic biological process

Leaving aside questions involving the morphology, systematics and evolution of marine organisms, most of the specifically marine biological problems are closely concerned with the basic biological process of producing organic matter. A more detailed treatment of this is given later (Chapter VI), but in order to understand the following section it seems to me essential at this stage to give a brief statement of the principles involved.

All animal life in the sea, as on land and in fresh waters, is dependent upon the synthesis of organic matter from inorganic substances; this synthesis is carried out in autotrophic plants by the process of assimilation. Chemosynthesis is totally insignificant by comparison with photosynthesis in which organic substances (carbohydrates, fats, proteins) are synthesized from water, carbon dioxide and mineral nutrients (nitrogen, phosphorus, sulphur compounds and other elements) under the influence of the energy of sunlight. The breakdown of these organic substances releases the imprisoned energy and allows both plants and animals to carry out functions requiring the expenditure of energy. Animal life is inconceivable without previous synthesis by plants.

Of the four groups of factors required for assimilation, water and carbon dioxide are always present in the sea in sufficient amounts. The incident light, however, is quickly reduced by absorption as the depth increases; thus below a certain depth the radiant energy is insufficient for assimilation. This means that below this depth light becomes a limiting minimal factor in the production of organic matter. At the same time it follows that the existence of all animals including those in the great depths is dependent upon the production carried out in the thin photosynthetic layer.

Production may also be limited by the fourth group of factors—the mineral nutrients—because some of these substances are present in such small quantities that when plant growth is maximal they may be used up completely. In areas where this scale of consumption has taken place, further production can only occur following the entry of water containing nutrients. The chemical constituents of sea water and its dynamic relationships are therefore of prime importance for the basic biological processes in all parts of the sea.

It seems to me therefore that a brief summary of the most important ecological factors is required, in which emphasis is laid on their distribution in time and space and attention drawn to their more important physiological effects.

For the sake of clarity these factors must be dealt with separately. It should be strongly emphasized, however, that the interaction of several factors can be observed in many cases. Brief mention has already been made of the multilateral relationships of organisms.

B. Abiotic factors

1. *Light*

From the biological aspect light must be regarded as the most important ecological factor because plant assimilation, that is the build-up of new organic matter and thus the primary production of food for animals, depends ultimately on radiant light energy. A comparison with conditions on land shows many corresponding features but also some important differences. Light frequently serves animals for visual orientation and it is also known that light has a stimulating effect on locomotion. In addition there are other living processes that are controlled by light; for example Grave (1927) records that the duration of pelagic larval life in the ascidian *Symphlegma viridis* is shortened in light but lengthened in darkness. In many cases periodic light change is probably the external factor which stimulates physiological rhythms in plants and animals.

Both land and sea share the periodical change of night and day and also of the seasons. The precision with which these periodic changes operate can be seen particularly well, for example, in the migration and breeding cycles of birds. It can be shown that these cycles, which are independent of temperature, are influenced solely by corresponding changes in light. Land plants exhibit annual rhythmical changes in hardiness to frost which are determined less by temperature than by changes in illumination.

Non-periodic changes in light intensity due to variations in cloud cover are also common to both sea and land. For life in the sea, however, the decisive factor is the light energy penetrating the water and not the light that strikes the surface. As part of the light is reflected at the surface less light penetrates than is apparent. The amount of reflection is dependent upon the position of the sun and also on non-periodic fluctuations in the roughness of the water surface. A prolonged or temporary covering of ice will naturally have a great effect on the light conditions in the water below. Non-periodic changes in illumination are therefore greater and more frequent in the sea than on land.

Furthermore, the amplitude of the periodic changes in light is greater in the sea than on land. Clarke (1938) compared the average amount of daylight in summer and winter at the Blue Hill Observatory with values recorded at a depth of 15 metres in the neighbouring Vineyard Sound. The measurements on land gave a ratio of 3:1, those in

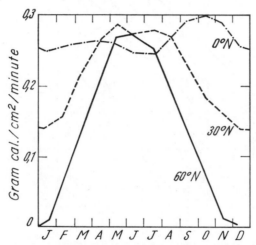

Figure 19. Seasonal variation in the total sunlight striking the surface of the sea in various latitudes (after Kimball 1928 from Moore 1958)

the sea 300:1. Figure 19 shows that the amount of incident light striking the sea surface during summer is almost the same in different latitudes, but that there are very marked differences in winter. This also applies, of course, to different areas of land and fresh water. The variations in the amount of light in water are significantly greater than on land; for instance with the sun at an angle of 50° or more the loss from reflection is about 3%, but with the sun at an angle of 5° this loss is 40%.

The light entering the water undergoes both qualitative and quantitative changes. Two components are involved in the extinction of light: scattering and absorption. Both are due to molecular processes in the water and the dissolved substances and also to the presence of suspended particles. The denser the population and the more dead particles floating in the water the less will be the depth of penetration

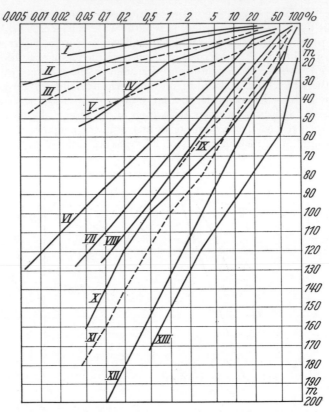

Figure 20. Depths to which light penetrates, expressed as percentage of the surface light, in various sea areas (after Clark 1933, 1937–39, Jerlov 1951). I, Woods Hole Harbour; II Mean curve for various coastal waters; III, VIII, X, XI, XII, XIII, Sargasso Sea; IV, Gulf of Mexico near mouth of the Mississippi; V, Gulf of Maine; VI, Coastal water; VII, Caribbean Sea; IX, Gulf of Mexico

of a given amount of light. In the vicinity of the coast, due to turbulence and land erosion, there is always a greater content of suspended matter than there is in the open sea; also, the population density of small organisms is usually greater. In fact there is a marked gradient in the depth of penetration from the coast to the open ocean, or, to express this in another way, at the same depth there is less light in coastal waters than in oceanic waters (Fig. 20).

Qualitative changes in the light occur because of differential absorption of the wave-lengths of sunlight. Figure 21 shows that in pure water the long-wave components and the ultraviolet are reduced to a greater extent than the blue-green component; the latter also decreases with depth so that at great depths no daylight is detectable. The UV component, and particularly the blue, are very strongly absorbed by the suspended matter and dissolved substances (e.g. Gelbstoff, see p. 101); as a result of this selective absorption the maximum transmissibility in cloudy water lies in the green. With particles from depths of 500 and 1,000 metres in the Sargasso Sea, Yentsch & Ryther (1959) found an absorption maximum in the blue.

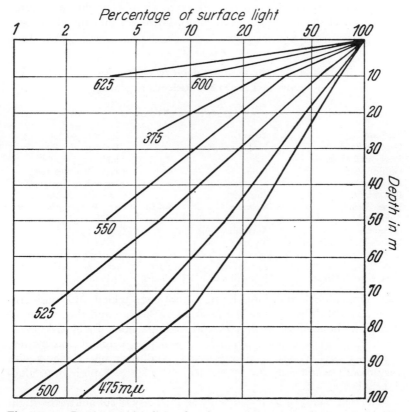

Figure 21. Depth to which light of various wave-lengths penetrates, expressed as percentage of the surface light at 11°25′S, 102°08′E (after Jerlov 1951)

These points are of fundamental importance in everything that happens in the sea, because they mean that assimilation can only take place in the surface layers that are sufficiently illuminated. It follows that in the final analysis all the animals, including those inhabiting great depths, are dependent for their nutriment on these productive

C—M.B.

surface layers. A detailed knowledge of the light conditions in the sea is therefore an important prerequisite for any assessment of the potential production; the available data on this subject, however, are still meagre.

By determining the gas exchange (oxygen production, carbon dioxide consumption) of phytoplankton in submerged flasks one can calculate the depth at which the assimilation and dissimilation of the plants are equal (the compensation point). The results of various authors have shown considerable differences in different areas as might be expected (Table 4). It should be borne in mind, however, that the glass flasks

TABLE 4

Depth of the compensation point in various localities

(Ketchum 1951 p. 337 and Richards 1957 p. 192)

Locality	Time of year	Depth in m	Author
Sargasso Sea	August	>100	Clarke 1936
Georges Bank	June maximum	59	Clarke 1946
English Channel	July	45	Jenkin 1937
Gulf of Maine	June	24–30	Clarke & Oster 1934
Loch Striven	Summer	20–30	Marshall & Orr 1928
Passamaquoddy Bay	Summer	17	Gran & Braaruud 1935
Puget Sound	Summer	10–18	Gran & Thomson 1930
Oslo Fjord	March	10	Gaarder & Gran 1927
Georges Bank	April minimum	9	Clarke 1946
Woods Hole Harbour	August	7	Clarke & Oster 1934
Loch Striven	March	5	Marshall & Orr 1928
Helsingør Sound	Annual average	4·5–7	Steemann-Nielsen 1937
Long Island, Great	July–August	0·8–2·1	Richards 1957

themselves absorb additional light, so that even taking into consideration the local differences in the position of the sun and the content of particles, the figures are not directly comparable. In clear oceanic water the compensation point lies at a depth of about 100 metres; in areas near the coast it is considerably shallower and in places with very cloudy water only the upper decimetre shows a positive production balance.

Petersson *et al.* (1934) found that the intensity of submarine daylight which prevailed at the compensation point was about 400 lux. This value will of course be less if the water contains pure phytoplankton and no animals. Moreover, it will be dependent upon the specific composition of the phytoplankton (see p. 313).

For many biological purposes the determination of the percentage of daylight which reaches a given depth is only calculable if at the same time the composition of the light is known; this is because assimilation is proportional to the light energy, but this is a function of the intensity and of the composition of the light (cf. the investigations of Stanbury (1931) on the growth of *Nitzschia closterium*). For this reason determinations of the depth of the compensation point have hitherto only been of academic value, because no account has been taken of the aborption by different flasks. Jerlov (1951) has calculated the light energy for a series of depths, and an example is given in Fig. 22. In this context it should be mentioned that in marine algae the position of the absorption maximum varies considerably and even 'for a given species of alga it may fluctuate markedly according to locality, depth and even the season' (Levring 1960 (see Chapter VI, 1).

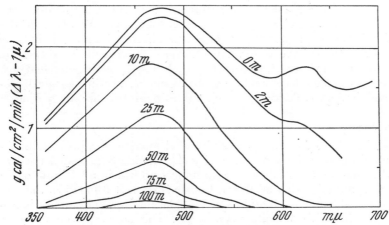

Figure 22. Spectral energy of the downward directed light at various depths (after Jerlov 1951)

Nevertheless, it is of biological interest to know the absolute depth to which light penetrates, because light is important in the orientation of many animals and also in stimulating their activity. Thus the extensive vertical migrations of many inhabitants of the pelagial are evidently determined by the light rhythm. Here it is worth noting that greater attention should be paid to moonlight, because living processes dependent upon the phases of the moon have been observed in many marine animals, in cases where experimental analysis has failed to show the influence of factors other than changes in light (see Chapter VI, 3).

There are only a few observations on the depth to which moonlight penetrates. Thus, Waterman (1954), for example, remarked that while diving in full moonlight, although he could still recognize details of the bottom at a distance of 1 metre the light through the analyser was of extremely low intensity (unfortunately the depth of the water was not given).

Finally it should be pointed out that polarized light is also present in water. Waterman (1954) recognized two components of this: the incident polarized light and a component originating from the water itself. For both components the intensity of the polarization is affected by the position of the sun relative to the point of observation. Various marine animals such as *Limulus, Eupagurus, Littorina*, etc., can detect polarized light and so orientation by this light is a promising subject for further investigation. This necessitates, however, a more detailed knowledge of the polarized light itself and of the physiology of light-sensitive organs in marine animals.

It is known, particularly from human and veterinary medicine, that in the presence of fluorescent substances light can cause fatal injuries in organisms. Pereira (1925) showed a 'photodynamic effect' of this kind in the eggs and larvae of the sea-urchin *Arbacia* on the addition of eosin to the culture water. Since natural fluorescent substances are known in sea water (cf. p. 101), Friedrich (1959) has suggested that such photodynamic effects might also be of importance in biological processes in the sea. Positive research results in this field are still lacking.

2. *Salinity*

The salinity is a measure of the total amount of mineral salts dissolved in unit volume of water. It is expressed as S ‰ and defined as 'the total amount of dissolved substances present in a kilogram of sea water, assuming that all the carbonate is converted to oxide, the bromide and iodide to chloride and that the total organic matter is oxidized' (Dietrich & Kalle 1957, p. 42). This customary definition in physical oceanography is only of limited value in biology, because it does not take account of the relative ionic composition and the pre-

TABLE 5

Salinity in adjacent seas (in ‰)

	Surface	Deep water
American Mediterranean Sea		
Caribbean Sea & Gulf of Mexico	34·73–34·88	34·98
European Mediterranean Sea		
Straits of Gibraltar	36·25	37·75
Eastern Mediterranean	>38–>39	>38
Red Sea (northern part)	40–41	40·5–41
Persian Gulf	38	38
Baltic Sea		
Belt	~16	~27
Central Baltic	7	<10–>20
Gulfs of Bothnia and Finland	1–3	

sence of trace elements. Let us consider first the total salinity, leaving the relative composition to be discussed later (see pp. 80 et seq.).

The average salinity of the sea is about 35‰. Deviations from this are found mainly in adjacent seas, in which the inflow of fresh water and the evaporation are not in the ratio of 1:1 and the exchange of water with the open oceans does not bring about any adequate compensation (see Table 5). Differences may also occur in the open ocean; thus in the tropics where there is a high rate of evaporation the surface waters show a higher average salinity, while in the polar regions considerable dilution may occur at the surface owing to the melting of large masses of ice. There are also vertical gradations in salinity. In the open ocean however these fluctuations are usually small; they are distributed over such wide areas that taken alone they scarcely have any biological significance. On the other hand, the picture is different in the vicinity of the coast, in sheltered bays, the estuaries of large rivers and so on. Here the salinity may fluctuate above or below the mean, to such an extent that the tolerance limits of many plants and animals are exceeded. It should be particularly noted that, in general,

TABLE 6

Freezing point and osmotic pressure in relation to salinity

(Dietrich 1952 pp. 435–36)

Salinity ‰	Freezing point depression	at 0°	Osmotic pressure (at) at 10°	at 20°
2	−0·108	1·30	1·35	1·40
4	−0·214	2·59	2·69	2·78
6	−0·320	3·87	4·01	4·15
8	−0·427	5·16	5·35	5·54
10	−0·534	6·45	6·69	6·92
12	−0.640	7·73	8·01	8·30
14	−0·748	9·04	9·37	9·70
16	−0·856	10·34	10·72	11·10
18	−0·965	11·66	12·09	12·52
20	−1·074	12·97	13·45	13·92
22	−1·184	14·30	14·82	15·35
24	−1·294	15·63	16·20	16·78
26	−1·405	16·97	17·59	18·22
28	−1·516	18·31	18·98	19·65
30	−1·627	19·65	20·37	21·09
32	−1·740	21·02	21·79	22·56
34	−1·853	22·38	23·20	24·02
36	−1·967	23·76	24·63	25·50
38	−2·081	25·14	26·06	26·97
40	−2·196	26·53	27·50	28·48

repeated fluctuations in salinity are more injurious than regular and gradual changes.

From the physiological and ecological points of view the significance of salinity lies in the osmotic pressure which it produces. The majority of oceanic invertebrates are isotonic, that is, the osmotic pressure of their internal fluids is the same as that of the surrounding water. Details are available in only a few forms. Sharks show isotonicity, but this is determined by the large amount of urea present in the blood.

Bony fishes are in general hypotonic, that is, the internal concentration of salt is less than that of sea water. This condition can therefore only be maintained by special regulatory mechanisms within the animal. Sea birds and marine mammals are also able to maintain the salinity of their body fluids at a level that is less than that of sea water. The excess salt taken in with food and water is excreted. In several animals that are isotonic the relative ionic composition of the body fluids is different from that of the surrounding sea water. In every case physiological work is necessary to produce and maintain this varying composition.

Osmotic pressure depends upon temperature (Table 6), and the temperature conditions under which marine organisms are exposed to varying salinities are therefore relevant. It is especially at the limits of tolerance that interaction between changes in salinity and temperature are to be expected. This will occur particularly in coastal areas and will naturally be most marked in tidal waters and in the spray zone. The copepod *Tigriopus fulvus* which lives in pools close to or above the high water mark (Ranade 1957) provides an example; the lethal

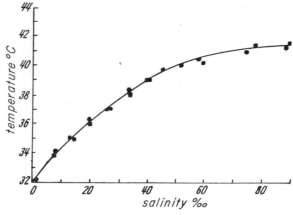

Figure 23. Lethal temperatures for the copepod *Tigriopus fulvus* at different salinities (2 experimental series VIII and IX 1956) (after Ranade 1957)

temperature rises with increasing salinity, thus this animal can live in shallow pools that warm up quickly and have a high rate of evaporation (Fig. 23).

As shown in Table 6, an increase in the salinity of a solution is correlated with a lowering of its freezing point. This may be fatal to homoiosmotic bony fishes in certain circumstances when the water temperature falls to below the freezing point of the blood, which has a lower salinity than the surrounding sea water.

The wide limits within which fluctuations in salinity can be tolerated are apparent from one of the tables given for copepods by Sewell (1948, p. 367) (see Table 7). In many other cases, however, the range is considerably smaller.

TABLE 7

Comparison of extreme salinities tolerated by various copepods

(from Sewell 1948 p. 367)

Species	Salinity ‰	Species	Salinity ‰
Acartia clausi Giesb.		Pseudodiaptomus serri-	
fresh water up to 36·00		caudatus (T. Scott)	10·85 to 43·80
A. bifilosa Giesbr.	0·30 to 32·00	Centropages hamatus	
Pseudodiaptomus tollingerae		(Lillj.)	13·5 to 23·90
Sewell	0·17 to 11·84	Oithona nana Giesbr.	13·68 to 43·80
P. annandalei Sewell	0·17 to 18·94	Labidocera pavo Giesbr.	14·12 to 32·00
P. binghami Sewell	0·17 to 18·94	Paracalanus parvus (Claus)	19·33 to 43·80
Temora longicornis Müll.	6·54 to 36·16		
Acartia longiremis (Lübb.)	6·72 to 35·32	Metridia lucens Boeck	28·10 to 35·40
Pseudocalanus elongatus		Calanus finmarchicus	
Boeck	7·25 to 35·30	(Gunn.)	29·00 to 55·30
		Acrocalanus gibber Giesbr.	32·95 to 35·30

On account of this physiological effect considerable attention has been paid to regions with low salinity (brackish water). Several attempts have been made to classify brackish waters according to their salinity and its influence on the fauna (see Chapter VII, 4). However, a number of other factors are involved in the reaction of organisms to reduced salinity, and so it seems to me that for the evaluation of the actual conditions we need an analysis of the other factors in addition to data on the salinity and its range of fluctuation.

3. Temperature

The living processes of organisms, being chemical by nature, are dependent upon temperature at least in so far as their rate is concerned, and therefore the temperature of the medium is an important ecological factor. As all physiological processes are tied to the fluid state of water it follows that the temperature range of living organisms must lie between the freezing and boiling points of water. In practice, however, it is restricted to a smaller range, and in many cases a species can live within a temperature range covering no more than a few degrees. Among the homoiotherms the range is very narrow as far as the internal temperature is concerned, but due to physiological thermoregulation the range of external temperatures that can be tolerated is relatively large.

In considering the temperature range for individual species it should be remembered that the various stages in a life history may have different requirements. Adults may survive at the limits of tolerance whereas the developmental stages are restricted to a much narrower range. In this context there is, however, a difference between the area in which a given species may occur and the area in which it is indigenous and maintains its population by reproduction. Here it is particularly important that the temperature necessary for reproduction and development (the critical spawning temperature) occurs at the time when the other factors upon which readiness to breed is dependent are also able to exert their influence.

It follows therefore that in discussing temperature, attention should be paid to the following points:

(a) Mean values in the long and short term;
(b) Annual range of fluctuation and the distribution of seasonal maxima and minima;
(c) The difference between the reproductive area and the area of sterile dispersal of the individual species.

The mean temperature of the seas of the world is 3·8°C, but this has little meaning from the biological point of view; even the fact that in equatorial waters the water column has an average temperature of only 4·9°C is only of interest biologically in so far as it indicates the great difference between the warm surface layers and the cold deep water.

There are differences in the regional distribution of the mean surface temperatures between the seas lying north and south of the equatorial region, not only in mean values but also in relation to the course followed by the isotherms. In the southern hemisphere these mainly lie parallel to the latitudes, whereas in the northern hemisphere, both in the Atlantic and Pacific, the isotherms fan out in such a way that they are drawn together more closely off the coasts in the west than in the east. This means, for example, that distribution limits determined by temperature are more sharply demarcated in the west than in the east, all the more so as the same principle applies to the isotherms at depths of 200 and 400 metres (Fig. 24).

In the open ocean the annual fluctuations in surface temperature have a characteristic distribution: in the polar regions and in wide areas of the tropics they amount to less than 2°C while in temperate regions between latitudes 30° and 40° they are about 10°C. In the western parts of the north Atlantic and north Pacific oceans that are influenced by offshore air currents the fluctuation may be over 18°C, and in marginal areas under the influence of the continents they may be over 14°C. In all cases, however, the annual fluctuation is relatively small in open water. It is also worthy of note that because of the specific heat of the water the extremes lag behind those on land.

The annual temperature fluctuations in a given region are not only

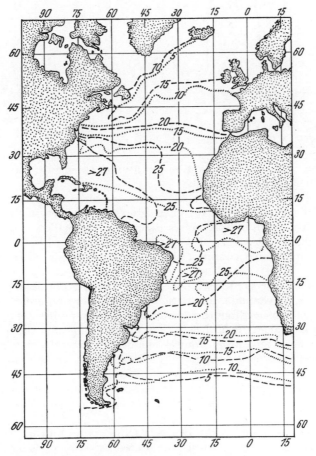

Figure 24. Isotherms for Atlantic surface waters in February (dotted lines) and August (broken lines) (after Hardy 1945 from Sewell 1948)

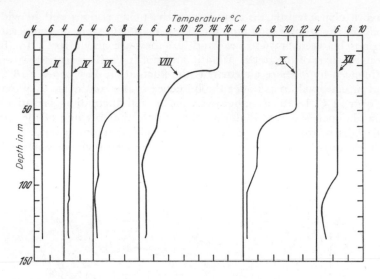

Figure 25. Vertical distribution of temperature (°C) during different months at 50°N, 145°W (N.E. Pacific) (after Tully *et al.* 1960)

dependent upon the illumination but also on changes in the currents which often occur periodically.

The annual range of temperature decreases sharply with depth. Thus Helland-Hansen (1930) gives figures for an area in the Bay of Biscay: at a depth of 100 metres the annual fluctuation is not much greater than 1°C, whereas at 0 metres it is about 8°. Tully *et al.* (1960) found similar conditions in the north-east Pacific (Figs. 25 and 26).

The upper layers of the sea floor, where they are continually covered with water, show annual fluctuations which correspond approximately to those of the neighbouring water layers. The position is different however in those parts of the sea floor that are subject to the influence of tides; here very large annual fluctuations have been measured, depending on the geographical position. Thus for the uppermost bottom layer in the shallows off the North Sea coast of Germany, Linke (1939) found that when the tide is out the temperature varies from +35°C in summer to —5°C in winter, an annual range of about 40°C.

In terrestrial habitats the range of temperature during a 24-hour period is often very large but in the open ocean it is very small (about 0·2 to 0·3°C) and even in shallow coastal waters it is scarcely more than 2°C. On the sea floor, provided it is continually covered with sea water, no diurnal fluctuation would be apparent. If, however, it is only covered periodically then considerable diurnal fluctuations may occur; on the bottom in shallow water Linke measured a daily temperature range of 15°C at the height of the summer, up to 10° in spring and autumn and about 5° in normal winters.

Those parts of the littoral that are subjected to daily and annual temperature fluctuations can only accommodate animals that are truly eurythermal.

Furthermore it is worth noting that in certain areas temperature differences occur within a short distance, namely where cold and warm sea currents come in contact with each other, as for example the Gulf Stream and Labrador Current, the Irminger and East Greenland Currents in the north-west Atlantic, or the Kuro Shio and Oya Shio Currents in the north-west Pacific (cf. Fig. 193, p. 345).

In assessing biological problems it is also significant that the annual fluctuations in temperature continue into deeper water, although in the open ocean an annual fluctuation of over 0·5°C would only be recorded down to about 140 to 300 m. Depths beyond this can generally be regarded as thermostable from the biological point of view. In areas of shallow sea the conditions are of course quite different, particularly when disturbance is caused by currents.

The large-scale vertical distribution of temperature shows some characteristic features. In the open ocean, between latitudes 50°N

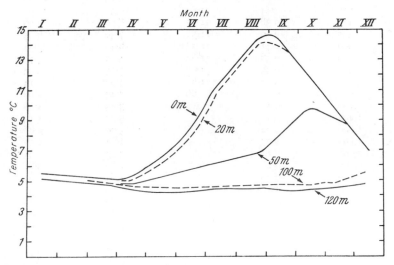

Figure 26. Annual curve for temperature at different depths at 50°N, 145°W (N.E. Pacific) (after Tully *et al.* 1960)

and 45°S, there is a warm-water sphere close to the surface which gives place towards the poles to a cold-water sphere via a relatively thin layer characterized approximately by the isotherms 8 to 10°C. This boundary layer, which in the tropics lies at 300 to 400 m and in the subtropics at 500 to 1,000 m, ascends towards the poles and reaches the surface at the oceanic polar front, which represents the boundary between different types of water.

In most regions the temperature decreases more or less regularly— or sometimes irregularly—with depth (see Table 8, columns 3 and 4),

TABLE 8

Four examples of the vertical distribution of temperature

from Dietrich 1952 pp. 472–474)

Depth in m	Weddell Sea 62°32·5′ S 24°32′ W	Red Sea 15°52′ N 41°43′ E	Philippine Trench 9°40·7′ N 126°51·3′ E	Equatorial Atlantic Ocean 3°30′ S 22°26′ W
0	−0·78	26·70	28·80	25·49
25	−0·84	26·10	28·30	25·57
50	−1·79	25·99	28·20	25·54
75	−1·88	24·25	27·50	24·52
100	−1·89	22·51	25·90	14·97
150	−0·73	21·84	20·58	13·47
200	−0·02	21·80	15·15	12·42
300	−0·44	21·60	10·50	10·91
400	−0·38	21·57	8·50	8·96
500	−0·40	21·57	7·30	6·95
600	−0·41	21·58	6·48	5·54
700	−0·38	21·60	5·80	4·77
800	−0·33	21·62	5·35	4·35
900	−0·29	21·63	4·90	4·14
1,000	−0·25	21·66	4·45	4·06
1,200	−0·17		3·85	4·28
1,400	−0·10		3·35	4·15
1,600	−0·02		3·00	3·86
1,800	−0·07		2·61	3·57
2,000	−0·14		2·25	3·32
2,500	−0·26		1·80	2·87
3,000	−0·30		1·64	2·68
3,500	−0·43		1·58	2·58
4,000	−0·50		1·60	2·01
4,500	−0·50		1·65	0·89
5,000			1·72	0·68
6,000			1·86	
7,000			2·01	
8,000			2·15	
9,000			2·31	
10,000			2·48	

but in the Weddell Sea, for example, the coldest ocean area, there is only a small difference between the surface and the deep water (column 1). In very large areas of the seas of the world the temperature at the bottom is less than 2°C, that is, the cold-water sphere is continuous from pole to pole and is only overlain by the warm-water sphere in the tropics and subtropics. The morphology of the bottom, of course, plays an important role in water exchanges. For instance, it prevents an exchange between the deep waters of the Red Sea and of the Indian Ocean and this results in a concentration of warm, high-salinity water at all depths in the Red Sea. Here therefore tropical stenothermal animals will find adequate temperatures even at depths of 1,000 m. In the same way the depths of the Mediterranean are cut off from those of the Atlantic by the Gibraltar sill.

In addition to the daily and annual temperature cycles, nonperiodic deviations from the mean values are known for many oceanic regions, and these usually last for several years; Rodewald has reported on this in a series of papers published between 1952 and 1959. This may result in displacement of spawning areas and changes in the occurrence of certain species, which often have serious effects on fisheries.

The formation of a thermal discontinuity layer in areas with marked annual fluctuations of surface temperature is of significance from the biological point of view. It may mean that a homothermal warm top layer is separated from the cold water below by a transition layer that is often only a few decimetres thick. This marked change may have a direct influence, as it will involve a gradient in density, thus affecting the buoyancy of living organisms and also of dead particles. The formation of the discontinuity layer will also cut off the deeper water from the surface until such time as the discontinuity layer breaks up as the temperature drops in autumn. If at this time the surface water is cooled to below the temperature of the deeper water a vertical exchange of water (vertical thermal convection) will then take place, resulting in an exchange of gases and nutrients between the surface and the deeper layers.

Numerous physiological and ecological investigations have already been carried out in relation to the occurrence of varying temperatures in different localities. The details cannot be discussed here, but the following results may be mentioned as they appear to be generally applicable. At about the upper temperature limit of their environment, polar animals, although they live continuously at low temperatures, have approximately the same metabolic intensity, measured in oxygen consumption, as tropical animals of the same group (Thorson 1936 in lamellibranchs, Vernberg 1959 in fiddler-crabs, Wohlschlag 1960 in fishes). Relative to body size the oxygen consumption of antarctic benthonic fishes living at about 0°C, for example, is almost as great

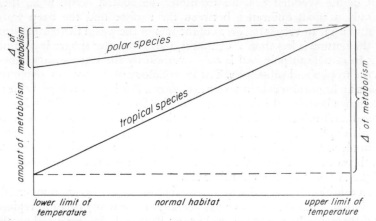

Figure 27. Diagram to show relationship between the metabolic intensity of tropical and polar species and the extreme values of temperature in their localities (original)

as that of tropical fishes at 30°C. On the other hand with decreasing temperature the oxygen consumption of warm-water animals falls much faster and lower than that of cold-water animals; thus the metabolic intensity of tropical animals at the lower temperature limit of their habitat is much less than that of polar animals, although the actual temperature range for the tropical animals is considerably higher. This is shown diagrammatically in Fig. 27.

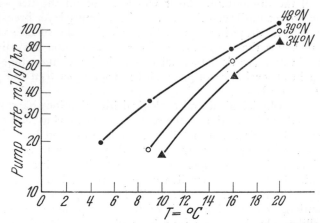

Figure 28. Rate-temperature curve of samples from three localities at different locations on the coast of western North America, of the mussel *Mytilus californianus* (after Bullock 1955)

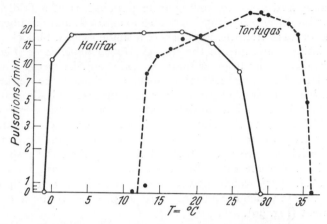

Figure 29. Rate-temperature curves for northern and southern samples of the medusa *Aurelia aurita*. The rate measured is that of swimming movements of the bell. The habitat temperatures are Halifax 14°C and Tortugas 29°C (after Bullock 1955)

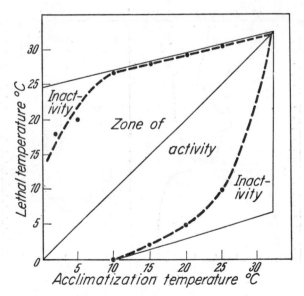

Figure 30. Relation between acclimation temperature and temperature tolerance in the lobster *Homarus americanus*, showing zones of activity and of inactivity where continued existence would not be possible (after McLeese and Wilder 1958 from Kinne 1963)

By comparative studies on separate populations of a fiddler-crab (*Uca pugilator*) Démeusy (1957) was able to establish that this phenomenon also occurs within the members of a single species. Bullock (1955) found similar conditions in the mussel *Mytilus californianus* (Fig. 28), and this also applies to other functions (Fig. 29). In species from the middle latitudes where seasonal temperature changes are marked the relation between metabolic intensity and temperature is also variable. This means therefore that physiological processes may undergo adaptations which are not primarily genetic in character. These adaptations may be partly due to an earlier conditioning during the development of the individual (see pp. 5, 6), and this may limit the response of populations. On the other hand several examples are known (cf. Precht *et al.* 1955) which show that the reactivity depends upon the temperature to which the animals were adapted before the start of the experiment (Fig. 30).

4. *Specific gravity*

In addition to salinity and temperature, sea water (depending on pressure) has an attribute generally known in physical oceanography as specific gravity and denoted by the symbol σ. Whereas distilled

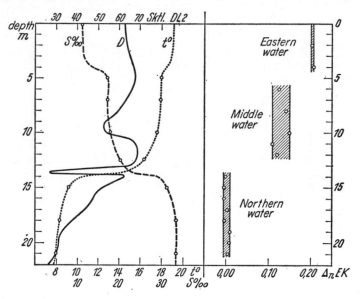

Figure 31. Vertical distribution of light transmission (D), temperature (t) and salinity (‰) at one station between Copenhagen and Elsinore, showing three water masses one above the other. A turbidity layer has been formed at the discontinuity layer at 13–14m (after Joseph 1949)

water has its greatest density at +4°C (specific gravity = 1), the density of sea water increases with decreasing temperature and reaches its highest value at temperatures below 0°. Increase of salinity, as for instance by evaporation, likewise causes a rise in specific gravity. Temperature and salinity therefore affect the specific gravity reciprocally when both factors are increased or decreased at the same time.

The specific gravity of sea water is of direct and indirect importance in biological processes. Differences in specific gravity between neighbouring water masses with unstable stratification result in currents which lead to the transport of organisms and to significant exchanges of substances in solution. Thermal and haline discontinuity layers constitute density discontinuity layers, which function as barriers to the vertical exchange of water, thus preventing the entry of heat and also the transport of gases and nutrients between the water masses separated by them. Sinking bodies and substances with a specific gravity less than that of the water in the discontinuity layer may collect in it. Accumulations of this kind have often been observed and described as plankton or turbidity layers (Figs. 31 and 32); on account of their

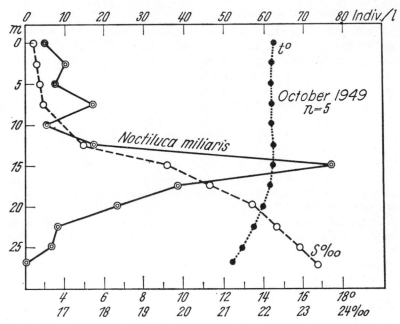

Figure 32. Mean values for the vertical distribution of *Noctiluca miliaris* in October 1949 at the lightship *Flensburg;* n = number of vertical plankton hauls (after Gillbrucht 1954)

nutrient content discontinuity layers may cause concentrations of planktonic animals. However, the reaction of zooplankton to discontinuity layers has still not been much investigated and no broad generalizations can be made; Banse (1955, 1957), for instance, observed a stratification of plankton caused by the thermohaline stratification of the water, whereas Harder (1957) confirmed the active immigration of copepods and larvae into discontinuity layers.

In a preliminary note on this problem Merz & Behrens (*Internationale Revue der gesamten Biologie und Hydrographie* for 1911) recorded that in the Sakrower Lake at Potsdam the small planktonic crustaceans, particularly copepods, prefer the limits of homogeneous water layers. Since then there have been several similar observations from the sea. However, the physiological basis for this is by no means clear. It is possible that the density as such or the density difference could act as an adequate stimulus, or it could be due to the thermal and chemical differences between the neighbouring water masses. It is also, however, possible that the concentration of nutrient matter could be a more important factor in the aggregation of animals at discontinuity layers than the presence of purely physical factors.

5. *Temperature–salinity relations*

When the temperature of a homogeneous water mass is plotted against the salinity, one obtains a single point on a t/S graph. Several such points from measurements taken in a large water mass will lie on a curve, which represents graphically the conditions in the vertical or horizontal plane and expresses the possibility of mixing between different water masses. The curve shows the relationships between temperature and salinity for those water masses from 200 metres down, which are not subject to surface influences and only change through the process of mixing.

Numerous observations have shown that some species are confined to certain types of water (pp. 74, 142, 356), and such correlations may be recognized by plotting the position of the species concerned on t/S graphs of the water masses; hence one can infer the degree of correlation and the type of water mass concerned. Figures 33 and 34 show examples of this. Cassic (1959) found correlations between the t/S relationships and the logarithms of the numbers of some zooplankton species in New Zealand waters, and Beklemishev (1960) writes of the copepod *Calanus pacificus*: 'Thus, in the case of *C. pacificus* its habitat is characterized by a part of t/s curve which shows what becomes of the biotope of the boreal community in this peripheral part of the *C. pacificus* range' (l.c., p. 299). Bary (1959) gives further examples, and Brinton (1962) shows similar relationships for euphausiaceans in the Pacific.

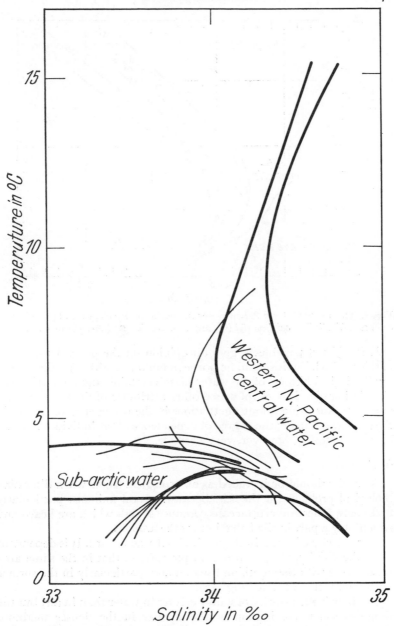

Figure 33. t/s curves for the localities of the pelagic worm *Poeobius* (thin lines) compared with those for sub-arctic and central Pacific water according to Sverdrup (after McGowan 1960)

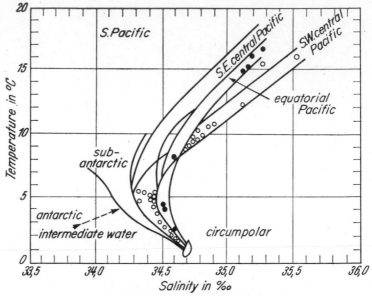

Figure 34. t/s graph for different water masses in the Pacific, showing t/s plots for catches of *Chauliodis sloani dannevigi* (O) and *C. s. secundus* (●) (after Haffner 1952)

It is still not possible to give an opinion on the general usefulness of this method, but it should be borne in mind that although the changing 'biological history' of the water masses certainly does not affect the t/s relationships it does influence other attributes of the water masses and may thus bring about fluctuations in the occurrence of species in the same water mass; the degree of dependence of the individual species on such factors is, however, very variable.

6. *Acoustics*

Acoustics should be included as an ecological factor when discussing biological problems in the sea, because several animals, particularly mammals, fishes and crustaceans, produce sounds which are heard and which play a part in the behaviour of the animals.

The velocity of sound in the sea is about 1,500 m/sec. It is dependent upon the salinity, temperature and pressure, so that in the same area it fluctuates with the depth and season and, particularly in the surface layers, with the latitude (Fig. 35).

Not only is the velocity of sound greater in water than in air, but the range of sound conduction is also greater in the denser medium. Therefore sound is very easily transmitted in water. Since the absorption of sound increases with the square of the frequency, relatively weak

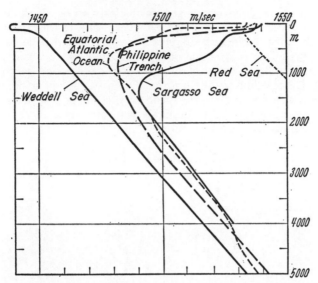

Figure 35. Examples of the speed of horizontal waves in the world's seas; with little or no thermal stratification the minimu is less marked, e.g. Weddell Sea and Red Sea (after Dietrich & Kalle 1957)

sounds, provided they are of low pitch, are audible at fairly great distances. It is not only the sounds made by members of the same species that play a part in the behaviour of animals. According to Bull (1957) fish subjected to repeated fishing with echo-sounder and trawl may develop conditioned reflexes, so that the sound waves alone (echosounder, engine and propeller noises) have a scaring effect. More detailed investigations on this are still lacking (on bio-acoustics see pp. 106–108).

7. Currents

Marine currents are in many ways of great biological importance both generally and locally. It is particularly significant that both in surface waters and in greater depths there are currents with a mainly horizontal movement, but which may vary in direction and permanence and even have a vertical component. As an example of the size of these currents we may cite the Cromwell Current in the Pacific which was only discovered a few years ago: this current, which is about 300 km wide and extends from close beneath the surface to depths of 200 m, lies almost exactly along the equator and flows from west to east in a region in which the water elsewhere flows westwards. It therefore becomes extremely difficult to analyse the transporting activity of marine currents from observations taken at the surface or at individual depths. The discovery, based particularly on the work done during the Geophysical Year, that deep currents flow in the reverse direction to the surface currents above them is important in assessing the transporting effect because many organisms undertake vertical migrations (see Chapter VII) and may be moved now in one direction, now in another. According to the duration of their stay in one or other of the currents and according to the speed of the currents such organisms may therefore remain in a relatively restricted area. Attention had already been drawn to this phenomenon by Hardy & Gunther (1935) (Fig. 36).

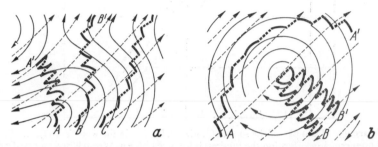

Figure 36. Examples, in plan, of the paths taken by plankton animals A–A[1], B–B[1], etc., migrating daily between surface waters (fine continuous arrows) and lower waters (fine broken arrows) which are moving in different directions. The night and day paths of the animals are shown in continuous and broken heavy lines respectively (after Hardy 1956)

The different currents are often characterized by striking differences in the temperatures of their water masses and they also show differences in salinity, thus providing conditions under which individual currents or current systems may be characterized by specific populations. It is possible to define the water masses of certain areas according to their populations and to deduce their origin from the presence of 'indicator species'; in addition the zones of contact and mixing between currents form very sharp limits to the distribution of many species.

Periodic and non-periodic fluctuations in the currents of certain regions may lead to far-reaching biological fluctuations. Thus the mass proliferation of certain plankton organisms, producing what are known as 'red tides', can be correlated with specific current conditions (see Chapter VI, 7, for the effect of red tides on populations). And in the area off the coast of Chile, for example, the periodic change in fish populations depends on changes in the course of the Peru Current which are therefore of importance to the fisheries. The inflow of high salinity Atlantic water into the Baltic Sea leads to far-reaching hydrographical changes particularly in the deeper parts and thus to biological fluctuations which are most easily recognized by fluctuations in fishery statistics (see p. 353). Similarly the fluctuations in the yield of the North Sea herring fishery have been, at least partly, correlated with changes in the inflow of colder water.

Vertical exchange of water masses may affect biological processes in two ways. Sinking surface water carries oxygen into deeper water. In an area where there is no vertical water exchange the oxygen will be consumed as a result of the decomposition of sinking organic substances. This will lead, as for example in the Black Sea, to the complete disappearance of oxygen in depths of over 250 metres and to the formation of hydrogen sulphide (Figs. 37 and 46). Animal life can therefore

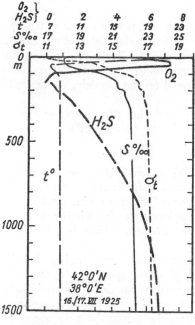

Figure 37. Vertical distribution of temperature (t°), salinity (S‰), density/specific gravity (σ), oxygen (O₂) and hydrogen sulphide (H₂S), both in cc/1 according to Soviet investigations in the Black Sea in July 1925 (after Dietrich & Kalle 1957)

only exist in depths far below the surface when surface water carrying oxygen sinks and becomes distributed in deep water.

Conversely, vertical exchange moves deep water to the surface. With the sinking of organic matter into deeper water the mineral nutrients forming part of such matter will accumulate in the vicinity of the sea floor. By the process of vertical exchange these nutrients are once more brought to the surface layers where there is active assimilation and thus retained in the chemical cycle. Thus, for example, the abundance of plankton and fish in certain areas off the west coasts of Africa and America is correlated with the upwelling of deep water.

In temperate regions with a seasonal temperature cycle in the surface waters convection currents bring up bottom water rich in nutrients and this forms the basis for the plankton outburst in the early spring. Nathanson (1906) drew attention to the great importance of vertical water exchange but did not, however, realize the connection with mineral nutrients.

It is also apparent that water which remains over long periods in the layers where assimilation takes place without the addition of deep water or of fresh water rich in nutrients, will become impoverished and will only support poor populations. The centres of the great anticyclonal currents in the northern and southern hemispheres are examples of this, whereas the currents in the cyclonal areas are rich in nutrients (Fig. 6b).

The movement of water plays an important role in the nutrition of many animals. Currents continuously bring fresh food to sessile animals and also hold particles longer in suspension. One could possibly find a correlation between the speed of currents close to the bottom and the density of the sessile animals which feed on the particles suspended in this water, even in deep-sea areas.

The individual currents and the circulation of the seas as a whole are therefore of primary importance in the biology of the sea.

8. Tides

The tides, that is the periodic half- or whole-day rises and falls of the sea surface, have several direct and indirect effects on biological phenomena in the sea. The movements of water associated with them, known as the tidal currents, may supplement or inhibit the transporting effect of other currents, a phenomenon which is particularly apparent in coastal areas.

The turbulence set up by tidal currents and their changing directions has indirect effects that are biologically important because they hinder the settlement of small particles thus causing increased cloudiness and preventing the formation of thermal stratification in the water.

In the lower reaches of tidal rivers there may be a shortening of the period of tidal rise and a lengthening of the period of fall; this phenomenon may be so marked that the fast tidal rise produces what is known as a bore.

The rise and fall of the water surface is naturally of the greatest importance in coastal areas where, depending upon the morphology of the coast, more or less extensive areas alternately dry out or become covered with water. Short-term fluctuations in temperature, insolation, salinity and degree of desiccation are associated with this change in water covering, giving rise to areas with extremely varied ecological conditions. Here one finds a fauna and flora poor in species but often rich in individuals which have special physiological and ecological adaptations. A distinct zonation can be recognized depending upon the period of flooding.

Further generalizations are not really possible because the configuration of the coast and the type of bottom have substantial effects on the populations, and furthermore the tidal ranges (the difference between the extremes of high and low tides) are extraordinarily variable. In the European Mediterranean Sea and in the Baltic Sea the tides are scarcely noticeable, but on the French Channel coast the tidal range may be more than 12 metres and in the Bay of Fundy ranges even exceeding 15 metres have been recorded (see pp. 78 and 79 for changes in water pressure associated with fluctuations in water level).

9. *Waves*

Waves appear at the boundary surface between two media of different density when these are moving at different speeds. The most obvious are the surface waves which are due to the action of wind. An analysis of the direct effects of surface waves on biological processes is still scarcely possible at the present time. Movement of the surface increases the uptake of oxygen by the water and it alters the reflection of the incident light so that changes occur in underwater illumination. Waves are associated with turbulence which, according to the circumstances, may move pelagic organisms down into deeper water. When water movements reach the bottom in areas of shallow sea the upper layers of the sea floor will be moved about and rearranged so that the habitats of benthonic animals suffer changes. Waves running up the coast often throw large numbers of organisms on to the beach, where they die; on steep coasts the extent of the spray zone is dependent upon the height of the waves running up and breaking.

Internal waves are not recognizable at the surface; they arise at the boundary between moving water layers of different densities. According to Krauss (1958) internal waves on the Norwegian coast can rise several hundred metres; as a result of this and depending on the direction in which the waves are propagated the sea bottom will be affected by movements whose direction and speed are naturally of immediate biological importance. Furthermore it has been found that storm centres can also give rise to internal waves, whose migration speed may be greater than that of the centre itself so that they may serve to indicate approaching weather conditions at distant locations. Schmidt (1958) developed a working hypothesis, according to which the fluctuating catch in the saithe fishery off Norway, said to be due to the fishes' 'sensitivity to weather', was in fact caused by the arrival of internal waves propagated by storms.

The term seiches is given to standing waves in which the water particles stop at a point and return along the way they have come; seiches occur particularly along coasts and are recognized by the changes they cause in water level. Sewell (1948) working in the Bay of Bengal observed periodic fluctuations in the salinity which were associated with changes in the numbers of planktonic organisms and the occasional occurrence of certain species which were normally lacking. 'This I attribute to a mixing of the water of the lower stratum with that of the surface layer by wave action when the lower stratum approaches sufficiently near to the surface to be affected. It thus seems possible that variations in the number of organisms present at any given level in any area may be due, not to active migration of the organisms themselves, but to purely passive movement brought about by changes in the level or the character of the stratum of water in which they are living' (l.c., p. 363).

10. *Hydrostatic pressure*

A column of water 10 metres high and 1 square centimetre in area exerts a pressure of one atmosphere, so that the pressure increases with depth and in the greatest depths reaches values of more than 1,000 atmospheres/cm². It was at one time thought that animals could not exist under such large pressures. Since water and therefore the fluids of the body can only be compressed to a small extent, organisms lacking gas-filled cavities scarcely suffer any damage when they move from deep water to the surface. The position is very different, however, in animals with gas-filled swimbladders, in which the lumen may change considerably with changes of pressure; they would be irreparably damaged if brought up from depths of a few hundred metres.

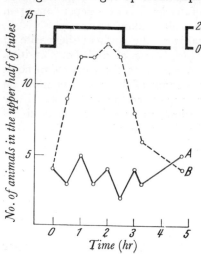

Figure 38. Number of zoea and megalopa larvae in the upper half of the experimental tubes. A = control, B = under increased pressure, equivalent to 20 m depth, for 2½ hours. Readings taken every half hour (see text) (after Hardy & Bainbridge 1951)

Relatively little information on the importance of pressure can be gleaned from the bathymetric behaviour of animals, because some species are tied to fairly restricted depths, whereas others are eurybathic, occupying a large depth range; in the latter, individual animals have, of course, the ability to adapt themselves to certain depths. At the same time, however, it is known that during their daily vertical migrations some fishes will move in the course of a few hours 400 metres up or down and some crustaceans 200 metres; that is, they will tolerate a pressure difference of 40 and 20 atmospheres respectively.

Earlier experimental investigations (summarized in Cattell 1936 and Schlieper 1963) were in general concerned with the use of very high pressures on littoral animals and these produced far-reaching effects which were frequently lethal. More recent observations (e.g. Hardy & Bainbridge 1951, Knight-Jones & Quasim 1955, 1957) have shown reactions to pressure changes, sometimes even with minimal changes, in both pelagic and benthonic animals (Fig. 38).

The physiology of pressure perception in animals without gas-filled cavities is so far unknown. Here perhaps there are two points which may help. At constant temperature and salinity the density of sea water increases with increased pressure; the same should also apply to the body fluids. Furthermore, pressure increases the dissociation constant of CO_2 and thus changes the solubility of calcium carbonate. This would also affect the metabolic processes. It would seem certain that even in animals without a hydrostatic gas organ pressure plays a not unimportant role, whether by directly influencing the physiology of the organisms or by changing other environmental factors such as density or the dissociation of CO_2. It is also possible to visualize a change in the protein molecules which might disturb such special functions as enzyme action or the permeability of membranes.

11. Salts

From the physical aspect the salinity of sea water is biologically important because it affects the density of the medium and thus the ability of the water to support floating and swimming organisms; it is also the main component determining osmotic pressure of the medium. The salts dissolved in sea water are also chemical factors which may be considered in two groups, according to their relative abundance.

The principal dissolved salts account for over 99% of the salinity (Tables 9 and 10). Their relative proportions to the total salinity are constant in all parts of the sea; estuarine brackish waters receiving a periodic influx of fresh water are exceptional.*

* The constant composition of the salts in sea waters eases the task of determining the salinity of a sea-water sample, because it is sufficient to determine one of the main

TABLE 9

Concentration of elements in sea water with salinity 33·34‰
(from Cox 1959)

	in mg/kg = p.p. million		in mg/ton = p.p. 1,000 mill.
Chlorine	18,980	Arsenic	10–20
Sodium	10,561	Iron	0–20
Magnesium	1,272	Manganese	0–10
Sulphur	884	Copper	0–10
Calcium	400	Zinc	5
Potassium	380	Lead	4
Bromine	65	Selenium	4
Carbon	28	Caesium	2
Strontium	13	Uranium	1·5
Boron	4·6	Molybdenum	0·5
Silicon	0–4·0	Thorium	0·5
Fluorine	1·4	Cerium	0·4
Nitrogen	0–0·7	Silver	0·3
Aluminium	0·5	Vanadium	0·3
Rubidium	0·2	Lanthanium	0·3
Lithium	0·1	Yttrium	0·3
Phosphorus	0–0·1	Nickel	0·1
Barium	0·05	Scandium	0·04
Iodine	0·05	Mercury	0·03
		Gold	0·006
		Radium	0·0000002

TABLE 10

Percentage of anions and cations in the salts of sea water
(from Harvey 1955)

Cl′	55·2	Mg··	3·7
SO₄″	7·7	Ca··	1·16
Br′	0·19	K·	1·1
H₃BO₃	0·07	Sr·	0·04
HCO₃+CO₃	0·35	Trace elements	0·02–0·03
Na·	30·4		

From the biochemical viewpoint the excess of the cations over the anions of strong acids is of great importance in relation to the capacity to form compounds. This gives sea water a weak alkaline reaction ($pH=8·10$ to $8·20$), and also a considerable buffering capacity towards additional acids and bases.

---

components and from this, by the use of constants, to calculate the total salinity. The chloride is determined by titration.

In spite of the constancy in the relative composition of sea water the principal components occur in different proportions in the body fluids of living organisms; in other words these organisms exercise a selective choice from the ions available to them in sea water.

TABLE 11

Molecular composition of sea water and cell sap expressed as % of Cl content

(from Osterhout 1933)

		Cell sap of	
	Sea water	Valonia macrophysa	Halicystis osterhouti
Cl′	100	100	100
Na·	85·87	15·08	92·40
K·	2·15	86·24	1·01
Ca··	2·05	0·285	1·33
Mg··	9·74	traces	2·77
SO₄″	6·26	traces?	traces

Table 11 provides a few examples of this phenomenon, which in some respects indicates the considerable physiological capabilities of the organisms concerned. It was particularly the accumulation of potassium in the cells of marine algae and the relatively small amounts of sodium that led Osterhout to his investigations on *Vallonia macrophysa* and *Halicystis osterhouti*; these were followed, amongst others, by the comparative investigations of Gross & Zeuthen (1948) on *Ditylum* and of Scott & Haywood (1955) on *Valonia* and *Lactuca*. The latter investigation especially showed that cation exchange is a physiological process dependent upon light and temperature. Large amounts of calcium extracted from the sea are deposited in the shells of foraminiferans, lamellibranchs and gastropods, in the skeletons of corals and many other animals, and to some extent also in algae.

We still know relatively little about the physiological significance of the principal salts in sea water and about the specific reactions of different organisms to them.

12. *Trace elements*

The other elements present in sea water account for only about 0·02–0·03% of the salinity. They are therefore only present as traces (Table 12) and consequently are known as trace elements.

In spite of the small amounts of these substances many play an extraordinarily important part in biological processes in the sea, because they are indispensable for biochemical processes. Almost all organisms show an accumulation of several trace elements; a number of instances

TABLE 12

Concentration of trace elements in sea water

Element	mg/m³ from Kalle 1952	mg/m³ from Harvey 1955
Deuterium	—	16,000
Fluorine	1,400	1,400; 1,300
Silicon	1,000	
Nitrogen		
(NH_3, NO_2, NO_3)	1,000	as NH_3–N < 5–50;
		as NO_2–N 0·1–50;
		as NO_3–N 1–600;
		as organic N 30–250
Rubidium	200	200
Aluminium	120	0–7; 27–270; 160–1,800; 1,900
Lithium	70	100
Phosphorus	60	as PO_4–P < 1–60; as organic P 0–16
Barium	54	30–90
Iron	50	3–21; 15–50; 0–30
Iodine	50	50
Arsenic	15	1·6–5; 2·4–3·1; 3–5; 15–35
Copper	5	1–25
Manganese	5	1–10
Zinc	5	14; < 8; 9–21
Lead	< 5	4; 5
Selenium	4	4
Tin	3	3; < 5
Uranium	2	
Caesium	2	2
Chromium	—	1–25
Titanium	—	1; 6–9
Molybdenum	0·7	0·3–0·7; 2; 0·4; 12–16
Gallium	0·5	0·5
Cerium	0·4	0·4
Thorium	0·4	0·01–0·001
Vanadium	0·3	0·2–0·3; 2·4–7
Yttrium	0·3	0·3
Lanthanum	0·3	0·3
Silver	0·3	0·3; 0·15
Antimony	—	0·2
Bismuth	0·2	0·2
Nickel	0·1	0·1; 0·5; 1·5–6
Cobalt	0·1	0·1
Scandium	0·04	0·04
Cadmium	—	0·032–0·057
Mercury	0·03	0·03
Gold	0·004	0·004; 0·008
Radium	0·0000001	0·07–0·58 × 10^{7}
Germanium	—	shown

of this are given in Table 13, which clearly shows the great physiological powers of the organisms in this respect. We are still, however, a long way from having anything approaching a complete picture of the situation in this field.

TABLE 13

Trace-element content in brown algae

(Ratio fresh weight: sea water) (from Black & Mitchell 1952)

Art	Nickel	Molyb-denum	Zinc	Vana-dium	Tita-nium	Chro-mium	Stron-tium
Pelvetia canaliculata	700	8	1,000	100	2,000	300	20
Fucus spiralis	1,000	15	—	300	10,000	300	8
Ascophyllum nodosum	600	14	1,400	100	1,000	500	16
Fucus vesiculosus	900	4	1,100	60	2,000	400	18
Fucus serratus	600	3	600	20	200	100	11
Laminaria digitata							
Thallus	200*	2	400	10	90	200	90
	200*	2	1,000	20	100	200	18
Stipes	300*	3	600	10	200	200	14
	400*	2	900	30	90	200	14

* From two different localities.

The accumulation of trace substances takes place selectively and in some groups of organisms or even in individual species it occurs in a specific way. Thus, in investigations on sponges Bowen & Sutton (1951) found titanium in *Dysidea crawshayi* but not in *D. etheria*. In *Terpios fugax* they found titanium only but in *T. zeteki* manganese, tin and vanadium as well as titanium.

The phenomenon of selective accumulation raises various problems. A number of workers have worked on the mechanism of the uptake of substances, particularly of metals. It appears that in some cases the ions, accumulate in the mucus produced by the animals and with which they are then consumed. Thus in tunicates (*Ciona intestinalis, Ascidia ceratodes*) Goldberg et al. (1951) showed the concentration of vanadium in the mucus produced by the branchial basket and its uptake by the gut. In *Mytilus edulis* Fretter (1953) considered that Sr^{90} was taken up directly by the ciliated epithelium of the gills and also by consumption of the mucous feeding sheets, after Korringa (1952) had suggested that Al, Cu, Fe, Zn, Hg and Mn might accumulate in the mucus of oysters. It would also be worth while investigating whether in such forms as the polychaete *Chaetopterus*, the gastropod *Olivella* and others the capture and filter nets constructed of mucus (see p. 237 et seq.) may also serve in the selective uptake of certain ions. The chemical reactions involved still require elucidation.

In many cases the suspended particles of organic and mineral origin in the seston (see Chapter VI, 1) evidently act as adsorbents for various elements. As a result these elements may be drawn into the biological cycle with the uptake of detritus by filtering organisms.

A second problem is that of the functional importance of the accumulated elements. Iron, copper and vanadium play an essential role in respiratory pigments; manganese is an important constituent of chlorophyll, cobalt forms part of the vitamin B_{12} complex, while calcium, strontium and silicon are very important skeletal substances. In many cases, however, the functional significance of trace substances is still not clear.

A further problem concerns the action and fate of the substances accumulated in certain organisms when these are transferred from one link in the food chain to another. It is possible that the selective accumulation of certain elements plays a definite role in the specialization of predators on certain prey organisms.

Finally it should be mentioned that the active accumulation of elements has great geochemical importance, because these substances are carried to the sediments in dead bodies or with sinking faeces from predators.

It is generally recognized that invertebrates store heavy metals in much greater quantities than vertebrates.

Quantitative estimates of most of the trace elements have been available for some time but they still do not allow us to give any opinion as to whether their accumulation can bring about a shortage, whether in time or space, which might restrict the distribution of certain organisms or groups of organisms. The fluctuations given in Table 12 suggest that something of this sort may occur in the case of certain elements. Thus Ryther & Guillard (1959) considered that the restricted production in the Sargasso Sea was probably dependent upon various metals. From the results of culture experiments (Ryther 1961) it has been sugggested that iron may be a factor in the distribution of oceanic and neritic planktonic algae, to the extent that the neritic algae make higher demands on the available iron than the oceanic algae. It is particularly phosphorus, nitrogen and to a certain extent silicon which have been shown to influence growth. Nitrogen and phosphorus are used by all organisms in the synthesis of protein, while silicon plays an essential part as a skeletal substance, e.g. in diatoms and some protozoans. A deficiency in one of these elements will affect the total biological production, hence particular attention has been paid to them following the work of Brandt at the beginning of the century (cf. also Wattenberg 1957). They have been termed minimal substances, since biological production will be held up if the concentration of one of them falls below a minimal level.

Phosphorus occurs either in organic compounds in living organisms and detritus or dissolved in inorganic form as P_2O_5'' and PO_4''. Organic phosphorus compounds may also be dissolved in the water.

The formation of protein compounds uses up the dissolved inorganic phosphorus, but the phosphorus is again freed from organic combination by the processes of metabolism and remineralization and thus taken back into solution. As a result of the continual alternation of build-up and breakdown, determinations of phosphorus content made on a single water sample have very limited value.

However, from the great number of determinations that have been made it is possible to drawn some conclusions of general validity on the spatial distribution of phosphorus. These are based on PO_4''. In the

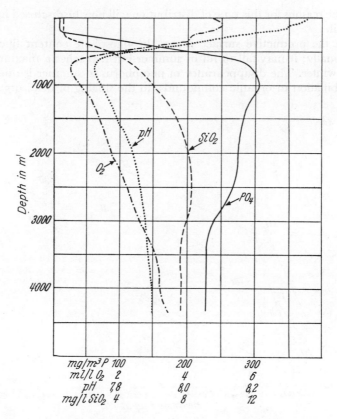

Figure 39 Vertical distribution of PO_4, SiO_2, O_2 and pH at 24°22′N, 145° 33′W (Pacific) (after Graham & Moberg 1944)

D—M.B.

oceanic region the following vertical distribution very frequently occurs:

1. an impoverished surface layer, whose vertical extent will be determined by the thickness of the active photosynthetic layer and by movements due to convection (cf. Graham & Moberg 1944);
2. a transition zone, in which the values for phosphorus increase considerably;
3. a maximal zone which
4. is then followed by a zone of more or less uniform decrease, leading to almost constant values at depths of 2,000 to 4,000 metres (Fig. 39). Considerable variation again occurs in the bottom water.

The reasons for this type of distribution will now be discussed in more detail.

In the productive surface layers the phosphate content fluctuates seasonally; it may fall to nil in summer but it reaches a maximum in late winter. The disappearance of phosphorus in summer is due to its combination in organic compounds, to the sinking of dead organisms

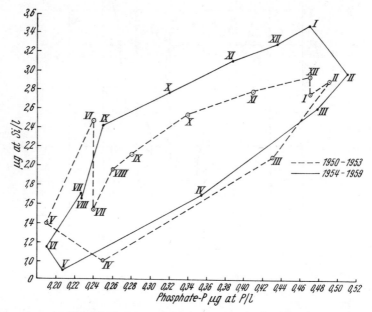

Figure 40. Si/PO₄ graph for 2-year groups in the sea area off Plymouth from the values given by Armstrong 1950–1959 (Original) (I–XII denote the months)

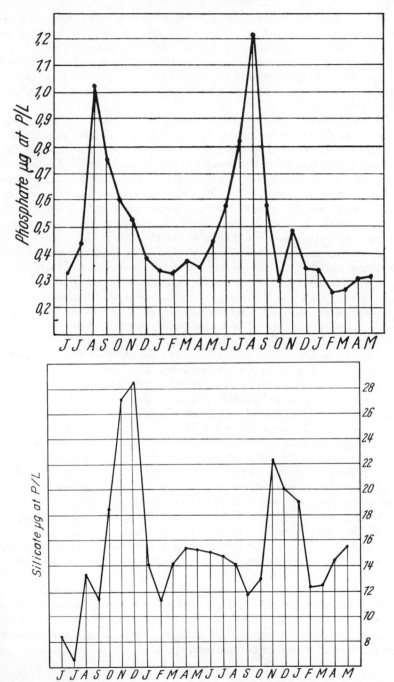

Figure 41. Annual cycle in the content of phosphate and silicate in the coastal waters off Madras (after Thirupad & Reddy 1959)

and faeces to the bottom layers, to the vertical migrations of zooplankton and to the absence of water mixing by convection owing to
the formation of a discontinuity layer. The breakdown of this layer in
spring and the increased turbulence brings about a return of remineralized phosphorus from the deeper layers into the surface waters.
There is therefore a seasonal cycle in the content of phosphorus (Figs.
40, 41 and 52).

Bruce & Hood (1959) found a marked daily cycle in the phosphate
content of the water in various bays in the Gulf of Mexico and thus
confirmed similar findings already made elsewhere. The fact that the
concentration was lower by day and higher at night shows that there is
evidently a correlation between phosphate consumption and photosynthetic activity. The extent to which phosphate is utilized is therefore
dependent upon the density of the population, and the close correlation
between the daily cycle of phosphate and the rhythmic daily fluctuations
in pH and oxygen content is then understandable.

The large-scale regional distribution of phosphate and its relation
to the plankton were first worked out by Hentschel & Wattenberg
(1930) from the results of the Meteor Expedition in the South Atlantic
Ocean (Fig. 6b). These investigations showed that in higher latitudes
the top 50 metres contain significantly larger quantities of phosphate
than they do in lower latitudes. The coast of West Africa is
characterized by particularly rich areas, and north and south of the
equator there are regions with relatively high phosphate values which
extend westwards in the form of tongues. This distribution is due to
exchange between the surface and deep layers; along the coast of West
Africa deep water upwells to the surface and is carried westwards by the
North and South Equatorial Currents; in higher latitudes convection
and the Atlantic deep water reaching the surface bring about an
enrichment in phosphate (see p. 74). The Atlantic deep water flowing
in a north–south direction acquires its high content of phosphate by
the addition of organic matter from the surface layers and by mixing
with the water of the bottom current. Harvey (1955) in fact regarded
the South Atlantic Ocean as a reservoir of plant nutrients.

Surface water which remains a long time without an intake of
phosphate must gradually become impoverished, because any phosphate
left tends to move downwards. This kind of aged surface water is found,
for example, in the Sargasso Sea, where the content of plankton is therefore always low (see p. 76).

Nitrogen is available in the form of three types of inorganic com-
pound, namely ammonia (NH_3), nitrite (NO_2'') and nitrate (NO_3').
In oceanic deep water the inorganic nitrogen is almost exclusively
present as NO_3', whereas the surface layers also have varying
proportions of the other compounds.

This fluctuating composition is dependent on the season and is the result of metabolic processes (Lagarde 1963).

Nitrogen is also present in sea water in dissolved organic form. This component can presumably only become available to other organisms by the action of bacteria. As in the case of phosphorus an annual cycle can be recognized in temperate regions.

Silicon plays an important role as a skeletal substance in diatoms and silicoflagellates. In many areas of the sea the diatoms show very marked seasonal maxima and this may lead to extensive reductions in the content of SiO_2 in the surface layers, so that less than 10 mg Si/m^3 has been found. Below the photosynthetic layer remineralization takes place, leading to marked concentrations of silicon. The vertical distribution of silicon shows similarities to that of phosphorus, but because mineralization is slower the maximal zone is less strongly marked and lies at a greater depth (see also Fig. 39).

The seasonal curve for silicon content at any one place is also similar to that for phosphorus.

In Figure 40 an attempt has been made to represent the annual cycle of phosphate and silicon in the form of a Si/P graph. This is based on the observations made during the period 1950–1959 by Armstrong in the sea area off Plymouth. The monthly averages of two periods each of four years have been plotted. Both curves are surprisingly similar, and from this one can conclude that in this area the annual cycle of biological and dynamic processes is of the same type taken over a period of several years. From data collected over a number of years one could in this way plot a standard curve; comparison with the curves of individual years would then indicate the fluctuations that take place from year to year. In areas in which, for example, the silicate consumers (mainly diatoms) are less abundant than in the sea off Plymouth the form of the corresponding curve would be different.

On the other hand the biological cycle may be overshadowed by hydrodynamic events. In the coastal waters off Madras, Thirupad & Reddy (1959) found silicon maxima coinciding with the minima for salinity and phosphate (Fig. 41 a and b). The high silicate content was caused by the inflow of fresh water rich in silicate during the rainy season, whereas the phosphate maxima were brought about by upwelling of deep water brought about by offshore winds.

13. *Oxygen content*

Living processes require energy, and for this oxygen is necessary (except in the case of anaerobic organisms), because the energy is produced by the oxidation of organic matter. Ever since the start of modern marine research considerable attention has therefore been devoted to the content and distribution of oxygen in sea water; in the early days, of course, it was thought that the depths of the oceans were unpopulated, owing to the supposed complete absence of oxygen. Numerous investigations have now shown that there is a sufficiently high content of oxygen for the existence of animals, even at great deapths; exceptions are the Black Sea (see pp. 75 and 96), parts of the Baltic Sea and, at certain times, fjords in deeply indented coastal areas.

The oxygen dissolved in the water is taken up at the surface from the atmosphere until saturation is reached; the capacity of the water to absorb oxygen depends upon temperature and salinity. These relationships and their importance in the assessment of biological processes have been discussed by Kinne (1962 a and b, 1963 a) (Fig. 42). An increase in temperature and/or salinity causes a decrease in the saturation value for oxygen: at 0° and 0‰ the oxygen content is 12–13 ml O_2/l, at 15° and 0‰ 7·2 ml O_2/l, at 15° and 30‰ 6 ml O_2/l, and at 15° and 55‰ only 5 ml O_2/l. Fluctuations in salinity and temperature therefore not only influence osmoregulatory activity and the speed of metabolic processes, but they also cause changes in basic respiration. Reactions to either salinity or temperature or to both changing at the same time need not therefore be caused by these factors themselves, but may in certain circumstances be due to a change in oxygen content caused by them; Kinne has shown the effect of this

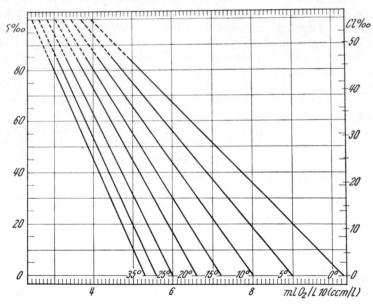

Figure 42. Relation of oxygen saturation to temperature and salinity (after Kinne 1962, 1963)

on the speed of development of the very euryhaline fish *Cyprinodon macularius*. It remains to examine in detail the extent to which the living processes investigated are in fact dependent upon the partial oxygen pressure in the medium.

Water may become supersaturated by the absorption of oxygen

TABLE 14

Formation of the oxygen minimum at three stations in the Pacific Ocean

(from Plunkett & Rakestraw 1955)

Station	Date	Depth	O_2 (ml/L) content
110.70	22.3.1953	0	5·53
28°40′ N		104	5·17
118°20′ W		381	0·72
		470	0·43
		566	0·33
		759	0·32
		955	0·52
		1,147	0·74
		1,487	1·10
T–P54	8.9.1953	0	5·67
40°34′ N		10	5·34
170°2′ E		273	4·94
		366	4·56
		371	4·54
		464	3·70
		469	3·68
		753	1·35
		758	1·37
		2,020	1·87
		3,340	3·40
T–P 99	26.10.1953	0	4·58
31°55′ N		24	4·56
142°12′ E		153	4·41
		257	4·61
		406	4·21
		506	3·96
		607	3·65
		803	2·24
		1,190	0·94
		1,995	1·76
		2,504	2·34
		3,220	3·03
		4,177	3·44
		5,254	3·67
		6,290	3·71

formed during plant photosynthesis (up to 311% according to Kuhl 1952). This phenomenon, of course, only occurs at the surface of pools where there is little movement of the water; its physiological and ecological significance have not been analysed in detail.

The vertical diffusion of oxygen from the surface into deep water is a very slow process, and since oxygen is being continually consumed at all depths by oxidative processes it follows that the movement or circulation of water must be of supreme importance in providing the deeper layers with oxygen. In areas where a bar or sill prevents circulation of the deeper water either completely or for certain periods of time, a prolonged or sometimes complete disappearance of oxygen may occur. Under these conditions anaerobic decomposition, especially by bacteria, may take place and this leads to the formation of hydrogen sulphide and thus to poisoning of the water. In the Black Sea, for instance, oxygen decreases sharply with depth and reaches zero at about 200 metres; from there downwards the content of hydrogen sulphide increases with depth (Fig. 37).

In sediments too, the necessary circulation is often lacking below

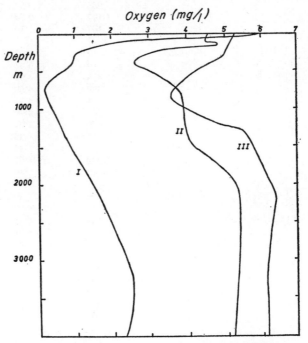

Figure 43. Vertical distribution of oxygen; I, south of California; II, in eastern part of South Atlantic; III, in the Gulf Stream (after Sverdrup 1938)

the oxygen-containing top layer and this may lead to the formation of hydrogen sulphide. This process is dependent upon the type of sediment, on its content of dead organic matter and on the rate of deposition. On bottoms in shallow water a few centimetres of black coloration in the sediment is in many places a characteristic sign of the absence of oxygen. On bottoms of this kind the only animals which can penetrate deep are those which are able to obtain their oxygen direct from the surface of the sediment (see Chapter V, 6).

In lower latitudes, between approximately 20°N and 20°S, the curves for oxygen content show a striking minimum at about 400–500 metres, i.e. below the actively assimilating upper layer; below this the oxygen content rises again and then decreases once more at greater depths (Table 14 and Fig. 43). The poverty of the intermediate layer must be

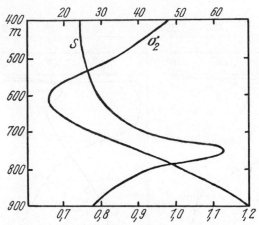

Figure 44. Oxygen minimu (O²) and particle mazimum (S) in medium depths (400–900m) from mean values at 13 stations in 0°–15°N in the Pacific (after Jerlov 1953)

due to the oxidative processes (respiration of animals and hetero-trophic plants, enzymatic decomposition) outweighing any intake of oxygen by lateral exchange (Fig. 44).

In certain areas the surface layers also show a lower oxygen content than would be expected from the temperature and salinity. This is due to the upsurge to the surface of deep water poor in oxygen.

From the biological viewpoint it is important to know whether this intermediate oxygen minimum has a restricting effect on the distribution of living organisms. Physiological investigations carried out over a long period have shown that the respiration of most marine animals is to a large extent independent of the partial pressure of oxygen and is not reduced until the oxygen content is reduced to a relatively low level (in bacteria for example not until it is below 0·34 ml O_2/litre according to Zobell 1941). In swimming animals, of course, there is probably a relation between the intensity of movement and dependence on oxygen pressure. In general, however, there have been no observations that the oxygen-poor layer influences either the regional distribution or the population density, and even the vertical migrations of planktonic animals are apparently not influenced by gradients in oxygen content.

This, however, does not apply completely in extreme conditions, such as have been found in the Arabian Sea. Here Vinogradov & Voronina (1962) found that the plankton biomass decreased as the oxygen content fell, but that a secondary maximum of plankton occurred in the zone of minimal oxygen (Fig. 45). These authors seek to explain

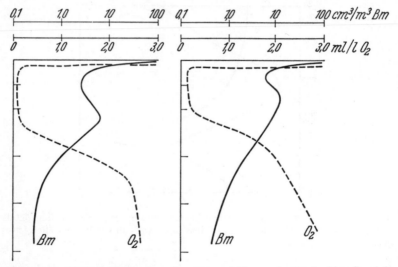

Figure 45. Vertical distribution of oxygen and biomass in the Arabian Sea (after Vinogradov & Voronina 1962)

this by postulating that some pelagic invertebrates, particularly copepods, are capable of acquiring energy anaerobically and can thrive here owing to the rich supply of food resulting from decomposition and also to the absence of enemies.

Confirmation of these phenomena may be expected from the Indian Ocean Expedition. In addition physiological experiments must elucidate the extent to which the life cycle of pelagic copepods, including their reproduction and development, can be anaerobic.

14. *Carbon dioxide and pH*

As in the case of oxygen, the content and distribution of carbon dioxide in sea water depends on absorption at the surface and on water exchanges, and also on the activities of living organisms, because CO_2 is removed from the medium in the process of assimilation and returned to it by respiration, that is by the oxidation of organic matter. Intensive assimilation in the euphotic zone can lead to a marked reduction in the carbon dioxide content and conversely zones with a high rate of oxidation will show an accumulation of carbon dioxide.

Carbon dioxide, however, is not only dissolved physically in the water but is also present in chemically bound form, as undissociated carbonic acid H_2CO_3 and also as carbonate (CO_3'') and bicarbonate (HCO_3') ions. According to the law of mass action there is an equilibrium between the free carbon dioxide, the hydrogen ions and the ions with combined carbon dioxide, thus:

(1)
$$\frac{[H \cdot] \cdot [HCO_3']}{[H_2CO_3]} = K_1$$

(2)
$$\frac{[H \cdot] \cdot [CO_3'']}{[HCO_3']} = K_2$$

A change in the content of carbonic acid due to the removal or addition of carbon dioxide only results in a relatively small change in the content of hydrogen ions; this means that sea water is well buffered, its pH range normally being from 7·0 to 8·5. Atkins (1923 a) found seasonal fluctuations of pH which were correlated with the development of the phytoplankton.

In the vertical scale a carbon dioxide maximum or a pH minimum corresponds to an oxygen minimum.

The dissociation of carbonic acid is largely determined by salinity, temperature and pressure. This means that the solubility of the biologically and geochemically important calcium carbonate will also be affected; the surface layers of the sea are supersaturated with calcium, and more so in the warm regions than in the cold. The solubility of calcium increases with depth, because the dissociation of carbonic acid increases under the hydrostatic pressure and thus a shift of pH to the acid side takes place. From the biological viewpoint this means that calcium is precipitated more easily in warm regions than in cold and in the surface layers than in deep water. Hence one can understand the regional distribution of some groups of living organisms and the fact that calcium skeletons are often poorly developed in deep-sea forms (see Chapter V, 8).

Hitherto no one has investigated the way in which the greater dissociation of carbonic acid may perhaps influence the respiration of living organisms or whether the physiologically important calcium economy is affected by the differing solubility of calcium in deep water.

15. *Other gases*

In addition to oxygen and carbon dioxide, nitrogen and rare gases such as helium and neon are also absorbed by sea water. Whether these have a biological role cannot yet be said; the possibility cannot be excluded in the case of nitrogen because nitrogen-fixing and nitrogen-producing bacteria and Myxophyceae have been found in the sea. Their influence is evidently small because it has not yet been possible to demonstrate any biological effects due to the content of gaseous nitrogen (Lagarde 1963).

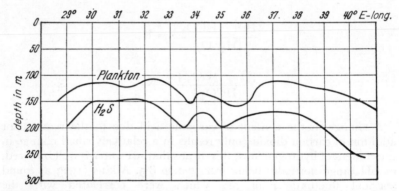

Figure 46. Lower limit of plankton and depth of the 0·5 cc $H_2S/1$ layer in a longitudinal section through the Black Sea at 43°N (after Neumann 1944)

Mention has already been made of the poisonous hydrogen sulphide (Fig. 46). This is formed by facultative or obligatory anaerobic bacteria during the decomposition of organic matter in the absence of oxygen; the bacteria use the oxygen combined in sulphates for their oxidation processes, so that the sulphate is reduced to H_2S.

16. *Polymers*

There is a further characteristic of water which has not so far been regarded as an ecological factor, and indeed its possible biological significance cannot yet be assessed. T. C. Barnes (1932 et seq.) reported that in water the proportions of the polymers mono-, di- and trihydrol vary according to the temperature. He found, for example, that in *Spirogyra* growth was stimulated when there was a greater proportion of trihydrol in the medium, that is, in water which had shortly before been melted from ice and brought up to the temperature of the experiment. Dunbar (1951 a and b) returned to this finding and suggested that their was a difference in the relative proportions of the polymers in water of polar and Atlantic origin. He considered it likely that the great productivity of the areas in which both water types mix could be traced back to the relative amounts of the different polymers. It is natural to extend these ideas to regions of upwelling (see p. 75); but this needs further verification, particularly on an experimental basis.

According to Friedman (1957) the surface waters of the Pacific off Hawaii contain a greater proportion of dihydrol than is present in rain and in the water of rivers and lakes and in that melted from ice and

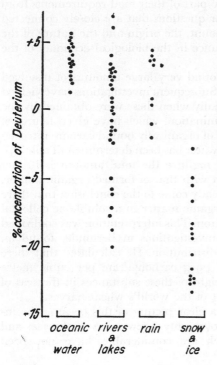

Figure 47. Percentage concentration of deuterium (D/H) in different water types from Hawaii (after Friedman 1957)

snow in a mountain area (Fig. 47). In view of the possible biological significance of such polymers it would be very valuable to have comparative determinations from different regions.

C. Biotic factors

Biotic factors are those produced by organisms and which have an influence on living processes in the sea; in many respects they are not easy to distinguish from the abiotic factors, because they are themselves dependent upon them. It seems to be more practical to retain such a division, but one should bear in mind their interdependence.

1. *Dissolved organic matter*

Discussion of this factor leads us deep into a general discussion of biological phenomena in the sea because the amount of dissolved organic matter is largely dependent upon the physiological activities of the living organisms. One problem of organic matter in sea water was raised by the theory of Pütter (1908 et seq.), who postulated that aquatic animals obtained a large part of their food requirements from dissolved matter. This raises four questions that are closely connected with each other, namely the amount, the origin and the nature of the dissolved matter and its importance in the biological economy of the sea.

Pütter believed that he had found very large amounts of dissolved carbon compounds in the water. Subsequent investigations have yielded significantly lower values, especially when these were obtained by the use of specific methods of determination, which have given figures of only a few mg/litre. The content of organically bound carbon, nitrogen and phosphorus in ultra-filtered water has been determined (Table 15). In spite of the small values in mg/litre the total amount is in fact surprisingly large in comparison with that of formed organic matter. Gaarder & Gran (1927) had already come to the conclusion that there was five to six times as much organic matter in a soluble or colloidal form as was present in rich plankton. This interpretation was confirmed by Krogh (1933) who, in his investigations in Bermuda, found an extensive and uniform vertical distribution. He calculated that there were about 1·5 g protein and 3·9 g carbohydrate per cubic metre and estimated that the total weight of these substances in the seas of the world was 20,000 times that of the world's wheat harvest.

We are still far from having a clear picture of this because the investigations so far undertaken cover only a few areas of the sea, and the seasonal fluctuations, which are considerable, have not been properly assessed.

Naturally this dissolved organic matter can only be derived from organisms. Broadly speaking there are a number of possible sources. There is a continual influx from the land (or from fresh water) which must be particularly important in coastal areas. The decomposition of dead organisms leads by autolysis to the formation of soluble sub-

TABLE 15

Examples of dissolved carbon, nitrogen and phosphorus in organic combination

Author	Area	mg C/l	mg N/l	mg P/l
Krogh 1934	Atlantic Ocean (Bermuda)	2·35	0·244	
Robinson & Wirth 1934	Pacific		0·09–0·32	
Dazko 1937	Black Sea	2·4		
Redfield et al. 1937	Golf of Maine 1–10 m in 5 Mon.			0·000–0·016*
v. Brand & Rakestraw 1941			0·105–0·239	
Kay 1954	Baltic	2·0–4·6		
Plunkett & Rakestraw 1955	Pacific	0·6–2·7		
Skopintsev 1955†	North Atlantic	1·04–1·97		
Harvey 1955	English Channel			0·002–0·008
Ketchum et al. 1955	Subtrop. Atlantic			0·002–0·009
Skopintsev 1959†	Atlantic	2·40–2·48	0·24–0·26	0·021–0·001
	Pacific	0·98–2·68	0·07–0·11	
	Greenland Sea	2·00–2·1	0·03–0·38	0·009–0·029
Duursma 1960	North Sea 0–10 m	1·03–1·38	0·11–0·28	
	North Atlantic 5–50 m	0·23–1·31	0·00–0·37 (0·61?)	

* Expressed as mg PO_4/m^3, calculated by division by 3000.
† Quoted from Duursma 1960.

stances; in this context it is indeed remarkable that fishes and other marine animals have a large percentage of free amino acids in their musculature, which passes into solution relatively quickly after death. A great variety of organic substances present in sediments (see p. 116), including many amino acids, are transported into the open water by turbulence. Dissolved matter is also derived from animal excrement and urine; here it is significant that, for example, copepods living with a plentiful supply of food only utilize a relatively small proportion of what they actually eat; this was shown by Fuller (1937), amongst others, by an analysis of the faecal pellets of *Calanus*. It is known that bacteria, yeasts and flagellates release enzymes, vitamins and organic metabolic products into the medium. Kreps (1934) observed the catalytic breakdown of organic substances in sea water that had been rendered germ-free. He considered it probable that this was due to the action of enzymes released from dead organisms.

Kain & Fogg (1958–1960) found that bacteria-free cultures of *Asterionella japonica*, *Isochrysis galbana* and *Prorocentrum micans* showed little or no growth; they suspected that organic growth factors such as cobalamin or thiamin were lacking. Pütter (1924) had already realized that planktonic algae release about 95% of their assimilates into the water.

This subject has been investigated on several occasions, using algal cultures. In bacteria-free cultures of *Carteria* and *Chlamydomonas*, Braarud & Føyn (1930) found that 'about 30% of the oxidizable organic matter which the algae had formed by their photosynthesis during the experimental period (the excess production) was passed into the liquid' (l.c., p. 20). The release of assimilates had also been confirmed in other culture experiments, e.g. in Myxophyceae by Fogg (1952), in *Chlamydomonas* by M. Allen (1956), in *Chlorella* by Roberg (1930) and by Tolbert & Zill (1956). In diatoms, on the other hand, the release of assimilates has not been established. In this context the littoral benthonic algae deserve a thorough investigation. Ericson & Lewis (1953) and Starr *et al.* (1957) were able to confirm the presence of substances with vitamin B_{12} activity in the culture liquids of certain marine bacteria; it was also found that the intracellular levels of vitamin B_{12} were significantly higher than the extracellular, indicating the part played by the bacteria in the production of these

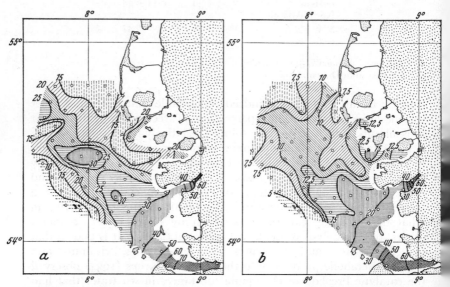

Figure 48. Distribution of (a) yellow substance ('Gelbstoff'), and (b) fluorescence at 0 m in the inner German Bight in August 1954 (after Kinne 1956)

substances. It is also worthy of note that Siepmann & Höhnk (1962) found that marine yeasts release riboflavin into the culture medium.

It is still not possible to provide a final assessment of the importance of this original source of dissolved organic matter in the sea. It should be mentioned that by determination of the amount of Gelbstoff (yellow substance) and of the fluorescence of dissolved organic matter Kalle (1956) was able, for example, clearly to distinguish the influence of the freshwater inflow from the River Elbe into the North Sea (see Fig. 48).

Kalle regarded both the Gelbstoff and the fluorescent substances as extremely stable metabolic products of living organisms or as decomposition products of organic matter, whose chemical nature is not very clear. The presence of free enzymes in sea water has been shown by several authors since H. W. Harvey (1925); carbohydrates have been demonstrated on many occasions. By adsorption on carbon and extraction with acetone or benzene Johnston (1955) found the following substances in North Sea water:

(*a*) an insoluble white substance,
(*b*) coloured material, possibly carotenoid,
(*c*) fairly large amounts of brownish waxy or fatty matter,
(*d*) yellowish orange matter fluorescing bright blue in ultraviolet light.

Water-soluble vitamins such as the B_{12} complex have been demonstrated repeatedly (e.g. Cowey 1956, Daisley & Fisher 1958). The summaries by Vallentyne (1957) and Duursma (1960, 1961) of the literature on organic substances in sea water have been used, with additions, in the preparation of Table 16.

The use of modern analytical methods, particularly paper chromatography, will doubtless provide more detailed knowledge in this field (thus, for example, B. M. Allen (1956) found non-volatile organic acids in *Chlamydomonas* cultures).

It will be of great biological interest to undertake comparable analyses in bodies of water containing a large concentration of a single planktonic plant, because under such conditions there have been several reports of far-reaching effects on interspecific relationships (cf. Chapter VI, 7), which have been traced back to the presence of poisonous substances.

At this preliminary stage in the investigations little can be said about the role of dissolved organic matter in biological processes. The selective absorption of light with a maximum in the blue has been mentioned. There is no doubt that dissolved substances, often adsorbed on suspended particles, represent an important source of nutrient for heterotrophic bacteria, yeasts and fungi. It has been shown experiment-

TABLE 16

Constituents of the dissolved organic matter

(from various authors)

Author	Substance
Belser	Uracyl, Isoleucine, Glycine
Burkholder	B_{12}
Creac'h 1955	Citric and malic acids
Droop et al.	B_{12}-Analogues
Kalle	Chlorophyll
Kishniae & Riley	B 12, Thiamin
Lucas 1947	Carotenoid
Slowey et al.	Fatty acids: lauric, myristic, palmitic, stearic, myristoleic, palmitoleic, linoleic
Tadashiro Koyama	Acetic, formic, glycollic and lactic acids
Wangersky 1952	Rhamnoside, Vitamin C

ally that some flagellates and diatoms take up vitamins or growth factors from the medium (H. W. Harvey 1939, Provasoli & Pintner 1953, Droop 1954, Lewin 1954, Braarud 1961), hence it can be concluded that the content of such organic matter in sea water is a prerequisite for the success of these organisms: 'it is a reasonable assumption that if an organism requires a growth factor *in vitro*, then this metabolite or its physiological equivalent should be found in significant amount in the environment' (Provasoli & Pintner 1953, p. 845). It is still an open question whether any specific biological significance can be attributed to the blue fluorescence, which is very widespread (see p. 56).

Broadly speaking, substances which may promote or inhibit biological processes can be termed biotics and antibiotics; in a long series of papers Lucas (1936–1955) has referred to them as external metabolites and has drawn attention to their role in the interspecific relationships of planktonic organisms. This subject will be discussed in more detail in Chapter VI, 7. The investigations of Wilson (1948–1953) on the settlement of the pelagic larvae of benthonic animals point to the action of certain substances in the substrate. In this connection it should be mentioned that littoral algae produce amounts of antibiotic substances.

It is highly probable that dissolved matter does not supply a source of food for pelagic or benthonic animals. On the other hand Zobell & Grant (1943; *cit.* Zobell 1946) reckon that about 70% of the organic matter dissolved in sea water is remineralized by the activity of bacteria, of which about 30% is transferred into cell substance and is therefore once more available as food.

2. *Seston (suspended matter)*

In addition to the salts and dissolved organic matter sea water also contains suspended particles which collectively form the seston. The amount of seston is determined gravimetrically or optically by measurement of the extinction or of the Tyndall effect. By these methods one can even take into account colloidal particles of the order of size of 2 μm.

TABLE 17

Quantitative relationship between inorganic detritus (sand grains 2-300μ) and flagellates in various parts of the Mediterranean (from Bernard 1959)

Mean number per ml		Ratio
Sand grains	Flagellates	Sand grains : Flagellates
0·52	660	0·0008
76	314	0·24
831	464	1·8
982	326	3·01
1,004	383	2·7
1,428	307	4·7
1,700	344	5
2,100	326	6·4
27,400	275	99
73,000	229	320

The seston has two main components, namely the living plankton and the mass of non-living suspended matter—the tripton or detritus. This second group contains two types of substances: the dead organisms (and fragments) and the inorganic constituents such as clay particles and sand grains. Little is known about the relative proportions of the different components in the seston, mainly because the relevant investigations have mostly been restricted to sea areas near the coast. In the vicinity of the coast the inorganic constituents, derived from the land or by turbulence from the sea floor, are more abundant than in the open sea; the ratio between living plankton and detritus fluctuates with the stages of development of the plankton.

As an environmental factor the seston is biologically important in several different ways. Mention has already been made of the absorption of light by suspended matter, and this is not only quantitative because there is a maximum in the blue and ultraviolet region. The organic constituents of the seston represent the sole source of food for many pelagic animals. Some components of the seston serve as a substrate for micro-organisms. The vital processes of micro-organisms and zooplankton and also free oxidation processes cause a noticeable consumption of oxygen in a high proportion of the organic components that lies below the euphotic layer. Krey (1959a, p. 79), for example, states: 'although there may be differences in many details, it is a general rule that in the surface layers there is an unmistakable correlation between high seston values and periodically very low oxygen values'. On the other hand a high seston content with a predominantly inorganic component has no effect on the content of oxygen.

According to Armstrong & Atkins (1950) the dissolved silicon taken up by diatoms is partly restored by solution from suspended silicate compounds.

There is another way in which a high seston content may hinder organisms mechanically. Thus, Bernard (1959) found that after winds blowing off the desert the seston content in the eastern Mediterranean was increased by the addition of sand grains of 2–30 µm and that this reduced the production of flagellates (Fig. 17); in the North Sea, herring avoid areas with dense concentrations of the colony-forming flagellate *Phaeocystis*; it appears that one of the main reasons for this is that the flagellates clog the gills.

These few examples will show the importance of the seston as an environmental factor and we may now turn to some of the other problems involved. On the basis of material from the Swedish Deep-sea Expedition on the *Albatross*, Jerlov (1953) drew attention to certain interesting points, of which only two will be mentioned here. In the deep sea a high content of seston is often present in the water layers close to the bottom, and this is presumably the result of the transporting action of currents. This may account for the presence and distribution of certain forms of life in deep sea areas (see Chapter V, 8). At one station (No. 128) a seston maximum due to living and dead plankton was found at a depth of 150 metres. The upper 75 metres showed supersaturation with oxygen and low phosphate values; from this one concludes that shortly before there had been a rich outburst of plankton, which subsequently gave rise to a seston maximum in deeper water.

Using water samples from depths of 300, 500 and 1000 metres in the Sargasso Sea, Yentsch & Ryther (1959) found that, on passing about 10 litres through a filter with a pore size of about 0·80 µm, there was a partial stoppage of the filter and a visible coloration appeared. The quantity of this particulate matter was not ascertained, but absorption spectra from acetone extracts gave maxima between 400 and 500 nµ, although the absorption peaks between 665 and 670 nµ due to chlorophyll and so characteristic of surface plankton were absent. If it is assumed that the formed matter consists of dead and decaying phytoplankton, then the pigments present were evidently carotenoids. These are coloured and absorb particularly in the blue, and thus contribute to the reduction of the light.

As regards the vertical distribution of the seston, maximal accumulations have frequently been observed in relatively thin layers; these are the so-called turbidity layers (cf. Krey 1954). By the accumulation of sinking particles these form a density discontinuity layer (see p. 104 and Fig. 49) which will be secondarily augmented by the concentration of particle-feeders. It is also possible that the phenomenon of scattering layers, the sound-reflecting intermediate layers in the

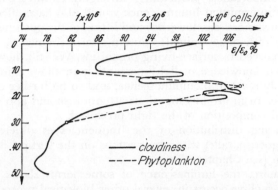

Figure 49. Correlation between number of phytoplankton cells and cloudiness at a discontinuity layer off the island of Hokkaido (after Koblenz-Mishke 1960)

open ocean (see pp. 336). can be traced back to such collections of seston, although it is more likely that the reflection is caused by the animal consumers rather than by the seston itself.

3. *Bioluminescence*

The light-producing ability of numerous marine organisms, particularly those in the pelagial, has long been known and research on this subject has been summarized, for example, by Hardy (1952). Any assessment of the biological significance of bioluminescence must be sought in the behaviour of the individual species—that is, in their autecology (see p. 169). This synecological factor of biological origin, which may possibly have a more general significance could not be properly assessed until quantitative measurements with a very sensitive bathyphotometer (Clarke & Hubbard 1959, Clarke & Breslau 1959) provided data on the frequency and intensity of the light produced. At depths of 100 metres south-east of New York these authors found maximum frequencies at night of more than 160 light flashes per minute while higher and lower values were recorded at other depths in the same locality and also in the Mediterranean. Each individual flash lasted from 0·2 to 1·0 second, but by the overlapping of flashes the period of 'sustained luminescence' could extend to 10 seconds. There was great variation in the intensity; in the Mediterranean the maximal values recorded corresponded with the intensity of the natural light present at depths of 350 metres.

If one reckons the average duration of a single flash as 0·6 second, with a frequency of 100 per minute and a uniform distribution, one will obtain a sustained luminescence which must be of considerable ecological importance, particularly at night. However, there is no doubt that intermittent luminescence would also have effects, especially as disturbances (e.g. lowering and lifting of the photometer at sea, movements of a single animal) cause a rise in frequency, which can even be observed in surface-living organisms. We still, however, lack any detailed knowledge of light perception in those animals which may possibly react to bioluminescence, and so no further conclusions are possible. In future investigations attention should also be paid to the spectral composition of the light produced.

The varying distribution of the frequencies at different depths provide a good parallel with observations on the vertical distribution of plankton (see Chapter VI, 2).

Furthermore, the luminescence of some forms shows a diurnal periodicity such as occurs in several other biological processes. In the flagellate *Goniaulax polyedra*, according to Hastings (1959), this periodicity persists during constant illumination and also in periods of fluctuating light and darkness. Backus *et al.* (1961) found that in May the bioluminescence at the surface off Woods Hole was 700 times greater at night than during the day, but in November only 100 times greater.

Kalle (1960) analysed the very striking luminescence which suddenly appears and then disappears, in order to find out how it is caused; here the light produced is very bright and shows 'changes that are almost mathematical, with a precision that cannot be influenced by any external phenomena'. He suspected that these were caused by subsidiary phenomena associated with local sea-quakes, the organisms being stimulated by seismic waves to produce luminescence.

4. *Bio-acoustics*

It has been known for a long time that some aquatic animals produce sounds. The extent to which this occurs in marine animals was not however realized until the wartime investigations, particularly by the Americans, in connection with defence against U-boats (Knudsen *et al.* 1948, Griffin 1955, Fish *et al.* 1952). The principal producers of sound are crustaceans, fishes and toothed whales.

Among the crustaceans the genera *Palinurus*, *Crangon* and *Synalpheus* and various crabs have long been known to produce sounds. In most cases stridulation noises are produced by rubbing one part of the body against another, but the species of *Synalpheus* produce snapping sounds by pressing a knob on the moveable joint of the claw into a depression in the fixed joint (Fig. 50).

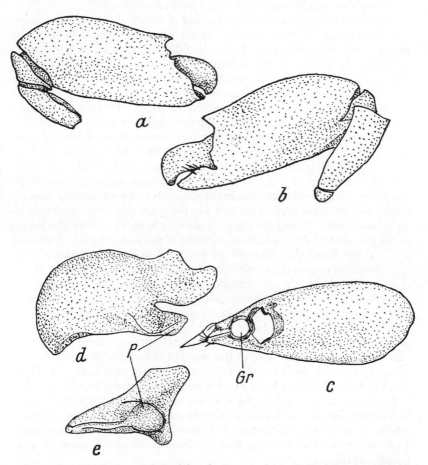

Figure 50. Pistol pincer of *Synalpheus laevimanus*: (a) and (b) from the side; (c), non-movable joint showing the roundish pit (Gr) after removal of the movable joint. The 'pollex' of the movable finger P, in (d) and (e), is pressed into the pit (after Volz 1938)

Sound production among fishes is very widespread and may be produced either by stridulation (rubbing of fin rays and skeletal parts against each other, e.g. in *Balistes* species), by grinding the teeth against each other, particularly those in the pharynx (e.g. *Caranx*, triggerfish) or by resonance produced by the rapid vibration of special muscles associated with the swimbladder (e.g. in Sciaenidae, *Trigla*, *Opsanus*, *Gradus* and others). In the case of sounds produced by stridulation or grinding the swimbladder may serve as a resonator.

There are several ways in which sound production may be of biological significance. It may function as a recognition signal or serve to attract and stimulate during the breeding season (Tavolga 1956), or it may be used as a threat signal which may also serve to mark a territory. Kellogg (1958, 1959) has been able to show convincingly that bottle-nosed dolphins (*Tursiops truncatus*) orientate themselves to obstructions with the help of sound reflections. There is a parallel here to the bats, but in the dolphins the pitch is much lower, and in them we have no knowledge of any directional sound production or reception. It remains an open question whether this orientation plays a part in keeping the animals in schools and in finding prey; it is possible that it functions in both ways. It has been suggested that deep-sea fishes may be able to orientate themselves by producing sounds which are reflected back from the sea floor, on the same principle as an echo-sounder (Marshall 1954). This idea is supported by the range of the sounds and the confirmed presence of resonator organs. Snyehiro *et al.* (1957) found that fishes of the genera *Scomber* and *Trachurus* were very sensitive to sounds, to which they reacted by changing the direction of swimming in the observation tank and making flight movements.

In *Bathygobius soporator*, Tavolga (1956) found that visual, chemical and sound stimuli played a part in sex discriminatory behaviour.

When we have more detailed knowledge on the part played by sound production in the autecology of individual species, sound recordings will become an effective tool in the study of behaviour patterns. In this context one may mention the analyses of sound production in birds and insects which have been carried out so extensively in recent years.

In the meantime we do not know how and to what extent the various individual sounds or the total repertoire of sounds affect those animals which do not come within the framework of species-specific reactions. In other words the synecological significance of sound production is largely unknown.

The same applies to ethology in general, and Bull (1961) has pointed out the importance of this subject in the field of marine biology. However, he holds the view that the ethologist 'should be encouraged to direct his research interests towards the behaviour of animals of major economic importance and not be allowed to divert effort to smaller and perhaps more easily kept and studied species.' (This utilitarian point of view appears rather out of place here, because most of the economically important fishes require larger installations for maintaining them than many others and doubtless only offer a restricted opportunity for gaining new knowledge.)

5. Competition

In a given habitat the members of one species will be affected by the demands of the other species present, if their ecological niches overlap. Where two species make the same demands on the available food, differences in other biological activities will have an effect on the food competition between the two species (see Chapter VI, 7). A good example of this is provided by the work of Kinne (1954, 1956 b) on the amphipods *Gammarus duebeni* and *G. salinus*. Both species have the same kind of physiological demand on abiotic factors, particularly salinity, but one of them, namely *G. duebeni*, is displaced by the other. Under the same conditions *G. salinus* shows a faster rate of growth and reproduction; owing to this *G. duebeni* is displaced into the less suitable marginal areas, where it is able to live because it has a greater physiological tolerance towards salinity, temperature and the oxygen content of the medium.

An analogous example, relating however to the nutrient content of the substrate, has been given by Altevogt (1957) for fiddler-crabs: 'There is an ingenious ecological separation between the species *Uca marionis* and *U. annulipes*, for the latter still inhabits substrates where *U. marionis* no longer lives, but can also thrive in the biotope (which is richer in food) of *U. marionis*.' There is a similar situation in the case of *Dotilla blanfordi* and *Dotilla myctiroides*.

Presumably here too the species with the greater ecological potentiality (*Uca annulipes* or *Dotilla myctiroides*) will be forced to move into less suitable areas by the species with the lower ecological potentiality. The progressive displacement of the north-west European barnacle *Balanus balanoides* by the invading species *Elminius modestus* has been analysed by Crisp (1958) and Barnes (1960), amongst others. *Elminius* not only has a greater reproductive capacity but also has greater powers of ecological adaptation. This enables it to colonize areas in the eulittoral in which *B. balanoides* no longer occurs; from there the species then moves into the settlement area of *Balanus* and displaces it (cf. also Connell 1961 a and b).

On a gastropod shell occupied by a hermit-crab Warburton (1953) found a large colony of each of the hydroids *Hydractinia echinata* and *Podocoryne carnea*. Throughout their whole extent each colony remained at a distance of 8–12 mm from the other. The most probable explanation of the existence of this 'no-man's land' is that the marginal polyps kill each other off to a distance corresponding to the length of their tentacles, so that competition for space leads to a reciprocal exclusion from this zone.

Plaine (1952) working at Woods Hole on the polychaete *Polydora ciliata* which lives commensally in the shells of *Pecten irradians* found two areas with significantly different population densities: in one population in which the lamellibranch *Anomia simplex* was attached to the *Pecten* by calcified byssus threads, *Polydora* occurred in much fewer numbers than in the other in which *Crepidula* had settled on the *Pecten* shells; evidently *Anomia* more or less excludes *Polydora*.

Noctiluca miliaris feeds mainly on diatoms and in calm weather it may show a mass outburst. This causes a simultaneous reduction in the zooplankton, particularly the copepod population; a contributory factor in this is competition for food because diatoms form the principal food supply for the copepods. The result is that the fisheries are adversely affected because plankton-eating fishes such as pilchard and anchovy disappear (Prasad 1958).

Ivlev (1961) has carried out experimental investigations on the ecology of feeding in fishes and has shown that lack of food or hunger is a special ecological factor. The build-up of a population is largely dependent upon the available food, particularly during the early stages of development; the degree to which hunger is tolerated varies considerably between the individual species and it is also dependent upon abiotic factors such as temperature and salinity. In some invertebrates, e.g. prosobranch gastropods, the course of pelagic and non-pelagic development is correlated with the feeding conditions (see Chapter V, 6).

In the course of investigations on the ecology of feeding it must be remembered that accumulated trace substances are also taken up with the food; in some cases these may be of great importance. Medusae, which have a water content much greater than 90%, can scarcely have any nutritional value, expressed in calories. But they are relatively rich in mineral substances, e.g. arsenic and zinc in *Cyanea* (J. & W. Noddack 1940), and this may possibly explain why they are important to those animals that eat them.

Competition occurs not only for food but also for space, light, oxygen, etc., but little is known about these aspects in marine biology. In every case competition acts as a regulator, for example in the relative density of settlement in the different habitats. The predator-prey and host–parasite relationships which are present in every biocoenosis will also have this kind of regulating effect. As was shown in the case of the gammarids cited above, biotic and abiotic factors as well as species-specific physiological tolerance are closely concerned in the co-existence of several species.

Competition is, however, not only an interspecific matter but also takes place between the individuals of a species, since these naturally occupy the same ecological niche. Within an ecological niche each individual requires a certain amount of space for feeding, settlement, respiration, breeding, etc. As the numbers of individuals increase the available space may, in some circumstances, fall below a critical value; this will especially be the case when a biocoenosis is disturbed and one species breeds faster than normally; intraspecific competition will then have a destructive effect (Schäfer 1951). Knight-Jones & Moyse (1961) have given a summary of examples of intra-specific competition among sessile forms.

D. The sea floor as an environment for life

In the biology of the sea, the sea floor plays just as important a role as the open water; in the euphotic regions along the coasts the sea bottom is settled by numerous plants which may in places form dense growths, while animals, of course, live on the sea floor right down to the great deep-sea troughs. In addition there is a considerable chemical interchange between the bottom and the water above it.

The suitability of the sea bottom as a substrate for settlement will depend upon its composition and structure. On these factors will depend whether animals can live only on the surface of the bottom or be able to penetrate into it. The epifauna living on the substrate differs in character from the infauna living within it; the infauna will also vary depending upon the different types of bottom, such as sand or mud (cf. especially Remane 1940) (Chapter V, 6).

The epifauna on hard bottoms (the epilithion) shows considerable variety whereas the infauna (the endolithion) is understandably less diverse. Chalk substrates are inhabited by such organisms as boring sponges (*Cliona*), polychaetes such as those in the genus *Polydora*, boring lamellibranchs of the genera *Saxicava* and *Petricola* and also the echinoid *Strongylocentrotus*. Here the substrate only serves as a protective home while the animals rely on the surrounding water for feeding and respiration. The epifauna contains numerous sessile forms, for which the hard surface acts as a substrate for attachment, e.g. sponges, coelenterates, bryozoans, barnacles, tunicates. In littoral coastal regions hard bottoms provide the habitat, whether preferred or exclusive, of numerous green, red and brown algae; some of these have special attachment organs and form dense growths, as in the species of *Fucus* and *Laminaria*, while others cover large areas with incrustations. Sessile animals, such as hydroids, polychaetes of the genus *Spirorbis*, bryozoans and tunicates often settle on algal thalli. Sediment accumulates in among the epifauna and also in clefts and provides a habitat for a rich soft-bottom fauna.

Chalk cliffs and rocks are generally less densely settled than granite or sandstone; here it is probable that mechanical rather than chemical factors are involved. Rocky coasts exposed to strong swell often remain unsettled, except for crevices and sheltered bays.

In the littoral the settlement on hard bottoms shows a very clear zonation which is closely correlated with the period during which the shore is covered with water and with the fluctuations which this causes in salinity, temperature and light (see Chapter V, 2).

On soft bottoms scattered individual lamellibranchs, stones or boulders often provide a substrate for the settlement of representatives of the hard-bottom fauna.

Artificial constructions made of stone, cement and iron and to a certain extent timber are comparable to natural hard bottoms. The settlement occurring in such places leads to the same ecological pheno-mena, but it also involves the problem of damage to the materials by the organisms (see Chapter VIII). Protective paints may be damaged mechanically and thus rendered ineffective; growths on ships' bottoms reduce the capacity and speed of the vessel; corrosion of timber, iron and cement by micro-organisms is fairly well known, and in particular a number of timber-destroying fungi have been found (Höhnk 1955 b, Meyers & Renolds 1960). In the case of timber, the shipworms (*Teredo*) and gribbles (*Limnoria* species) not only use the wood as a home but also as a food substrate; here there are cases of symbiosis analogous to those found in timber-feeding insects (cf. the summary of Ray 1959, 'Marine boring and fouling organisms'). Fungi of the genus *Myce-lites* attack the spines of sea-urchins and the teeth of fish; they have also been found in fossil material.

Little is known about the settlement on hard bottoms outside the littoral region, because the usual dredges, bottom samplers and core samplers give disappointing results. We may expect significant results from underwater photography and television and from direct obser-vation by divers. The extent of the hard bottom decreases with dis-tance from the coast as it becomes covered with sediment, but there is much evidence to suggest that even at great depths there are un-covered hard bottoms with a settlement of animals. Laban *et al.* (1963), however, mention that 'in any case it is striking to confirm once more, that on hard substrates at depths in total darkness (within the confines of the abyssal and hadal regions), the amount of sea floor covered by epifauna is extremely small'.

Pure sandy bottoms frequently occur in coastal areas and in shallow seas. They were long thought to be particularly inimical to settlement, because only a small macrofauna is present. Careful investigations on littoral and sublittoral sandy bottoms, particularly by Remane, have now shown that these frequently accommodate a very rich microfauna, which lives in the system of spaces between the sand grains (mesopsam-mon). The size of the spaces between the sand grains is dependent upon the size of the grains themselves and as a result there is in fact a considerable difference between the populations of coarse and fine bottoms; the latter are often so closely packed that they contain practically no system of spaces and thus no mesofauna; they are then very hard and the infauna is also poorly developed.

The composition of the microfauna living in the system of spaces is very varied; Delamare Deboutteville (1960) has published an extensive account of this habitat. The general characteristics of this interstitial fauna include small body size, elongated body form, reduction or absence of pigment formation, development of attachment organs, and reduction of the number of eggs and larvae. The geographical distribution of interstitial organisms is surprisingly wide, as is already apparent from observations on a few species, such as *Protohydra leuckarti* (according to Wieser 1958); this will certainly be confirmed in the case of other species after more detailed investigations in more extensive areas. The connecting links between the infauna of marine sandy bottoms and the habitats in terrestrial underground waters and cave waters are of great importance and interest (cf. Thienemann 1950, Delamare Deboutteville 1960).

In addition to the microscopic interstitial fauna sandy bottoms have a macroscopic infauna, in which some forms merely lie in the sand, such as *Branchiostoma* and various lamellibranchs, while others burrow and are more or less mobile. Among these are the polychaetes *Nephthys*, *Glycera* and *Ophelia*, the gastropod *Natica* and in the upper tidal zone beetles of the genus *Dyschirius*. There are also tube-building forms such as the echinoid *Echinocardium*.

The animals living in sand rely for food partly on settled detritus

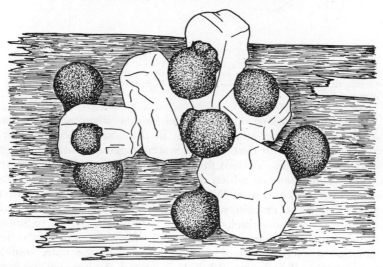

Figure 51. The fungus *Arenariomyces cinctus* on sand grains from eulittoral coastal sand (after Höhnk 1958a)

or particles in the water and partly on an autonomous source, since in shallow waters with sufficient light a flora of diatoms and peridinians develops which acts as a primary source of food. This is grazed off the surface of sand grains by small forms, while some larger animals such as polychaetes of the family Opheliidae consume the sand grains together with the food particles that are attached to them. Höhnk (1955 a) has also shown specific sand-swelling micro-fungi (Fig. 51), and there are probably also yeasts; these too provide an autochthonous source of food for sand-dwelling animals.

Physical and chemical reactions take place between the bottom and the waters above, particularly in places where muddy sediments occur; investigations in this field go back quite a time particularly for fresh waters (among others Ohle 1935, 1938, Einsele 1936, Bruce & Hood 1959), while corresponding work for the sea has been carried out in shallow inshore waters.

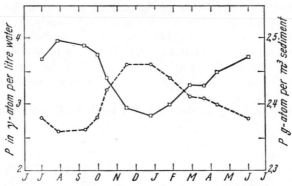

Figure 52. Annual fluctuations in the phosphorus content of the surface layer of the sediment (– – –) and in the ——water in Hurricane Harbour, Florida (after Miller 1952)

Miller (1952) working on the coast of Florida (Biscayne Bay) found that the curve for phosphate content in the bottom showed an inverse relationship with that in open water, suggesting that exchange takes place (Fig. 52). Various factors are involved in the mechanism of this exchange, among which the most important appear to be the content of oxygen and iron. When the oxygen content of the bottom layers is high, as occurs with a full circulation of water, insoluble ferriphosphate is formed, which will then accumulate in the sediment during the winter. If the oxygen content decreases owing to poor water circulation the ferriphosphate will be reduced to the more soluble ferrophosphate; if hydrogen sulphide is formed the iron will become combined as iron sulphide and the phosphate ions will be released.

Rochford (1951) and Jitts (1959) have investigated the absorption and desorption of phosphate in the muddy sediments or river estuaries; here too they find that oxygen plays a decisive role. Jitts (l.c., p. 20) states: 'It follows then that under average conditions the silt would act continuously as a trap for phosphate at the expense of the overlying water, particularly during the fresh run-off cycle when highly oxygenated waters with low salinity are bringing down large quantities of terrestrial phosphate and silt'. This is shown diagrammatically in Fig. 53.

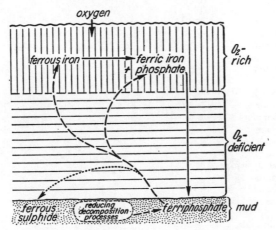

Figure 53. Diagram showing the processes involved in the exchange of phosphate between the sea floor and the open water (drawn by K. Schmidt)

The cycle of biologically important trace substances such as phosphate is therefore not only dependent upon the relationship between the formation and breakdown of organic matter and the vertical exchange of water but also on the chemistry of the sea floor. Observations have not yet been made on this kind of process at great depths in areas far from the coasts; they would certainly yield important information on the large-scale cycle of various trace substances; they would also be important from the geochemical point of view.

Analyses of organic matter in the sea floor are also urgently needed. The demonstration of free amino acids, fatty acids and carbohydrate (Plunkett 1957, see Table 18) in sediments and rocks may possibly bring into prominence the question of the nutrition of microbes and sediment-feeding animals—a subject which requires extensive experimental study. Table 18 summarizes the organic compounds found by Plunkett (1957) in an anaerobic and an aerobic sediment in the Caribbean Sea; quantitative determinations would be particularly valuable

TABLE 18

Organic substances in the sediment of the Cariaco Trench in the Caribbean Sea

(from Plunkett 1957)

| | in the anaerobic area | | in the neighbouring aerobic area | |
| | 10°37·5' N | 10°46·5' N | 12°36' N | 11°27' N |
	65°24' W	64°52' W	58°32' W	61°32' W
Sugars				
Sucrose	+	+	+	+
Glucose	+	+	+	+
Fructose	+	+	+	+
Galactose	−	+	−	+
Arabinose	+	+	−	−
Xylose	+	−	−	−
Amino acids				
Leucine	+	+	+	+
Isoleucine	+	+	−	−
Glycine	+	+	+	+
Phenylalanine	+	+	+	+
Valine	+	+	+	−
Aspartic acid	+	+	+	+
Arginine	−	+	−	−
Tyrosine	−	−	−	−
Glutamic acid	−	+	−	+
Organic acids				
Propionic acid	+	+	−	−
Fumaric acid	+	+	−	+
Lactic acid	+	+	−	+
Tartaric acid	+	+	+	+
Citric acid	+	+	+	+
Oxalic acid	+	+	+	−
Glutaric acid	+	+	+	+

for an assessment of the physiological and nutritional importance of these substances.

Biocoenotic, autecological and ethological investigations should also help in assessing the part played by the sea floor in the behaviour of organisms. It may provide: a resting place; a refuge and hiding-place; a spawning site; an ambush where roving predators lie in wait for prey; an attachment substrate for sessile forms which obtain their food by whirling or by catching it with tentacles; food for sediment-feeders. The sea floor also provides a habitat for the infauna and mesofauna, and in some places also a place to feed; in other cases on the other hand food is taken from the water above the bottom. These numerous examples of the dependence of organisms upon the bottom could easily lead to generalizations which could well be erroneous, and it is essential that the relationships involved should be tested for each individual case (pp. 225, 251).

Attention should also be drawn to the fact that the bottom does not remain uninfluenced by the organisms that settle there. Tube-dwelling animals suck in water for respiration; the oxygen content diffuses into the surrounding sediment and this allows oxidation to take place even below the surface. Sediment-feeders such as the lugworms (*Arenicola*) move considerable amounts of sediment from anaerobic zones up to the surface and thus, like the burrowing forms, bring about a rearrangement. Tube-building forms such as many polychaetes, molluscs and crustaceans line their tubes with mucus, so that when the population is dense a soft bottom may be bound together. Calcareous rocks and coral reefs are penetrated by boring organisms belonging to various groups (algae, sponges, polychaetes, lamellibranchs) so that they succumb more quickly to the mechanical action of the waves (e.g. Bertram 1936). These effects on the structure of the bottom brought about by organisms are also of importance to geologists interpreting fossil sea floors.

To a considerable extent the material of the bottom sediment is largely supplied directly by organisms; extensive areas of the oceanic deep-sea floor are covered by the 'shells' of pteropods, foraminiferans (*Globigerina*) or diatoms. The material of coral reefs, including the encrusting calcareous algae, becomes broke up to form coral sand.

The great variety of relationships between organisms and the sea floor makes it necessary to consider this complex in greater detail both from the autecological and the synecological points of view.

CHAPTER IV

PLANTS AND ANIMALS OF THE PELAGIAL

1. Classification

Various systems of classification and subdivision have been proposed in an attempt to obtain in a general picture of the great variety of pelagic life or to bring together groups with common characters.

The collective term nekton is used for those animals which by powerful swimming movements are largely independent of currents and waves. Among these are the toothed and baleen whales, seals, sea-lions and fur-seals, many fishes, large cephalopods and a few crustaceans. Many of these undertake extensive migrations between their breeding grounds and principal feeding areas.

In contrast to the nekton, plankton is a collective term used to cover all those organisms whose powers of movement are insufficient to prevent them being moved about by currents and waves; plankton means 'that which drifts'. The plankton has been classified in a number of different ways.

1. The difference between plant and animal plankton—phytoplankton and zooplankton—is generally accepted and rightly so because phytoplankton is very largely made up of organisms capable of assimilation; these are the primary producers upon which the zooplankton depends for food.

2. Within the zooplankton one can differentiate between the permanent or holoplankton and the temporary or meroplankton. In holoplanktonic animals all the developmental stages belong to the plankton and no bottom-living stages occur. Almost all animal phyla (except sponges, bryozoans and phoronideans) are represented in the holoplankton at least by a few forms. In contrast meroplanktonic animals have stages in their development which are tied to the bottom or to firm substrates, so they do no carry out their full life cycle in the open water (this was termed transitory plankton by Johnstone 1908). In very many marine animals the eggs and larvae live pelagically while metamorphosis leads to a benthonic way of life. This method is by far the commonest;

it occurs for example in many polychaetes, crustaceans, bryozoans, echinoderms, lamellibranchs, gastropods and fishes. Species with alternation of generations frequently have a benthonic and a pelagic generation, particularly in the coelenterates. In other cases benthonic animals lay their eggs in clutches or they protect the eggs, and only the larvae are planktonic. In a few cases only the eggs are present on the bottom, while the larvae and adults belong to the pelagial, as for example in the herring.

The role played by the open water as a medium in the life of such meroplanktonic animals is therefore extraordinarily variable and has to be analysed for each individual case. In addition the behaviour varies within a group or even within an individual species according to depth or geographical position, that is, it is determined by ecological factors (see Chapters IV, 6, and V, 6).

3. Hentschel (1939) introduced the terms kinetic (for metazoans, protozoans and peridinians) and akinetic plankton (for diatoms). He analysed plankton samples into four main groups, metazoans, protozoans, peridinians and diatoms, and arranged the samples according to the increasing numbers of the peridinians, plotted these values in order on the abscissa and the corresponding values for the three other groups on the ordinate. The curves for the protozoans and metazoans were similar to the peridinians in so far as they ran more or less as a straight line and increased with increasing values. From this, Hentschel concluded that these groups showed similar ecological behaviour, which is expressed in the number of individuals, whereas no correlation appeared from the curves for diatoms. It certainly seems reasonable to doubt whether the different numerical relationships of the two groups are in some way correlated with active movement or its absence, as the terms imply.

4. The size of the organisms is reflected in the following classification:

megaloplankton	> 10 mm	} only animals
macroplankton	1–10 mm	
mesoplankton	0·5–1 mm	larger diatoms, small zooplankton, many larvae
microplankton	>60 μm	} most phytoplankton
nannoplankton	>5 μm	
ultraplankton	>5 μm	bacteria, small flagellates

From the viewpoint of biological processes these much-used subdivisions are only significant in so far as one can expect high productivity from abundant micro- and nannoplankton, because the majority of planktonic plants belong to these orders of size. When there is a rich mesoplankton one must first differentiate between the zoo- and phytoplankton components before drawing further conclusions. A rich population of megalo- and macroplankton presupposes the rich development, either simultaneously or in the very recent past, of the groups of lesser size order, which serve as a food source. This system of classification is of broad practical use in considering problems of biological productivity, but the extent to which the different groups function as producers or consumers will depend upon their species-specific composition.

The terms net, filter and 'centrifuge' plankton refer to the method of capture, but the mesh size of the nets, the pore diameter of the filters and the speed and duration of centrifugation have not been standardized. It is therefore not always possible to make comparisons between the quantitive results of different investigations.

The following terms, based on the body form of planktonic organisms, have been coined and used particularly in the older literature: chaetoplankton (with marked flotation processes), discoplankton (disc-shaped), physoplankton (more or less bladder-like), rhabdoplankton (elongated, rod-shaped forms). This system, based on the characterization of form, does not in itself express very much, although the form of an organism may in some cases be important.

7. Haeckel (1890 and 1892) used the terms monotonous, prevalent, polymictic and pantomictic plankton to express the relative composition of plankton according to the species. In monotonous plankton about 75% of the individuals belong to a single species, and in prevalent plankton about 50%; in polymictic plankton the appearance of the samples will be determined by several species, while in pantomictic plankton no single species is prominent by reason of the number of its individuals. The species that predominate in monotonous and prevalent plankton are known as characteristic species, and in some cases whole plankton communities may be named after them.

8. The following terms have been used in connection with the spatial distribution of planktonic organisms:

(a) Haeckel (1890) used the terms superficial or pelagic for the surface plankton, and abyssal or bathybic for the plankton of the deep sea. Zonaric plankton is tied to a certain horizon while pamplanktonic or euryplethar organisms occur in all water layers.

(b) Fowler (1898): epiplankton, 0–100 fathoms depth, mesoplankton, 100 fathoms above the bottom, hypoplankton, the lowermost 100 fathoms.

(c) Lo Bianco (1903) based his classification on light and differentiated:
Phaeo- = light plankton, 0 to about 30 m

Knepho- = shade or twilight plankton, 30 to about 500 m
Skoto- = darkness plankton, which occurs in the greatest depths.
Haecker (1908) designated the zone below 1,500 m as the 'night'
zone with nyktoplankton.
Lo Bianco's classification corresponds roughly to the subdivision of the pelagial into euphotic, mesophotic and aphotic zones given on p. 44.

(d) The terms neritic and oceanic plankton refer respectively to the area of shallow seas and the region of open sea far from the coasts.

9. Steuer (1933) used the term allogenetic for those forms which were occasionally introduced from the main oceans into the adjacent seas, where after some time they eventually disappeared. In my opinion the meaning of this term could be enlarged so that it is not only restricted to the occasional occurrences of oceanic forms in adjacent seas but could also be used for oceanic forms in the neritic, for neritic forms in the oceanic and for the sterile distribution areas of planktonic species. Correspondingly one could also speak of allogenetic benthos.

The multiplicity of terms and the different ways in which they have been defined make it difficult to study the older literature and also some of the modern work. It would therefore be desirable to have a nomenclature that was in common use. In the present book I propose for the vertical distribution to use a classification which has been evolved in recent years and which takes account of the results of the expeditions carried out during recent decades. It has the following divisions:

Epipelagial   = photic zone.
Mesopelagial   = uppermost aphotic zone, whose lower limit coincides approximately with the 10° isotherm and therefore varies between about 100 and 700 m. The 10° isotherm is chosen because this is the surface isotherm at the limits of the polar regions.
Bathypelagial   = middle aphotic zone, which extends between the 10° and 4° isotherms; the lower limit varies in the great oceans according to the differing depth of the isotherms.
Abyssopelagial = deep aphotic zone, which extends to depths of 6,000 m.
Hadopelagial   = region of the deep-sea troughs.

As in any classification of a continuous medium this system can only provide certain guiding principles; it is still not possible to classify the distribution and functions of the organisms diagrammatically, even though one can recognize certain important biological phenomena within the individual zones.

2. Phytoplankton

Plant plankton is of great importance not only in the biology of the pelagial but also of the benthal, particularly in areas distant from the coasts, because it provides the principal source of primary nutrition. It is composed of various groups, mainly of unicellular organisms.

Diatoms play a significant role in all parts of the sea, and they are dominant in the temperate and cold regions. The diatoms are characterized by having a bipartite siliceous shell, in which one part (the lid or epitheca) fits over the other part (the hypotheca) as in a box. The shells are characteristic of the species in their general form and also in the arrangement of their numerous pores and slits; they surround the protoplasts, the exchange of matter with the environment taking place through the perforations. Chromatophores with yellowish, green or brown pigments are present singly or in numbers. In many cases threadlike processes are developed; a number of species form more or less extensive colonies.

The reproduction of diatoms is of considerable importance in ecological processes owing to the speed at which they multiply by the continuous division of each cell into two. This is, however, accompanied by a simultaneous reduction in the average size of the individual cells. When a diatom divides the smaller hypotheca becomes the epitheca of one daughter cell; it follows that after three divisions only one individual of the original size will remain, in addition to three whose size will be determined by that of the original hypotheca and two even smaller size groups one with three and one with a single individual. This means that an increase in the number of individuals is not accompanied by a proportionate increase in the biomass. In the production of food, however, it is the mass that is important and it is this aspect of reproduction in diatoms that is of ecological significance.

The reduction in size does not, however, go on indefinitely; once a certain minimal size is reached there is either a rejuvenation and restoration of the original size by auxopore formation or the cells may gradually degenerate and die off.

Many diatoms store assimilates in the form of oil instead of carbohydrates, and this must be regarded as important in biological metabolism. This content of fat in the primary producers is largely responsible for the accumulation of fat in the consumers. At the same time the presence of fat droplets provides the diatoms with a degree of buoyancy.

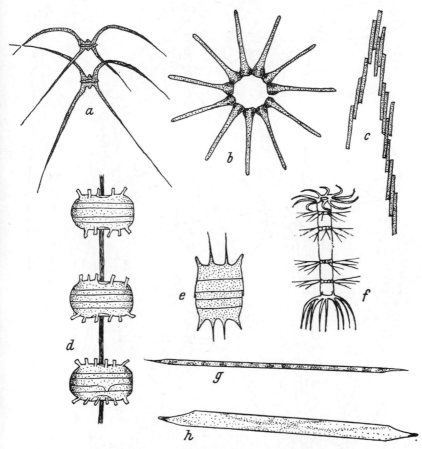

Figure 54.   Some cell forms among pelagic diatoms (based on Tregouboff & Rose 1957) 2, (a) *Chaetoceras didymus* var. *anglica*, (b) *Asterionella japonica*, (c) *Bacillaria paradoxa*, (d) *Thalassiosira nordenskioldi*, (e) *Biddulphia mobiliensis*, (f) *Bacteriastrum elegans*, (g) and (h) two species of *Rhizosolenia*

Some of the main types of diatoms are shown in Fig. 54.

As producers the dinoflagellates (Fig. 55) take second place to the diatoms, both as regards the number of species and their average abundance. Only some of them have chromatophores and are capable of photosynthesis, others live as heterotrophs and therefore rely on preformed organic matter; of these some feed saprophytically on dead material, others are parasitic. The Ellobiopsidae are often considered to be parasitic dinoflagellates (but see Boschma 1949); these organisms

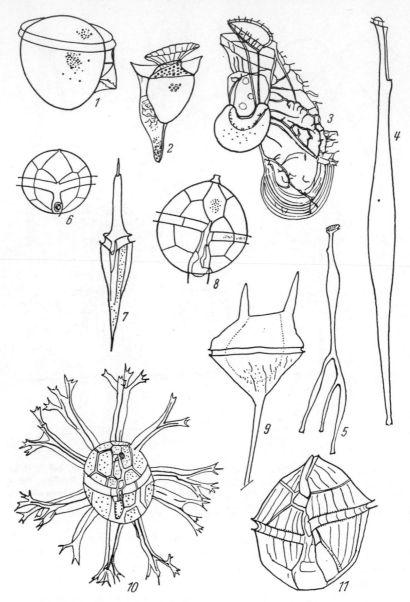

Figure 55.   Various forms of dinoflagellates (after Baleck 1962): 1, *Phalocroma ovum*; 2, *Paraphistioneis para*; 3, *Histioneis dolon*; 4, *Amphisolenia schroederi*; 5, *A. thrinax*; 6, *Diplosalopsis sphaerica*; 7, *Oxytoxum subulatum*; 8, *Peridinium subpyriforme*; 9, *Certium pentagonum subrobustum*; 10, *Cladopyxis brachiolata*; 11, *Goniaulax fragilis*

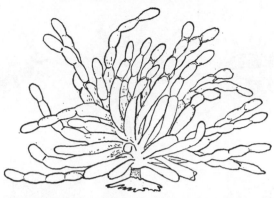

Figure 56. *Amallocystis racemosus* from the abdomen of *Pasiphaea tarda* (after Dahl 1951)

have mainly been observed in crustaceans in which they produce bushy processes which hang from the host to a length of several millimetres (Fig. 56). Infection of copepods by other parasitic forms may be very serious and of considerable significance to the populations, because in most cases, according to Jepps (1939), the infected animals are castrated. Nothing is known about the ethology of parasitic dinoflagellates. Other dinoflagellates are important because they undergo mass multiplication which may seriously affect the other inhabitants of the area.

Dinoflagellates prefer the warmer parts of the sea, although a number also occur in temperate latitudes. In contrast to the diatoms they can move by means of two flagella and are provided with a cellulose membrane (except in those that are naked); they are extraordinarily diverse in form (Fig. 55).

The Coccolithophoridae are very small autotrophic organisms in which the protoplast is enclosed by small disc-shaped or spiny calcareous platelets (coccoliths) (Fig. 57). They are known particularly from the warmer seas but little work has been done on the group as a whole (according to Bernard & Lecal 1960 the identification of fine structure in the hard parts requires the use of an electron microscope). In spite of their small size they play a considerable part in the food economy in some areas, in so far as can be judged from their occurrence in the gut of filter-feeders. In preserved plankton samples the hard parts often dissolve quite quickly and the soft parts are no longer identifiable. Coccolithophores may occasionally show mass outbreaks (according to Gran 1912, 5–6 million per litre in Oslo Fjord).

Coccolithus fragilis, a species that prefers lower temperatures, has been found in great numbers (7,000–240,000 cells per litre) at depths down

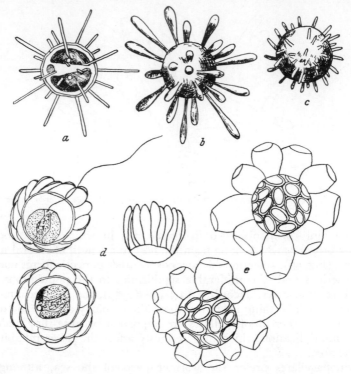

Figure 57. Some coccolithophores from the open sea (after Lohmann 1912): (a) *Rhabdosphaera stylifer,* (b) *R. claviger,* (c) *R. hispida,* (d) *Deutschlandia anthos*

to 4,000 m by Bernard (1953). He therefore regards them as 'une source de vie et de nourriture essentielle dans les couches profondes des mers chaudes et tempérées du globe' (l.c., p. 46). The presence of these small cells is not in itself proof of their importance as food, because their mass is small and it has not been confirmed whether the individuals found in deep water are living, growing and reproducing.

The coccoliths are important micro-fossils (Fig. 58).

In contrast to the coccolithophores the equally small silicoflagellates, which have delicate siliceous shells, do not play any great role as producers.

A special position is occupied by the members of the genus *Phaeocystis.* These are naked, brown flagellates which are principally neritic and may be of particular importance owing to the fact that they form jelly-like colonies during mass outbreaks. Areas with dense concentrations of *Phaeocystis* are avoided, for example, by herring (cf. Wimpenny (1938) and see Chapter VI, 7).

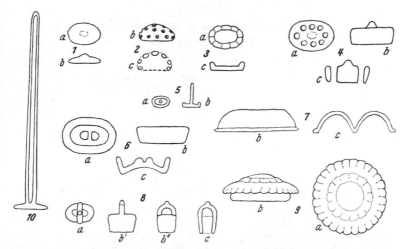

Figure 58. Coccoliths of: 1, *Acanthoica rubus*; 2, *Calyptrosphaera tholifera*; 3, *Pontosphaera huxleyi*; 4, *Syracosphaera dalmatica*; 5, *S. molischi*; 6, *S. binodata*; 7, *S. pulchra*; 8, *Zygosphaera wettsteini*; 9, *Tergestiella adriatica*; 10, *Ruginiaster longistylis*: (a) = from above, (b) = side view, (c) = vertical longitudinal section (after Kampter 1941).

There are several other naked flagellates which sometimes occur in great numbers. They preserve very badly and so it is scarcely possible to estimate the part they play in the production of organic matter; Gross (1937) regarded them as of considerable importance. Many of the remains in plankton samples that are regarded as flagellates are probably swarming spores of protistans, sperms, etc.; these can certainly be regarded as biomass but have no function as producers.

Green algae are only represented by very few species in the marine plankton, the most important being *Halosphaera viridis*. In *Meringosphaera*, according to Schiller (1916), the shell wall and its bristles are distinctly siliceous.

The Myxophyceae or blue-green algae are commoner in the marine benthal than in the pelagial but it seems that their ecological importance has been little investigated. The species *Trichodesmium erythraeum* which is characterized by a red pigment, sometimes occurs in enormous, quantities in the Red Sea, and is in fact responsible for the name given to this area of sea. The 'olive-green cells' mentioned by Hentschel (1936) are probably *Microcystis* and other members of this group; Bernard & Lecal (1960) mentioned a *Nostoc* (3–5 μm long) as the dominant form in some samples from the Indian Ocean.

The olive-green cells mentioned above must be heterotrophic, like other Cyanophyceae, since they have been found at depths with no light. From his investigations Hentschel (1936) reached certain far-

TABLE 19

Vertical distribution of 'olive-green cells'

(from Hentschel 1936)

1. Depth in m	2. Individuals per litre	3. % of the total Nannoplankton
0	4·5	0·1
50	1·8	0·1
100	31	0·7
200	108	31
400	95	59
700	46	69
1,000	34	74
2,000	19·6	75
3,000	11·8	73
4,000	11·1	73
5,000	8·6	58

reaching conclusions. Table 19 gives the mean numbers of individuals found by him and their percentage in the total nannoplankton at different depths in the Atlantic Ocean. In relation to the other components of the nannoplankton the olive-green cells are particularly abundant in deeper water, and therefore occupy an important place in Hentschel's order of precedence based on numbers. From this Hentschel concluded (l.c., p. 212) that 'the distribution of olive-green cells in the deep sea . . . is evidently of fundamental importance for the understanding of oceanic life in general. This life is only based in part on the food provided by autotrophs; the extraordinary importance of dissolved organic food, which is incessantly carried from the photic to the aphotic water layers, can no longer be contested.'

Any assessment of the part played by a group of organisms in the production of food in a given volume of water depends less on their relative numbers than on their absolute mass and the speed at which they multiply. Hentschel's highest values were 108 individuals per litre. Schiller (1931) and more particularly Hasle (1959) found significantly larger numbers. If one reckons on a diameter of 4 μm it would require about 16 million cells per litre to produce 1 cubic millimetre of cell substance. The actual biomass present is therefore very small, and we have no indication of the speed of reproduction. This means that any assessment of the actual nutritional role of the olive-green cells must still be regarded as provisional.

The same can be said of the other heterotrophs such as bacteria, yeasts and fungi, to which increased attention has been paid in recent years (e.g. Zobell (1946), Kriss & Novoshiloff (1954), Kriss et al. (1960), Höhnk (1958)); there are still very few data on the absolute

numbers and relative amounts of these organisms. Thus, Kriss *et al.*
(1960) found the following proportions in cultures from the tropical
Indian Ocean:

  700 non-sporogenous motile and immotile rods
  200 polymorphic cells, probably mycobacteria
  100 cocci
   40 sporogenous rods
   11 yeasts
    5 actinomycetes
    3 fungi

It is also unknown whether the forms found actually exist free in
the water and independent of firm substrates, that is as true plankton,
or whether they live attached to other particles so that their occurrence
is tied to special conditions. In many respects the vertical distribution
curves for bacteria in the upper layers resemble those of the true phytoplankton and it is assumed that the bacteria to a great extent exist
epiphytically.

The nutritional value of yeasts, for example, is in itself undisputed,
as has been shown for instance by Kriss & Novoshiloff (1954) in feeding
experiments on the polychaete *Nereis succinea*. But as in the case of the
olive-green cells the amount is small (in individual cases, 1,000 or
more yeast cells per litre) and so it appears questionable whether these
could be of any considerable importance in providing carbohydrates,
fats and proteins for planktonic animals. Their role as distributors of
vitamins in aphotic depths can scarcely be estimated but may well be
more significant. It is also beyond doubt that they play a part in the
removal of dissolved organic matter.

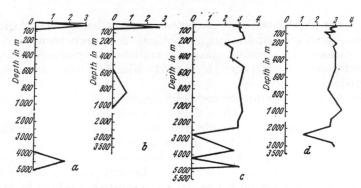

Figure 59. Vertical distribution of heterotrophs in the Indian Ocean (number of
bacteria per 40cc water) (after Kriss *et al.* 1960) Positions: (a) 60–70°S 20°E; (b) 60–
70°S 40°E; (c) 10°–0°S 95–97°E; (d) 10–20°N 95–97°E

From vertical profiles plotted along meridians in the Indian Ocean Kriss *et al.* (1960) established maxima and minima in the bacteria counts (Fig. 59), from the distribution of which they were able to draw conclusions on the route of water exchange between the Antarctic and the tropics. In the absence of any detailed analysis of what is required for the development of different population densities I am not convinced of the validity of these conclusions; we also need to find out whether there are sources of error in the methods of culturing different population densities *in situ* which may be misleading. The same applies to the investigations on micro-organisms living under the ice in the Polar Sea (Fig. 60).

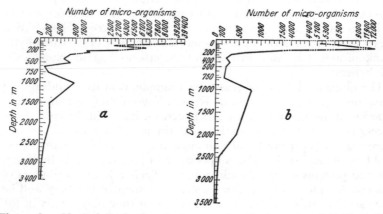

Figure 60. Vertical distribution of micro-organisms in the region of the North Pole based on counts from filters (a) in July, (b) in September (after Kriss 1960).

It is not possible to draw conclusions on the ethology or ecological importance of the parasitic fungi, which have frequently been found in both diatoms and animals, since the available observations are too restricted in time and space. This is a very wide field where research is still required.

Finally, mention must be made of Sargasso weed (*Sargassum natans, S. fluitans* and others) as part of the plant world of the pelagial. These are brown algae which are derived from attached neritic species. They sometimes occur in large masses, as in the Sargasso Sea of the Atlantic Ocean; they reproduce vegetatively and doubtless contribute to the vegetable matter in the pelagial. Ecologically, however, Sargassum weed is important not so much as a source of food but because its drifting plants provide a substrate for the settlement of numerous sessile animals (hydroids, bryozoans, barnacles); in addition motile forms that can attach themselves are able here to live in the open ocean (e.g. certain crustaceans, the Sargassum fish *Histrio histrio* and the pipefish *Syngnathus pelagicus*). There are also fish, even flying-fish, without any special adaptations, that find food, shelter and a place to spawn among this weed. Forms that drift at the surface such as the siphonophores *Physalia* and *Velella* or the gastropod *Ianthina* are frequently associated with this habitat (cf. Parr 1939, Woodcock 1950, Adams 1960). The animals living here are not pelagic in the sense that open water is their sole medium; they may therefore be termed pseudopelagic (see p. 198). So far as I am aware faecal material from macrophagous vegetarians has not been found.

3. Zooplankton

The following account of the animal groups occurring in the plankton is not, of course, intended to be exhaustive. It is hoped, however, that

TABLE 20

Frequency of pelagic tunicates in S.E. Australian waters from plankton hauls 1935-1941

(from Thompson 1948)

Copelata		Desmomyaria	
Oikopleura longicauda	269,912	*Thalia democratica*	1,421,652
Oikopleura fusiformis	57,485	*Ihlea magalhanica*	132,144
Fritillaria pellucida	46,654	*Salpa fusiformis*	25,057
Oikopleura rufescens	45,718	*Traustedtia multitentaculata*	6,531
Oikopleura dioica	15,159	*Brooksia rostrata*	2,765
Megalocercus huxleyi	11,294	*Salpa maxima*	1,561
Stegosoma magnum	9,751	*Salpa cylindrica*	1,184
Oikopleura albicans	8,897	*Iasis zonaria*	914
Oikopleura cophocerca	3,597	*Pegea confoederata*	682
Oikopleura cornutogastra	3,483	*Cyclosalpa pinnata*	236
Fritillaria formica	1,580	*Cyclosalpa bakeri*	110
Oikopleura parva	1,536	*Thetys vagina*	91
Oikopleura intermedia	1,263	*Ritteriella amboinensis*	41
Fritillaria borealis f. sargassi	815	*Metcalfina bexagona*	2
Fritillaria haplostoma	300	*Cyclosalpa affinis*	2
Fritillaria fraudax	211	*Cyclosalpa floridana*	1
Fritillaria megachile	150	*Cyclosalpa virgula*	1
Tectillaria fertilis	143	*Ritteriella picteti*	1
Fritillaria bicornis	100		
Althoffia tumida	4		
Kowalevskaia tenuis	3		
Bathochordaeus charon	1	Doliolida	
		Doliolum denticulatum	112,794
Pyrosomatida		*Dolioletta gegenbauri*	28,863
Pyrosoma atlanticum		Doliolidia	
(colonies)	12,815	(blastozooids)	13,717

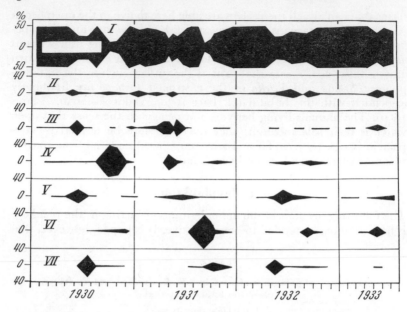

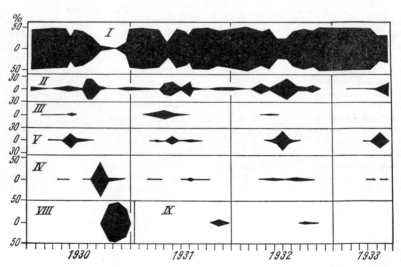

Figure 61. Relative percentages, by count, of the main groups of zooplankton organisms found from January 1930 to July 1933 at stations (a) outside (b) inside, Delaware Bay (after Deevey 1960).
I Copepoda, II bottom larvae, III *Sagitta*, IV Hydromedusae, V Fish eggs, VI Cladocera, VII Tunicata, VIII *Noctiluca*, IX Siphonophora

it will give an overall picture of the varied nature of the fauna and also show that certain animal phyla or classes are completely absent from the pelagial or are only represented by a few forms. Thus, for example, there are no pelagic nematodes, bryozoans or lamellibranchs; apart from their larvae the actinians and echinoderms have very few pelagial species. In addition to the special characteristics of the pelagic fauna it is interesting that the absence of the groups mentioned indicates that this environment provides a set of conditions to which certain animals could not become adapted, or if so only in a few cases.

In many instances our knowledge of the species and their specific and generic status is still insufficient, and many forms have only been described from inadequate material. Although a knowledge of the species should in no way be regarded as an end in itself, nevertheless their correct naming is one of the first and most important prerequisites on which to base further investigations and conclusions. The systematics of the fauna will therefore remain of fundamental importance even in the future, even although physiological, ecological and general biological problems remain of primary interest.

It is remarkable that in catches taken in a single area the species in individual classes or orders are present in very different numbers; frequently one species shows an absolute preponderance while the others are very much scarcer (Table 20). The relative numbers of the different species may also fluctuate during the course of the year (see Chapter VI, 7, Fig. 61). One can therefore only draw reliable conclusions on the plankton population of an area on the basis of long-term investigations with catches taken periodically.

In the pelagial, therefore, the different phyla are represented very unequally as regards the relative numbers of their species. In addition, the relative number of individuals of each group varies considerably, frequently fluctuating in space and time (Tables 21 and 22, Fig. 171).

TABLE 21

Relative frequency of six main groups of pelagic invertebrates in 47 surface hauls in the Pacific
(from Beebe 1926)

Crustacea	1,033	52·4%
Mollusca	367	18·7%
Coelenterata	207	10·5%
Annelida	191	9·7%
Insecta (*Halobates*)	89	4·5%
Urochorda	82	4·2%
	1,969	100·0%

TABLE 22

Percentage of individual animal groups in the zooplankton in the central tropical Pacific
(from Hida & King 1955)

Organisms	% Surface	Deep water
Copepoda	65·3	59·0
Foraminifera	11·7	5·6
Eggs	10·3	10·7
Tunicata	4·3	0·9
Gastropoda	2·7	1·9
Chaetognatha	2·0	2·3
Radiolaria	0·6	5·3
Crustacean Larvae	1·0	1·2
Ostracoda	0·1	7·2
Euphausiacea	0·9	2·6
Siphonophora	0·5	0·4
Amphipoda	0·1	0·4
Various	0·4	2·5

Owing to the wealth of available material it is not possible to give a complete account of the animals present in the plankton, and in some cases one can only deal with whole groups. Attention will be paid primarily to the holopelagic forms, that is to those which spend their whole life cycle in the open water, but the pelagic stages of meropelagic species will not be neglected.

Protozoans

Among the rhizopods, the heliozoans are often found in the plankton, particularly in the neritic region, but the number of individuals collected, and also the number of samples taken, is relatively small; mention may be made of *Heterophrys myriopoda*, *Raphidiophrys marina*, *Acanthocystis aculeata* and *A. pelagica*.

Foraminiferans with siliceous and calcareous shells are found in great numbers, both of species and individuals, in various benthonic habitats, but in the pelagial they are only represented by few forms with calcareous shells. Ellis & Messina (1940) recorded a ratio of 1,200 recent benthonic forms to 26 pelagic species, but more recent work suggests that this ratio is probably even more in favour of the former.

The commonest genus among the pelagic foraminiferans is *Globigerina*, of which the more or less spherical shells are also abundant in oceanic sediments in some areas (Fig. 62). The foraminiferans belong primarily to the warm-water region, but there are also a few species in colder areas. An ecological analysis is scarcely possible because even in the material from the Meteor Expedition (according to Hentschel

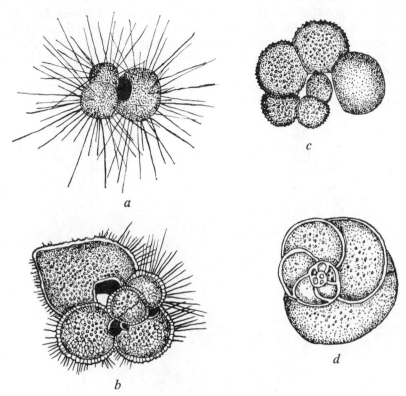

Figure 62. Pelagic Foraminifera (after Tregouboff & Rose 1957); (a) *Globigerina bulloides;* (b) *Globigerinoides sacculifera;* (c) *Globigerinella aequilateralis;* (d) *Tetromphalus bulloides*

1933) *Globigerina* was so rare and its distribution so scattered that it was not possible to discern any regular pattern in its distribution.

In contrast to the foraminiferans, the radiolarians are almost exclusively inhabitants of the marine pelagial; they do not have a calcareous shell, but build their skeleton out of organic substances, silicic acid or in some cases out of strontium sulphate (*Acantharia*). The siliceous skeletons are particularly abundant in oceanic sediments.

The radiolarians occur mainly in the warm-water sphere and principally in the upper water layers; the presence in them of symbiotic algae (zooxanthellae) is noteworthy as these are presumably important in the nutrition of these animals. There are also, however, many deep-water forms, from the distribution of which Haecker worked out a vertical classification of the pelagial.

The diversity and beauty of the radiolarian skeletons have been largely responsible for stimulating systematic investigations. Brandt (1895) considered that the variable vacuoles in the outer plasma have a hydrostatic function enabling the animals to remain at certain depths. Little is known of the role of the radiolarians in the general ecology of the pelagial.

Among the only ciliates occurring in the pelagial are the heterotrich tintinnids; these are sometimes very numerous in the neritic region and

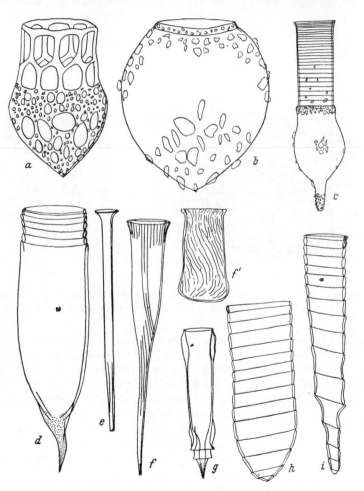

Figure 63. Forms of tintinnid loricae (after Jörgensen 1924): (a) *Dictyocysta elegans,* (b) *Stenosemella ventricosa,* (c) *Codonellopsis orthoceras,* (d) *Favella ehrenbergi,* (e) *Salpingella acuminata,* (f) *Rhabdonella spiralis* var. *elongata,* (f¹) *R. spiralis* f. *hydria,* (g) *Xystonellopsis paradoxa,* (h) *Coxliella annulata,* (y) *C. fasciata*

are regarded as good ecological indicators owing to their sensitivity to fluctuations in salinity, temperature, etc. (see e.g. Gillbricht 1954). They are characterized by the possession of an armour (lorica), from which they can be identified in preserved material. The loricae have an alveolar structure: 'the alveoli contain a low gravity fluid which may contribute to the ability of these organisms to float' (Reichenow 1949, p. 321). Over 700 species are known, and they are principally characterized by their loricae; they were classified into fifty-one genera by Kofoid & Campbell (1929) (Fig. 63). More recently Halme & Lukkarinen (1960) have shown the great variability of the loricae in *Tintinnopsis* in Finnish waters, thus suggesting that the use of these structures in identification may not be entirely reliable.

The tintinnids have seldom been observed in large numbers in the oceanic pelagial; they have been found at great depths but it is not certain whether these are living animals or sunken loricae.

Among other protozoans there have been occasional records of epizootic suctorians, and with suitable methods these ought to be found more frequently.

Coelenterates

The hydrozoans are represented in the pelagial by the Hydromedusae, which are usually small and are particularly common in the neritic region, since they have sessile polyps in the sexual generation. On the other hand, the orders Trachymedusae and Narcomedusae are largely holopelagic; their polyp stage is either completely suppressed or is in the form of a stolon-like structure on the mother medusa or even on other meduae (e.g. in *Cunina proboscidea*). A transition to the holopelagic habit by a different kind of modification of the alternation of generations has also made it possible for the Hydromedusae to invade the oceanic region, where they occur both in the upper layers and also at great depths. Russell & Rees (1960) for instance describe the hydromedusa *Bougainvillia platygaster* as a species which is in the process of becoming an oceanic form: in this species the immature germinal tissue gives rise directly to stolons or polyps from which new medusae are budded off. From the ecological point of view the mother medusa provides a substrate for its own reduced polyp generation.

Pelagic hydranths have been evolved in the family Margelopsidae; a peculiar form of float is produced in the subfamily Pelagohydrinae. According to Dendy (1902) and Garstang (1946) the aboral end of the pelagic hydroid *Pelagohydra* is regarded as the stalk region or hydrocaulus with its axial parenchyma and peripheral labyrinth of canals. Rees (1957) regarded the float as the dilated hind end of the hydranth in which the posterior whorl of tentacles has become scattered due to

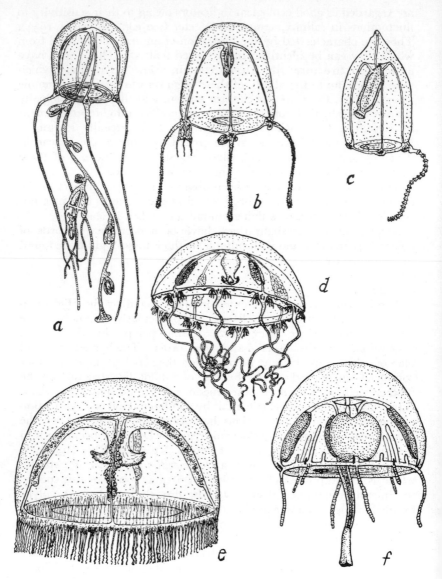

Figure 64. Examples of Anthomedusae (a, b, c), Leptomedusae (d, e) and Trachymedusae (f) (after Tregouboff & Tose 1957, somewhat modified: (a) *Sarsia gemmifera*,
(b) *S. prolifera*, (c) *Corymorpha nutans*, (d) *Eucheilota cirrata*, (e) *Laodice undulata*, (f)
Liriope tetraphylla

the swelling of this part of the hydranth into a float. On this interpretation the float is the basal part of the hydranth in which the parenchyma supporting the diaphragm is enormously developed, eliminating not only the posterior (aboral) chamber but also the posterior part of the oral chamber (Fig. 65).

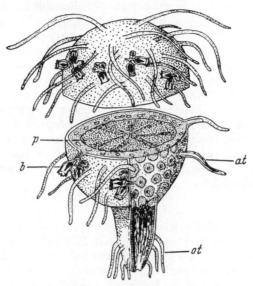

Figure 65. *Pelagohydra mirabilis* (after Garstang from Rees 1957, somewhat modified); *at* = aboral tentacle, *b* = blastostyle with developing medusa, *ot* = oral tentacle, *p* = parenchyma

Similar evolutionary tendencies also occur among the Scyphomedusae. According to Russell & Rees (1960) the deep-sea species *Stygiomedusa fabulosa* has cysts which develop from the germinal tissue and lie in endodermal pockets in the roof of the brood chamber. These authors regard the cysts as atentacular scyphistomae from each of which a single medusa is budded. At first, however, young medusae retain their connection with the central lumen of the scyphistoma. Tubular outgrowths develop from the cyst and eventually project freely into the stomach, allowing food to be taken from the gastrovascular cavity of the mother animal. The cyst wall can therefore be regarded as a chorion for the scyphistoma and the young medusa. This is therefore an example of viviparity.

In siphonophores, which are hydrozoans, the evolution of a holopelagic way of life has been made possible by the formation of colonies and by the marked division of labour between individuals. The chondrophorans *Velella* and *Physalia* live drifting at the surface of the sea; several siphonophores are particularly abundant in warm seas.

Among the actinians there is one group, the Minyadidae, which has become free-swimming. In these animals the base is curved round to form a cavity which is filled with a fibrous or lamellar mass of chitin which acts as a float. It is not certain whether this has a hydrostatic function which presupposes that the gas content can be regulated. The ethology of these animals deserves further investigation.

Among the ctenophores there are only a few benthonic forms, the majority—some 80 species—living holopelagically. In general they are voracious predators which can literally cause havoc in the plankton. Some species such as *Pleurobrachia pileus* and *Beroë cucumis* are markedly euryoecous and may therefore occur in various geographical latitudes and at different depths, whereas others are adapted to more limited conditions and are more restricted in their occurrence: *Mertensia ovum* is arctic, *Callianira cristata* antarctic while *Cestus veneris* and *Beroë forskali* are restricted to the warm-water sphere. A few species are inhabitants of the bathypelagial.

Vermes

The pelagic turbellarians, *Planocera gaimardi* Blainville 1828 (= *Planaria pellucida* Mertens 1832) and *Planctoplana challengeri* Graff 1892, have only been occasionally recorded and then only in small numbers. According to Prudhoe (1950) the synonymy is not at all clear and nothing is known of the habits of these animals.

Among the nemertines some of the Polystilifera are pelagic; they are almost exclusively bathypelagic and have usually only been found as single individuals. In contrast to the benthonic nemertines the pelagic species are small, usually somewhat flattened and in certain cases they have tentacles or short fins. Owing to their rare occurrence there is no detailed information on their habits.

Articulata

Among the annelids only a few polychaete families have evolved holopelagic forms, namely the Tomopteridae, Typhloscolecidae, Alciopidae, and some of the Phyllodocidae. Most species live in the warm-water region but they often occur sporadically and solitarily, so that the role of these predatory forms in the biology of the pelagial is of secondary importance in comparison with that of the benthonic polychaetes which play a very important part in the biology of the benthos. A number of polychaetes produce epitokous forms with swimming bristles during the breeding season and they may then form an important part of the plankton for a short period of time and in a relatively restricted area. The best-known example is the palolo worm.

The crustaceans have numerous pelagic species and this class in fact provides the major part of the zooplankton.

Cladocerans are particularly associated with the neritic region. The genera *Podon*, *Evadne* and *Penilia* sometimes account for a considerable proportion of the plankton. Thus in Block Island Sound, Deevey (1952) sometimes found that *Podon intermedius* accounted for up to 25% of the total number of organisms. The genus *Penilia* appears to require a minimum summer temperature of 21°C to produce a population whereas its resting eggs will probably tolerate temperatures even below 0°C.

Among the ostracods a large number of genera and species are pelagic, but they seldom form an important part of the plankton. A particularly striking form is the very large *Gigantocypris* which has a diameter of 20 mm.

The copepods are represented in the plankton by a large number of species (about 800) and they often constitute the major part of organisms present (see Table 22). Their vertical distribution extends from the surface down to great depths, and geographically they occur both in high latitudes and in the tropics; their ecology is marked not only by the wealth of individuals but also by their ethological differentiation. A great number of them are primary consumers and thus constitute an important intermediate link in the food chain; others live semi-parasitically or have gone over to a completely parasitic existence as parasites on fishes and invertebrates.

Among the Eucarida the order Euphausiacea, containing a single family, is exclusively pelagial. They occur in all the oceans in almost all regions; they are frequently very abundant and in some areas they provide the major part of the food of baleen whales and some fishes. They have therefore been the subject of a great number of morphological, systematic, ecological and physiological investigations. The most highly differentiated genus is *Euphausia* which has 30 species; the genera *Thysanopoda*, *Stylocheiron* and *Thysanoessa* have 11, 10 and 9 species respectively, while a few genera such as *Meganyctiphanes*, *Bentheuphausia* and others are monotypic.

Among the decapod crustaceans the Natantia have a large number of holopelagic forms; there are also some species that are only found sporadically in the pelagial, so that they can be regarded as meroplanktonic. It is noteworthy that in the pelagic genera *Gennadas*, *Funchalia* and *Hymenopenaeus* (family Penaeidae) the statoliths consist of a chitinous secretion and not of foreign matter that has been picked up. Whereas the species of *Sergestes* live in various ecological conditions, the genus *Lucifer*, which has few species, is endemic in the surface layers of warm seas. In the caridean families Pasiphaeidae and

Hoplophoridae the majority of the species appear to be bathypelagic.

According to Siewing (1951) the Peracarida can be separated into two groups: 1. Lophogastridae, Mysidacea, Amphipoda. 2. Cumacea, Tanaidacea, Isopoda.

In the Mysidacea and Amphipoda (Lophogastridae not known) the young hatch in a fully developed condition, whereas in the second group the young lack the last pair of thoracic extremities at hatching. In the first group (Lophogastridae again unknown) the marsupium is retained between two brood periods, but in the second group the oostegites are lacking in the period between two broods. 'The lophogastrids, amphipods and mysids are united by the common possession of a pyloric funnel and the pyloric "bristle-mill" ' (l.c., p. 168). It is interesting that a large number of pelagic forms occur in the first group, whereas in the second group only a few isopods live in the pelagial. The reasons for this are not yet clear.

The Arachnoidea are not represented in the pelagial and of the insects only the gerrid genus *Halobates* is pelagic. These bugs live at the surface and have even been found in areas far from land. Imms (1936) recognized five species.

The Chaetognatha or arrow-worms are almost exclusively holopelagic; the genus *Spadella*, which has only a few species, is benthonic. Large hook-like jaws and fast movements enable these animals to catch living prey, and since they are extremely voracious and sometimes occur in large numbers they play a very important part in the ecology of the pelagial; they themselves also serve as food for other predators. Most chaetognaths are oceanic and a few are bathypelagic. Some species serve as good indicator forms because they are tied relatively closely to certain types of water.

Mollusca

Squids and certain gastropods are pelagic, whereas the Amphineura, Scaphopoda and lamellibranchs are exclusively benthonic, although in many cases their larvae occur in the plankton.

Ecologically the pteropods are undoubtedly the most important group, for some species occur in great numbers and thus provide an important source of food for consumers higher up in the food chain. The Gymnosomata, which have no shell, live as predators and often seize their prey with suckers and hooks. In general they are not very abundant (see e.g. Tesch 1950) although in arctic regions *Clione limacina* may become an important source of food for baleen whales, as also does *Limacina helicina* among the Thecosomata. The latter group has either spirally twisted or symmetrical calcareous shells, or internal shells of a gelatinous or cartilaginous consistency. The Thecosomata feed on micro- and nannoplankton, are found both in the epipelagial and bathypelagial and most of the species occur in warm-water regions.

Among the nudibranch opisthobranchs there are a few aberrant pelagic forms. *Glaucus* is characterized by the bush-like arrangement of the cerata; these animals are bright blue and belong to the epipelagial or pleustal in warm seas. They appear to feed mainly on the polyps of *Velella*. The family Phyllirhoidae has several genera with a few species, most of which have only rarely been found; mention may be made of the genera *Phyllirhoe* and *Cephalopyge* (cf. Dakin & Colefax 1936, Stubbings 1937).

The heteropods are strikingly modified prosobranchs, in which the front part of the foot is laterally compressed to form a fan-shaped disc which executes undulatory movements during swimming. The hind part of the foot forms a tail-like structure which in *Atlanta* carries an operculum. The shell may sometimes be large enough to contain the whole animal, but very thin (*Atlanta*), or much smaller than the animal (*Carinaria*) or completely reduced (*Pterotrachea*). These predatory forms occur almost exclusively in warm-water regions.

It is scarcely possible to give a brief summary of the pelagic cephalopods that would be useful to non-specialists, but Pickford (1949), for example, cited synonyms for *Vampyroteuthis infernalis* from seven different genera. Rees (1954) showed that the juvenile cephalopods described under various specific names in the genus *Macrotritopus* are only different developmental stages of the benthonic octopod *Scaeurgus unicirrhus* and may grow to a considerable size when transported into deep water.

Numerous cephalopod species occur in the pelagial, most of them evidently living at great depths when adult; they show very great diversity in form. Some are good swimmers with predatory habits, others are evidently sluggish animals which from their structure would appear to feed on plankton. In many cases it is scarcely possible to differentiate between planktonic and nektonic forms, although those in which the subcutaneous connective tissue is much developed to give the body a more or less gelatinous consistency must be regarded as planktonic (see p. 147 et seq.).

With the exception of the aberrant holothurians of the group Pelagothurida, the echinoderms are exclusively benthonic. These holothurians, which have rarely been found, appear to be dependent for food on sinking particles and to have only limited powers of movement.

Among the protochordates only the Larvacea (Copelata Appendicularia) and Thaliacea have evolved holopelagic forms.

The Larvacea are free-swimming tunicates with a laterally compressed tail; they have been regarded as primary neotenous forms but not the basic stock of the tunicates (Berrill 1950). They are mostly small and at the most a few millimetres long: *Bathochordaeus*, which appears to prefer deep water, occasionally reaches a length of 25 mm. To catch the nannoplankton on which they feed these animals build a cuticular 'house', part of which forms a filtration apparatus; the animal sits inside the house (family Oikopleuridae) or carries it in front of the mouth (family Fritillariidae). Most of the species, of which about eighty have been recognized, live in the epipelagial of the warm-water sphere down to depths of about 200 metres, but a few have occasionally been caught down to about 3,000 metres. The genera with the most species are *Oikopleura* and *Fritillaria* (see Table 20).

The class Thaliacea has three orders, the Pyrosomida, Doliolida and Salpida. In the colony-forming Pyrosomida the individual animals are united by gelatinous material to form a cylinder open at one end and with a common central cavity. The branchial openings of the individual animals lie at the outer surface, while the atrial openings open into the central cavity. The water currents produced by the individual animals unite and pass out from the open hind end of the colony, which therefore moves by a form of jet propulsion with the closed end forwards. The colonies, which have a gelatinous or carti-laginous consistency, may reach a length of up to 50 cm. They are characterized by particularly brilliant luminescence.

The Pyrosomida occur principally in the upper 500 metres of the world's warm-water belt, although a few catches have been made down to 2,000 metres.

The Doliolida (Cyclomyaria) and the Salpida (Desmomyaria) also prefer regions with warm water. Their bodies are more or less barrel-shaped, the water used for respiration passing in at one end and out at the other. The muscle bands form hoop-like rings which are closed in the Doliolida but usually incomplete ventrally in the Salpida. The sexual forms of the Doliolida live solitarily: their eggs produce larvae, which after metamorphosis develop into asexual oozoids which by budding produce new individuals and with these form colonies. In the salps, on the other hand, the asexual form is solitary and produces a tubular stolon which gives rise to the aggregated sexual forms by budding.

In general the Doliolida are smaller (*Dolioletta gegenbauri* up to 3 cm) than the Salpida which may reach a length of up to 15 cm (Ihle 1937 for *Thetys vagina* and *Salpa maxima*). Like the Pyrosomida both these orders feed on nannoplankton.

4. Form and function in pelagic organisms

We still know relatively little about the autecology of pelagic animals and their synecological relationships. Certain conclusions on these subjects can be drawn from analyses of form and function but it should be emphasized that these may have only limited validity. Form and function are very closely connected, but similar functions may be carried out by different kinds of structure. Form and function can only be understood in relationship to the environment and to the behaviour of the animals that live in it.

Form and size

Even in the early days of research on plankton it was realized that, in spite of their higher specific gravity relative to water, the animals and plants float within a restricted horizon and evidently only expend a small amount of energy in maintaining their buoyancy. Since the speed of sinking is partly determined by the frictional resistance and this is in part determined by the form of the sinking body, the shapes of pelagic organisms have been considered in relation to buoyancy and classified into groups.

Forms with the body broadened into a disc or parachute shape have been termed discoplankton. Examples from the plant plankton would include the diatom *Planktoniella sol*, the coccolithophore *Petalosphaera grani* and the cystoflagellates *Leptodiscus medusoides* and *Craspedotella pileolus*. This shape is found in animals of the most diverse phyla and classes: among the coelenterates many medusae and the siphonophore *Porpita*; in the worms the pelagic nemertines and tomopterids: the copepod genus *Sapphirina* and phyllosoma larvae; among the molluscs the opisthobranchs *Phyllirhoe* and *Cytopyge*, some pteropods such as *Clio cuspidata*, *Diacria trispinosa* and *Desmopterus papilio*, and cephalopods like *Mastigoteulhis cordiformis* and others. Several cephalopods have broad membranes between the long arms which may function as a parachute when the arms are extended. The small number of pelagic holothurians also have this kind of shape.

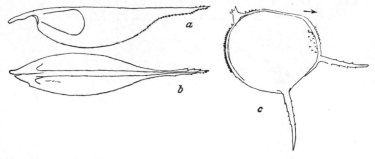

Figure 66. Contrasting shapes among pelagic ostracods; (a) and (b) *Conchoecia daphnoides* from the side and from below; (c) *Thaumatocypris echinata* (after Skogsberg 1920)

Elongation of one axis of the body gives long rod-like shapes or rhabdoplankton. Examples of this are the chaetognaths, the ostracods *Conchoecia daphnoides* (Fig. 66a, b) and *C. caudata*, the copepod *Macrosetella* and the amphipod *Xiphocephalus* (*Rhabdosoma*), and among the cephalopods some deep-sea forms such as the species of *Doratopsis* and *Toxeuma*.

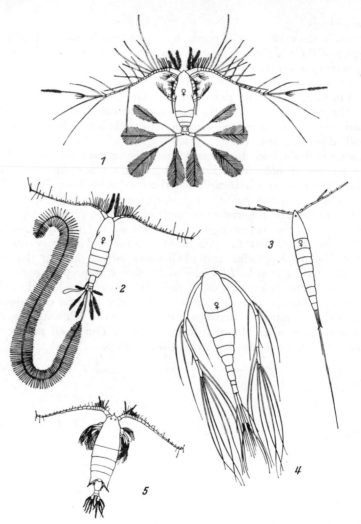

Figure 67. Types of form of female pelagic copepods: 1, *Calocalanus pavo*; 2, *C. plumulosus*; 3, *Macrosetella gracilis*; 4, *Mormonilla phasma*; 5, *Labidocera acuta* (1–4 from *Faune de France*, **26**: 5 after Delsman 1939)

Physoplankton comprises forms that are balloon-like or spherical. Examples are certain ctenophores such as *Pleurobrachia* and *Beroe* and also some medusae; in the pelagic hydrozoan *Pelagohydra* the 'float' is a bladder-like expanded corymorphid hydranth. The crustaceans *Mimonectes* (Amphipoda), *Thaumatocypris echinata* (Fig. 66c) and *Gigantocypris* (Ostracoda) are almost spherical. Other examples are the pteropods *Diacria quadridentata* and *Carolinia* species and the sac-like cephalopods like *Cranchia*, *Teuthowenia*, *Vampyroteuthis* and others. The Appendicularia form houses which are voluminous in relation to the body.

The forms known as chaetoplankton have elongated processes which give them a particularly bizarre appearance. There are examples of these among the ceratians, diatoms, foraminiferans and radiolarians; many crustacean larvae have this kind of grotesque appearance; copepods such as *Calocalanus pavo*, *C. plumulosus* (Fig. 67), *Oithona plumifera* and *Mormonilla phasma* show enormous development of feathery bristles; *Microsetella rosea*, *Aegisthus aculeatus*, and the larvae of *Mesorhabdus* and *Euchaeta* have very elongated single bristles, particularly on the furea. Long transverse spines on the shell occur in the pteropod *Clio cuspidata*. The long oar-like antennae of copepods may, like the long tentacle cirri in tomopterids, serve as flotation devices; many deep-sea crustaceans have much elongated antennae and extremities (Fig. 68). Certain meroplanktonic forms such as epitokous polychaetes (particularly in the syllids and nereids) as well as larvae

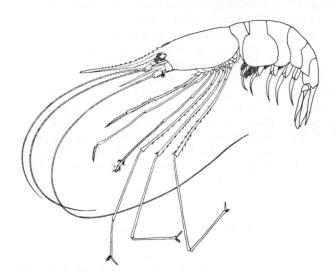

Figure 68. *Plesionika longipes* (after de Man 1920)

(e.g. spionids and mitraria) also have long bristles which are spread out when locomotion ceases.

In some cases as, for example, in the pelagic deep-sea crustaceans, the long body appendages may be regarded less as flotation mechanisms and more as taste organs or as carriers of other sense organs. In ceratians, however, it has been observed that the length of the horn-like processes varies with the time of year, and this can evidently be correlated with varying water densities caused by temperature changes.

The range of sizes in the pelagic world is surprisingly varied. Whales with a length of over 25 metres and a weight of 100 tons are only conceivable in a medium which supports the whole body, and cephalopods such as the Architeuthidae with a body length of 9 metres are scarcely imaginable as benthonic animals. There are also other classes of animals with representatives in both the pelagial and the benthos which have developed their largest species in the pelagial; examples are the copepods *Megacalanus principes*, *Bathycalanus richardi* and *Valdiviella insignis* with a body length of 12 mm, the ostracod *Gigantocypris* (20 mm), the mysids *Gnathophausia ingens* and *G. gigas* (142 and 115 mm, Fage 1941) and the amphipod *Thaumatops magna* (80 mm). The spatial distribution is also of interest for within the pelagic groups the largest representatives generally occur at greater depths. In addition to those already mentioned there is the appendicularian *Bathochordaeus* with a length of over 20 mm, which appears gigantic in comparison with its epipelagic relatives. Even within a species there is often a vertical size distribution, the larvae and young animals being found in higher water layers than the adults. This applies particularly to many fishes.

This summary of the shapes and sizes found in the pelagial, which is by no means complete, shows the great diversity that exists within individual groups; this can be contrasted with the relatively uniform nature of the abiotic environment; the diversity is all the more striking because it occurs not only in the pelagial as a whole but also within the epipelagial and bathypelagial. This diversity includes numerous cases of specialization and excessive development which must be regarded as special adaptations to living conditions in the pelagial.

The development of long body appendages and diverse body forms within a single epipelagial group may be explained by postulating that these increase the frictional resistance and decrease the speed of sinking. This view is supported by the frequent occurrence of such forms in warm waters. Deviation from the 'normal form' would therefore have a definite positive selection value. It appears that the specialized forms often occur only in small numbers and exist simultaneously alongside unmodified forms, that the deviations proceed in very different directions and that there is a significant increase in the specialization of epipelagial forms in tropical and subtropical regions.

This is particularly well shown in the copepods, in which elongated, flattened or bladder-like forms or those with extreme development of spines occur alongside 'normal forms'; there are similar examples among the dinoflagellates (Fig. 55) and the amphipods. Bathypelagic fishes show very great diversity in body form, development of fins, fin

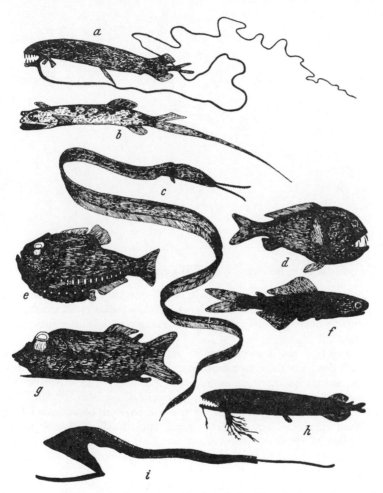

Figure 69. Variation in form among bathypelagic fishes (redrawn from various authors): (a) *Grammatostomias flagellibarba* (20–22 cm); (b) *Gigantura chuni* (about 13 cm); (c) *Nemichthys scolopacus* (up to 50 cm); (d) *Caulopacus longidens* (about 13 cm); (e) *Argyropelecus gigas* (8–10 cm); (f) *Lampanyctus elongatus* (about 15 cm); (g) *Opisthoproctus soleatus* (6–7 cm); (h) *Chirostomias pliopterus* (about 20 cm); (i) *Eurypharynx pelecanoides* (up to about 50 cm)

F—M.B.

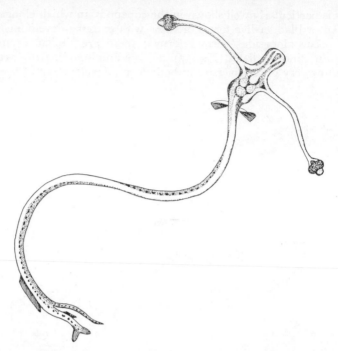

Figure 70. Larva of *Idiacanthus* with stalked eyes (redrawn from Marshall 1957)

components, appendages on the chin or gill-covers (Fig. 69). The development of long eye-stalks in the larvae of *Idiacanthus* is an example of an excessive development in the juvenile stage, which later disappears (Fig. 70).

Diversity in form is sometimes so heterogeneous that it appears to lack direction. Its existence suggests that it does not have negative selection value; but at the moment we cannot say that it has positive selection value, such as increasing frictional resistance, until we have more detailed knowledge of the habits and behaviour of the individual species. The relatively great uniformity of abiotic factors in the pelagial appears, however, to exert a low selective pressure on mutational changes; in other words the pelagial seems to allow great ecological licence.

Specialization and excessive development indicate a general uniformity of the environment in space and time; relatively small changes in the structure of the environment lead to the disappearance of specialized forms. The evolution of a large number of special forms within a given area suggests either that the environment has remained constant over a long period of time or that the rate of evolution has been particularly fast. For the bathypelagial at least there is nothing to support the second alternative; thus Carter (1961, p. 223) writes: 'The low temperature will slow the whole tempo of life and therefore, presumably, its evolution, but there is no reason to think it will alter the type of evolution' (cf. pp. 306 and 369). The apparently low rate of reproduction in bathypelagic animals is also at variance with an increased rate of evolution. Thus, one is led to conclude that the evolution of highly specialized forms in the epipelagial and bathypelagial of the tropics and subtropics has been made possible by a long period of constant environmental conditions.

The development of extreme body size in various animal groups has already been mentioned (see p. 148) as striking, and this is particularly so in some bathypelagic invertebrates. In the epipelagial this phenomenon is especially apparent in the baleen whales, whale shark, basking shark and swordfish. The evolutionary importance of body size has been discussed in detail by Rensch (1954) and the positive and negative growth allometry explained. Any discussion of the ecological importance or selection value of this development of size must involve the question of food. Baleen whales and the sharks mentioned above are filter-feeders which have to pass large volumes of water through the filtration apparatus provided by the baleen plates or gill-rakers in order to obtain the necessary amounts of food. In the baleen whales there is evidently positive growth allometry in the development of the

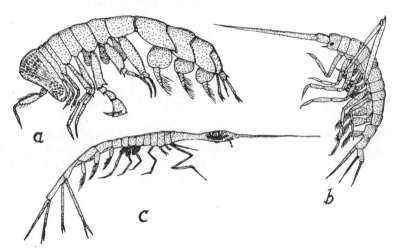

Figure 71. Three amphipods as example of forms showing growth allometry: (a) *Phronima sedentaria* male; (b) *Scina crassicornis* female; (c) *Rhabdosoma* (after Tregouboff & Rose 1957, modified)

lower jaw and skull, which leads to an enlargement of the mouth cavity in relation to the size of the body. It remains to be investigated whether similar positive allometry can be confirmed in the whale shark and basking shark.

In bathypelagic forms as mentioned above, increased speed and rate of filtration associated with a poor food supply could have led to an increase in body size associated with growth allometry. As far as I know there has been no detailed investigation of this in invertebrates, although this seems to be very desirable, because the differing vertical distribution of young fish and adults and the development of the mouth in many deep-sea fishes (see pp. 182 ff.) supports such a view. This problem should not, however, be approached solely from the point of view of obtaining food; in *Thaumatops* and other amphipods, for example, there is a positive growth allometry of the head, which is doubtless correlated with the development of enormous eyes (Fig. 70).

It is not really possible to draw any general conclusions because in many groups there is an increase in size with a decrease in water temperature. In the chaetognaths, according to Kuhl, 15 warm-water species have an average length of 18·1 mm, 8 deep-sea forms an average of 28·8 mm, whereas 8 cold-water forms have an average length of 40·2 mm; for *Sagitta bipunctata* the maximum length in the central Atlantic was 12 mm, but the animals were already sexually mature at 5 mm; in the arctic on the other hand they reached a length of 44 mm and sexual maturity was not attained until they were 30 mm long. The following table shows the position in some amphipods:

Species	Average body length (mm) Boreal	Arctic
Themisto abyssorum	9	17–21
Parathemisto oblivea	6	17
Hyperoche medusarum	5–6	up to 15
Hyperia galba	12–14	up to 24

Barnard (1932, p. 266) writes of the amphipod *Cyllopus magellanicus*: 'In size males vary from 9–15 mm, and ovigerous females from 8–16 mm, in both cases the larger sizes coming from higher latitudes'. In this context it should be mentioned that in the cold-water regions eggs with a large yolk content are often developed and this possibly presupposes an enlargement of the body with positive growth allometry of the gonads.

Günther & Deckert (1950) consider that the small amount of water movement in the deep sea, in comparison with the surface and swell conditions, is of decisive importance as a selection factor in the development of size: 'In the same way the stillness of the water in the deep sea occasionally allows the special development of size in vagile bottom-living groups and in planktonic families'.

There is no doubt that this factor should be considered for the deep sea but it should in no way be regarded as absolute or constant in size because even the deep sea is not free of currents. In the surface waters of polar regions the small amount of water movement cannot be considered as a factor contributing to the development of size. The development of large average length may be due to different kinds of factors or to combinations of these; so it must be assumed that the biological significance of an increase in size in the deep sea may differ from that in the polar regions.

The gelatinous nature of many pelagic animals is also a formative characteristic. Examples of this include: medusae and ctenophores; among the molluscs the heteropods and the pteropod *Cymbulia* with the pseudoconch, as well as some cephalopods such as *Benthoteuthis*, *Ctenopteryx*, *Chaunoteuthis* and others; the holothurians *Pelagothuria* and *Enypniates*; the appendicularians with their 'houses' as well as the salps and Pyrosomida. In many deep-sea fishes the skeleton is cartilaginous and the connective tissue is frequently gelatinous; the larvae of deep-sea angler-fishes (Ceratioidea) living at shallow depths in warm seas have a very gelatinous skin; the reduction of this during metamorphosis is probably responsible for the fast sinking to depths of over 1,000 metres in which the adults live (Bertelsen 1951). The same phenomenon therefore occurs in inhabitants of the epipelagial and in deep-sea forms.

The frequency with which this character is found in pelagic animals, as compared for example with littoral forms, suggests that the development of gelatinous tissue is adaptive. Gelatinous tissues have a high water content and therefore a relatively low specific gravity, which will facilitate buoyancy in open water (Table 23).

TABLE 23

Specific gravities of gelatinous material (up to 22°C)

(after Jacobs 1944)

Species	Specific gravity of the gelatinous material
Forskalia sp. (skin)	1·0260
Eucharis multicornis	1·0256
Olindias mulleri	1·0246
Cotylorhiza tuberculata	1·0268 in water with specific gravity of 1·0260
Salpa maxima	1·0275 measured one day after capture

However, some deep-sea benthonic forms also have this kind of connective tissue, so it cannot be assumed that this character is always dependent upon the same selection factors. Living conditions in the bathypelagial and bathybenthal ought to be favoured by the relative stillness of the medium, while for epipelagic forms other factors must be involved which are not yet clear.

Coloration

The development of gelatinous tissue is frequently, but by no means always, associated with marked transparency of the body in pelagic animals; thus, for example, the chaetognaths are transparent but not gelatinous. Extreme transparency occurs both in the epipelagial and in the bathypelagial and is seen in many meduase, siphonophores, ctenophores, turbellarians, nemertines, polychaetes, chaetognaths, copepods, amphipods, mysids, heteropods, cephalopods, appendicularians and salps and also in fish larvae. In transparent animals it is quite common for certain organs, such as eyes, gut, gonads or muscles, to be brightly coloured and to show black, brown, yellow, red or blue tones. When seen in an aquarium the form of the body will then appear to be broken up. This provides a certain protection against being seen and is surely the biological significance of this phenomenon. The transparency becomes most effective when observed against the light but it may be visually ineffective when viewed at certain angles. Thus, in diving experiments Bainbridge (1952) observed (p. 108): 'The clearest views could be seen by looking about 10° either to the left or to the right of the bright patch formed by the sun on the surface of the water. In this manner a sort of dark ground illumination is obtained and even the most transparent forms stand out very clearly.' Protection from being seen is therefore only relative and it can scarcely be of importance in animals living at great depths.

In many cases animals living close to the surface show blue tones, with the upper side usually bluish and the underparts white or silvery. Well-known examples among pelagic fishes are the herring and its relatives, the mackerels, tunnyfish, swordfish and dolphin-fish.

The 'disconants' (Ankel 1962) such as *Velella*, *Porpita* and the Portuguese man-o'war *Physalia* which float at the surface usually occur together with the gastropod *Ianthina* and the nudibranchiate opisthobranch *Glaucus*. All these forms are blue on the surface that faces upwards and white below, even when as in *Glaucus* the belly is turned upwards so that the animals hang from the surface in an inverted position. Ankel (l.c.) used the term 'blue fleets' for these associated forms, and they would include copepods such as *Anomalocera*, *Labidocera* and *Sapphirina*, the isopod *Idothea metallica*, the amphipod *Synopia ultramarina* and the hemipteran *Halobates*. The pelagic actinians of the family

Minyadidae are blue and so is the cephalopod *Tremoctopus* which swims close to the surface, and is usually pale blue-violet above and silvery-white below when in an unexcited condition. The drifting barnacle *Lepas fascicularis* also shows blue coloration.

We cannot yet say whether this widespread phenomenon is of general and valid significance. The blue uppersides may very well be regarded as camouflage coloration because when seen from above they differ little from the blue or blue-grey of the water. The pale underside seems to be less simple to explain; when seen from below towards the surface, even a pale object will appear more or less as a silhouette and therefore be clearly visible. When the animals move, whether passively by wave action in the case of drifting forms or actively by undulatory swimming, the pale underside will sometimes be lit up and sometimes disappear; in this way the moving body when seen from below or from the side will not offer any constant visual target but will appear broken up, thus providing a certain amount of visual protection.

Ankel disagrees with the idea that this type of coloration serves as camouflage and would ascribe 'a different reason common to all, perhaps even a sensory significance common to all' (l.c., p. 364), whereas Mertens (1962) suggests that the surface forms may be protected in this way from the eyes of birds; as in the case of transparency (see above) any protection of this kind would be only relatively but not absolutely effective.

In the deeper parts of the pelagial, blue and white more or less disappear and are replaced by duller colours, among which black, blackish-brown and violet predominate. Numerous invertebrates are coloured an intense red, as for example certain medusae, the ctenophore *Beroë abyssicola*, nemertines, polychaetes such as *Tomopteris* and the Typhloscolecidae, chaetognaths, several crustaceans, cephalopods like *Benthoteuthis megalops*, *Mastigoteuthis flammea* and also some fishes.

This coloration is generally uniform, with less of the mottling or patterning that occurs in related forms in the epipelagial. The dark colours of bathypelagic animals may be correlated with the darkness of the environment, because red deep-sea animals, which look so conspicuous in daylight, are not seen in their natural habitat where at least the red component of the light is completely lacking.

In some cases red coloration in the invertebrates is evidently correlated with temperature. In west Greenland waters Kramp (1942 a and b) found different intensities of red in *Beroë cucumis* and *Aglantha digitale*, the greatest intensity occurring in the coldest areas or in the colder parts of the year. Heegaard (1948) mentions that in *Meganyctiphanes* the red pigment spots disappear when the animals are kept at higher temperatures. It is still not clear whether there is a correlation between red coloration and low temperature in bathypelagic animals.

In contrast to the bathypelagial, the benthal and also those terrestrial habitats that lack light frequently have unpigmented forms. This difference suggests that the bathypelagial is not in absolute darkness and furthermore that the widespread bioluminescence is important (p. 365); the development of the sense organs also points in the same direction. In this context, attention should be drawn to the phenomenon emphasized by Zenkevitch & Bierstein (1956), namely that the hadopelagic animals in the deep-sea trenches are unpigmented; nothing is known about luminescence in this hadal fauna.

Hjort (1911) was the first to point out vertical gradations in colour among pelagic fishes; in the well-illuminated layers of the north Atlantic Ocean he found that blue colours were predominant down to about 150 metres, there were mainly silvery-grey tones in an intermediate zone between 150 and 500 metres, and in greater depths brown to blackish tones. Similar gradations have also been found in other sea areas, but the actual depth positions involved are dependent upon the latitude and the cloudiness of the water. It is of particular interest to note that Kemp (1917) found red or dark coloration in the fishes and crustaceans living in the very cloudy and thus poorly illuminated waters of the Indian river Matlah. Holthuis (1963) has given an account of red-coloured prawns from surface brackish waters in the tropics; some of these waters lay in complete darkness or in deep shade, others provided the animals with a chance to spend long periods in the dark among rocks and corals on the bottom. Darkness is in fact the common ecological factor that can be correlated with red coloration, because food provides no clue, some of the animals involved being detritus-feeders, some herbivores feeding on algae and others carnivores.

These points suggest that dark coloration can be correlated with weak illumination. In deep twilight, dull-coloured animals are the least visible so that the coloration can be regarded as an adaptive protective character (Marshall 1954). Motais (1961) is opposed to this view on the basis of two experimental findings. Julon (1954) observed an increase in visual acuity in poorly illuminated surroundings after injection with the melanophore hormone of the hypophysis, while Motais himself found a higher content of this hypophyseal hormone in dark-coloured deep-sea fishes than in pelagic fishes and the lowest content in the blind cave-fish *Anoptichthys*. From these results he concludes that in abyssal fishes with highly developed eyes this hormone helps vision in poorly illuminated places; according to this view the dark coloration would to a certain extent be a side phenomenon.

The two hypotheses of Marshall and Motais are not mutually exclusive but are complementary. The extent to which Motais's view can also be applied to dark or red deep-sea invertebrates and thus be generally valid must await experimental evidence. A relationship between pigment and visual acuity in weak light has been found in the case of the eyes of nocturnal birds and in night blindness. In this context closer attention should perhaps be paid to the findings of Fisher *et al.* (1955) on the concentration of carotenoids in the eyestalks of some crustaceans (Euphausiacea).

Movement and buoyancy

The movement of pelagic animals plays a role in preventing sinking, in obtaining food, in finding a mate, in covering large areas in horizontal or vertical migrations and in flight. In a given biological situation movement may in some circumstances differ quite considerably in the same animal, so that a more exact statement would only be of value if the total ecological picture were taken into account. Looked at in detail the movements of pelagic animals show considerable diversity, but certain basic types can be recognized.

Movement by flagella and cilia only occurs in small forms owing to the low degree of efficiency: this method occurs among the flagellates, ciliates and rotifers and in the numerous pelagic larvae of worms, molluscs, bryozoans and echinoderms. The fusion of cilia into bands with a co-ordinated beat has achieved a higher degree of efficiency in the ctenophores.

Paddling movements are often carried out with limbs acting as levers; here the effective stroke and the return to the starting position are associated with various rotatory movements. This sequence of movements is particularly beautiful to watch in the turtles. Paddling movements of this kind are seen at their best in the penguins, in which the fin-like wings make these birds the fastest swimming animals. Copepods make jerky movements by strokes of the first antennae, while in euphausians, amphipods and other crustaceans the paddling strokes of the extremities produce rather similar swimming movements. The same applies to the parapodia of some polychaetes such as *Tomopteris* and *Lopadorhynchus*. The large lateral flaps of some ctenophores (*Ocyropsis, Mnemiopsis, Leucothea*) also work as oars or paddles, as do the wing-like processes developed from the foot in pteropods (Fig. 84).

Swimming movements made by undulations of the body are widespread; here the whole body may make lateral or up and down movements (e.g. *Cestus*, some polychaetes, fish such as eels and leptocephalus larvae), or the movements may be restricted to the hind part of the body while the front remains facing the direction of movement.

This occurs in many fishes; in others forward movement is effected by means of the propeller-like twisting of the caudal fin, which may be attached to the body by a narrow peduncle as in fish like the mackerel. The horizontal tail fluke of whales works in a similar way. In the sunfishes (Molidae) reduction of the tail region (Fraser-Brunner 1951) is correlated with increased development of the dorsal and anal fins (Fig. 72) and a rearrangement of the musculature, but the speed of movement is much reduced.

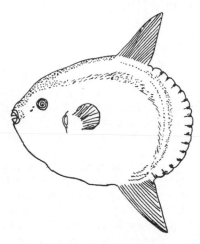

Figure 72. *Orthagoriscus mola* (after Apstein 1905)

In the normal position the ventral fin of the heteropods is turned to face upwards, and by undulating movements it not only supplements the snake-like movements of the body but can also produce an equivalent forward movement on its own. In cephalopods the lateral fins are sometimes highly developed and give the animals surprising speed. Jerking undulations allow the chaetognaths to execute jumping movements.

Movement by jet propulsion is very widespread. In this method a cavity is filled with water which is then expelled forcibly through a restricted opening. The force of the escaping jet of water drives the animal forwards in the opposite direction. This type of movement is well known among meduase, in which the pumping movements of the umbrella move the animal with the upper or aboral side of the umbrella facing forwards. This medusoid type of movement also occurs in the pelagic holothurian *Pelagothuria* and in the cystoflagellates *Leptodiscus medusoides* and *Craspedotella pileolus*. Cephalopods take water in through slits in the mantle cavity and then expel it in jets through the siphon. Normally the siphon is directed forwards so that the animals move with the hind end in front; elongated siphons can be positioned in different directions thus enabling the animal to move off sideways or even with the tentacle end forwards. In many cephalopods membranes are developed between the arms to form a funnel which can also carry out pumping or medusoid movements (Fig. 73).

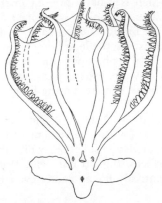

Figure 73. *Cirrothauma murrayi,* an example of a cephalopod with membranes between the arms (after Chun)

In the Thaliacea, the salps and doliolids use the contractions of the muscle bands to expel water from the atrial opening so that the animal moves with the branchial aperture forwards. In pyrosomid colonies each individual closes its branchial opening and by muscular contractions moves the water into the central cavity whence it is expelled in a single jet. The appendicularians have a structure which prevents the back-flow of the water sucked into the house by the

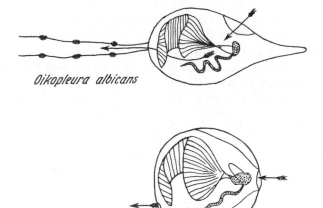

Oikopleura albicans

Oikopleura longicanda

Figure 74. *Oikopleura* showing direction of the water current produced by the undulating movements of the individual within the housing

undulatory movements of the animal (Fig. 74); the exit opening does not allow the water simply to flow out but rather to spurt out under pressure to give propulsion. Long-tailed crustaceans can move backwards by powerful flapping of the tail.

In connection with the asymmetry which is usually present in the body structure, steering organs and stabilizers are often developed for straight-lined movement. Examples are the fins of fishes, whales, seals, cephalopods, the lateral fins of chaetognaths and the long body appendages present in some pteropods, heteropods, cephalopods and appendicularians. Long appendages may also serve as taste organs, so evidence should be sought before any conclusion is made on the function of this type of structure.

Some animals that swim fast can propel themselves out of the water, as for instance the flying-fishes. The power for this movement comes from the caudal fin which has an elongated ventral lobe, while the broad pectoral fins serve as planing surfaces, but do not beat. With up-currents these fish can be carried relatively far and may even be lifted on to the deck of a ship (Fig. 75).

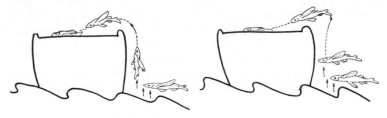

Figure 75. How flying-fish land on the deck of a ship with the help of an upstream of air (a) according to Möbius, (b) according to Mohr (from Mohr 1954)

The horizontal tail fluke of whales moves up and down so that if the animal is facing towards the surface it may be propelled into the air, as happens so frequently with dolphins. Verrill (cf. Abel 1916) gave the name flying squid to the fast-swimming cephalopod *Stenoteuthis bartrani* beause it can leave the water by powerful jet propulsion. Even copepods have been observed to leave the water at times. The biological significance of these aerial jumps is still not clear, although in some cases they may be escape movements with the air serving as a refuge.

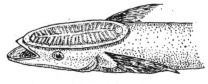

Figure 76. Shark-sucker, *Echeneis naucrates*. The sucking disc on the head is formed from the dorsal fin, the rays of which are modified to form the transverse ridges (after Norman & Fraser 1948)

The remora (*Echeneis remora*) which has its dorasl fin modified to form a suction organ (Fig. 76) may be termed haptic. It has been frequently recorded as a companion of sharks. Cirripedes in, for example, the genus *Lepas*, are sessile and attached to drifting objects, but *Lepas fascicularis* develops its own float and is therefore independent of foreigh bodies (Fig. 77 (see p. 171)). Empty houses of salps are used

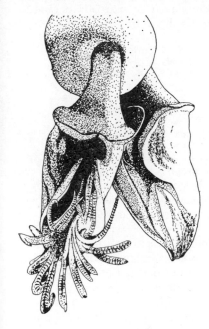

Figure 77. *Lepas fascicularis,* two animals on a common foam float; the animal on the right is closed (redrawn after Ankel 1962)

as living quarters and a means of transport by the females of the amphipod *Phronima sedentaria* and by the small males of the cephalopod *Ocythoe tuberculata* in which the larger females are benthonic. Many epizoic and parasitic animals in the pelagial also live sessile on their hosts.

Both phytoplankton and zooplankton are capable of compensating for the excess weight of their bodies, produced by the specific gravity of the protoplasm with its content of skeletal and other components, even without active movement; they can keep themselves afloat at a given depth. In the older literature on plankton much attention was devoted to problems connected with flotation. Ostwald (1903 b) summarized various observations and calculations in a single formula:

$$\text{speed of sinking} = \frac{\text{excess weight}}{\text{friction}(=\text{viscosity} + \text{resistance due to shape})}$$

The viscosity is affected by temperature and salinity and will therefore vary according to time and place. The resistance due to shape is mainly dependent upon the shape of the body, particularly on the relative area of the body surface and its shape, that is, on the form of the body circumference in the horizontal plane. Using a water column 25 cm tall at a temperature of 18·5°C and a salinity of 35·01‰, Gardiner (1933) determined the time taken by narcotized *Calanus finmarchicus* of different sizes to sink to the bottom:

Length of animal (mm)	Period of sinking (sec)	Length of animal (mm)	Period of sinking (sec)
2·1	181·3	3·2	70·3
2·3	110·6	3·4	58·9
2·6	114·9	3·6	57·9
2·8	106·8	3·9	43·9
3·0	74·3	4·0	31·0

Increase in the resistance due to shape can only slow down the speed of sinking, but cannot completely stop it unless the excess weight is less than a certain amount. Here, however, the action of water movements on bodies with a relatively large surface area and a varied shape will be of importance.

The speed of sinking can be considerably decreased by a reduction in the excess weight, so that true flotation occurs; sometimes the organisms can even become buoyant and float at the surface (overcompensated system, Jacobs 1935). In many pelagic animals heavy skeletal matter is reduced as compared with that of their benthonic relatives. Thus in the pteropods the shell, if present, is delicate and fragile, and the same applies to the heteropods. Among the cephalopods many forms lack shells or the rudiments of shells; the Loligonidae and Oegopsidae only have narrow, horny shells without calcium. Many deep-sea fishes have poorly ossified skeletons and this also applies to the species of *Orthagoriscus* living close to the surface. In deep-sea fishes one can presume that there is a basic correlation with vitamin supplies.

Reduction of skeletal substance is synonymous with reduction in ash content, as has been shown in some crustaceans.

In this context further investigations would be desirable before the general validity of this phenomenon can be finally assessed.

Attention has already been drawn to the development of forms with gelatinous tissues having a high water content (p. 153). The inclusion of water is important to flotation because it reduces the specific gravity of a given volume. The importance of water content can be illustrated by the behaviour of the benthonic sea-anemone *Tealia felina* which when conditions become unfavourable 'fills itself with water so that it takes on the appearance of a balloon. A slight water movement, such as that produced by a fish swimming by, is then sufficient for the anemone to be lifted up and to drift to another place' (Pax, T.N.O. III, e, p. 155). When the water content is over 90%, as in medusae, the ability to float is considerably enhanced. Gelatinous tissues with a high water content are found, for example, in medusae, siphonophores, ctenophores, nemertines, some crustaceans such as the copepods *Eucalanus* and *Haloptilus*, the prawn *Notostomus* and the amphipods *Sphaeromimonectes valdivia* and *Microphasma agassizi*, in heteropods, deep-sea cephalopods, tunicates as well as in some fish and fish larvae, particularly those from deeper water. Among the cephalopods the Cranchidae possess an extensive fluid-filled coelom, with the fluid accounting for about two-thirds of the total weight of the animal. Large amounts of cell fluid are also found in certain unicellular organisms, such as *Noctiluca miliaris*, some radiolarians and diatoms such as *Ditylum* and *Ethmodiscus*.

The degree of buoyancy cannot be explained solely by the content of water. Since the animal is isotonic with the surrounding sea water, a reduction in specific gravity can only be attained by an exchange of the heavy ions for lighter ones. Thus, for example, in *Noctiluca* and in the coelomic fluid of the Cranchidae there is a high concentration of NH_4 ions (up to about four-fifths of the total cation content); the

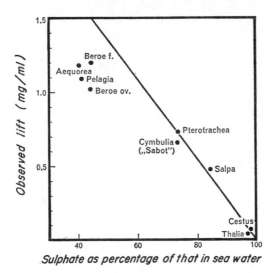

Figure 78. Observed lift of body fluids plotted against sulphate concentration (after Denton 1963)

elimination of sulphate ions appears to be of particular importance (Fig. 78). Denton (1963) cites the results of Beklemischew, Petrikowa & Semina (1961) according to which the diatom *Ethmodiscus rex* with a volume of about 1 cubic millimetre keeps afloat by having a reduced concentration of $Na^{\cdot}$, $K^{\cdot}$, $Ca^{\cdot\cdot}$, $Mg^{\cdot\cdot}$ and SO_4''.

We may now consider the problem of how these organisms move up and down at various depths; here it is possible that the uptake and release of pure water plays a role, thus involving changes in the osmotic conditions. When stimulated the Thalassicola (radiolarians) allow the ring of vacuoles in the outer protoplasm to shrink and the organism sinks (Figs. 79 and 80); since any subsequent ascent is associated with the formation of new vacuoles, we may presume that the contents of the vacuoles formed have a low specific gravity. Brandt (1895) ascribed special significance to the CO_2 expelled following respiration, but as far as I know there has been no detailed investigation of this mechanism.

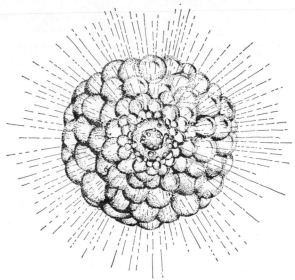

Figure 79. *Thalassicola pelagica,* a radiolarian without a skeleton but with foamy protoplasm around the central capsule (after Jacobs 1938, somewhat modified)

Davis (1953) has postulated that an organism will not use a single method to reduce its specific gravity but that in addition to ionic regulation it may also change the gas and fat content of its tissues.

In any case water uptake and the addition or removal of ions or ion groups are physiological processes associated with buoyancy; we do not yet know the site of the cells or tissues responsible for these activities.

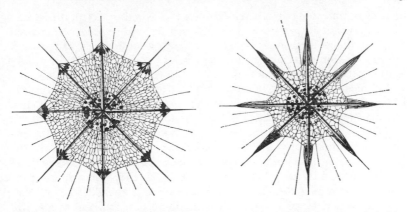

Figure 80. An acantharian (Radiolaria): *left;* outer mantle of plasma expanded by contraction of the myonemes: *right;* myonemes relazed, plasma mantle contracted (after Jacobs 1938, somewhat modified)

The answer might be provided by investigations such as those of Chapman (1935) on the histology of the jelly in coelenterates, taken in conjunction with physiological studies.

The physiological basis for the deposition of substances with a low specific gravity to reduce excess weight are more apparent; these include such substances as fats and gases. Fat deposition in many pelagic animals and the formation of fat as a product of assimilation in phytoplankton have been regarded as specific adaptations to buoyancy. This interpretation should however be treated with caution because the fat content fluctuates considerably dependent upon the nutritional conditions or on the amount of assimilation, without there being any demonstrable change in the ability to remain buoyant. Furthermore, the maintenance of buoyancy at a given depth requires a relatively rapid change in the content of the material producing buoyancy; it would therefore be desirable to have some evidence that this is possible in the case of fat content. In addition there is the fact that phytoplanktonic organisms remain within certain horizons of depth or illumination, according to the species, and that they will stay there even when assimilation is intensive; the formation of large amounts of fat or carbohydrate would move them either above or below their preferred horizon if their buoyancy was regulated solely by these substances. One must therefore presume that other mechanisms are involved in the regulation of gravity.

In considering the formation of gas as a flotation device it should be remembered that a given volume of gas can only keep a body at a given depth; if an organism with a content of gas is transported passively to above or below the depth to which it is adapted it will suffer increased buoyancy or increased speed of sinking. With decreased water pressure the gas expands and drives the body up, with increased pressure the volume of the gas decreases and buoyancy is reduced. Organisms that float by gas are therefore characterized by the ability to give off gas and to re-form it.

A typical 'gas-floater' is, for example, the siphonophore *Stephanomia* in which there is a gas bladder at the upper end of the stolon. When this bladder is emptied the animal sinks to the bottom (Jacobs 1937). If the intact animal is stimulated a gas bubble escapes through a fine opening at the upper end of the bladder and the animal sinks below the level where stimulation occurred. It later rises by the secretion of gas from a gland situated at the base of the bladder. The volume of gas can also be changed by muscles in the wall of the bladder, so that contractions will bring about a decrease in volume and the animal will sink.

Many fishes have a hydrostatic organ in the form of a swimbladder; this gives them so much buoyancy that in still water they only require to expend a small amount of muscular energy to maintain a buoyant condition. In the minnow, experimental investigations have been made on the effect of increased and reduced pressure, on the refilling of the swimbladder after the loss of gas and on the composition of the gas. This is so well known that it need not be discussed in further detail here; in principle the results are also valid for other fishes with swim-bladders. Deep-sea fishes with swimbladders are of special interest because, as a result of the water pressure, they have to produce larger amounts of gas in order to maintain a given buoyancy, than those species which live close to the surface. According to a calculation by Parr (1937), a fish living at a depth of 1,000 metres has to produce about 300 times as much gas as one at 10 metres, in order to attain the same degree of buoyancy. Marshall (1951) regarded this figure as much too high; he pointed out, however, that bathypelagic fishes have much larger *retia mirabilia* and gas glands than shelf fishes, although their swimbladders are much smaller in relation to the body cavity.

Observations on freshwater and littoral marine fishes have shown that the swimbladder is first filled by snapping up atmospheric air at the surface. Naturally this cannot happen in the case of fish living at great depths, and we do not know how this process is carried out if the young stages do not live near the surface. The composition of the gas in deep-sea fishes is also unknown. Trewavas (1933) reported that in *Opisthoproctus* (Fig. 69, g) the swimbladder lies immediately beneath the skin of the belly which is flattened and sole-like. In view of this it is to be assumed that these fish swim or drift back downwards and that the telescopic eyes which are directly dorsally in fact look downwards at objects which are illuminated from above.

Not all pelagic fishes have a swimbladder and we do not know whether some possess an equivalent organ in its place, because no morphological correlations can be recognized between the presence or absence of swimbladders on one hand and the development of other characters on the other. In certain cases the swimbladder has lost its function as a hydrostatic organ. Thus Kotthaus (1952) has described *Hoplostethus islandicus* in which the swimbladder is almost completely filled with fat. In contrast to its two related species with well-developed swimbladders *H. islandicus* may be regarded not as bathypelagic, but as benthonic or hypopelagic. Marshall (1951) reports the same in the pelagic forms *Gonostoma elongatum* Günther, *Cyclothone microdon* Günther *C. braueri* Hespersen & Taning, *C. livida* Brauer and *C. acclinideus* Garmon.

The small cephalopod *Spirula spirula* must also be reckoned among the 'gas-floaters'; in warm-water areas its shells are sometimes washed ashore in enormous numbers. Schmidt (1922) and Bruun (1943) have reported observations on living animals, which are positioned vertically in the water with the shell end upwards, so that a hydrostatic apparatus must be present in the shell. Denton *et al.* (1967) have recently shown that *Spirula* regulates its buoyancy in much the same way as *Nautilus* and *Sepia*. The exit chamber of the shell is more or less spherical, its opening being smaller than the diameter in the middle. It cannot therefore be a chamber to live in, but should be regarded as an inner shell functioning as a hydrostatic apparatus. According to Schmidt the same applies in *Nautilus* and he writes: 'In the Cephalopoda the counteraction of the weight of the shell has come about in the first instance not through its reduction but through its becoming buoyed up by gas secreted in its interior; . . . '*Nautilus* on this view possesses by far the most primitive Cephalopod shell. The metabolic rhythm, which finds its expression in the fact that the normal secretion of calcium carbonate to form shell becomes periodically replaced for a time by the secretion of gas has no longer, if it ever had, relation to the breeding season but extends to the early stages of the life history. What is the physiological meaning of this rhythm is quite unknown. The number of chambers would appear to preclude the possibility of its being seasonal' (l.c., p. 29). In the fossil belemnites according to Abel (1916) the chambered phragmocone may have served as a hydrostatic apparatus to compensate for the weight of the massive rostrum. As far as I am aware we do not know with certainty the extent to which the chambering of the hind end of the gladius in recent cephalopods, e.g. *Chiroteuthis*, has a similar function.

Denton and his co-workers (1961) have described a hydrostatic apparatus in the cuttlebone of *Sepia officinalis* which swims well and for long periods. The chambers of the cuttlebone, which are independent of one another, contain gas, the volume of which is regulated by the osmotically controlled intake and release of fluid into and from the cuttlebone. The changes in fluid content take place at the hind end of the cuttlebone where the siphuncular wall has a special system of drainage vessels.

Over-compensated systems drift at the surface, sometimes projecting above it. There are only a few forms that do this. *Glaucus*, a shell-less opisthobranch gastropod drifts upside down immediately below the surface. It regularly contains gas in its gut; this is probably air taken in through the mouth. We do not know any more about its biology and so cannot give any further information on the mechanism of flotation in this warm-water form.

Here we can include the siphonophore *Physalia* and the chondrophores *Velella* and *Porpita*, in which the pneumatophore is particularly large. In *Physalia* the pneumatophore is a simple air chamber with a pore through which the gas can be expelled by the contraction of the walls. On the other hand in *Velella* and *Porpita* the pneumatophore consists of a complicated system of chambers (Fig. 81). These are probably filled by secretion of gas, because after emptying the little chambers became filled again in a short time.

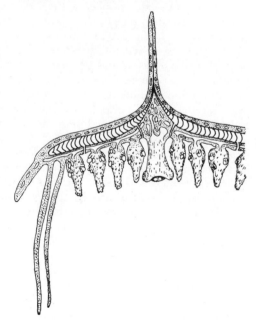

Figure 81. Section through *Velella* showing the air-filled chambers (after Hyman 1940, *The Invertebrates*, somewhat modified)

Gas bladders projecting above the surface catch with wind and act as sails, hence the terms 'by-the-wind sailor' and 'Portuguese man-o'war'. Woodcock (1944) reports observations according to which *Physalia* and *Velella* is so orientated to the wind as to cause the animals to sail about 45° to the left of the downwind direction. The deviation should be to the left in the northern hemisphere, but to the right in the southern: at 25–40°N some 421 animals were deflected to the left and only 9 to the right. The author correlates this with the fact that the wind-induced helical vortex pairs are asymmetrical with a larger clockwise vortex and a smaller anticlockwise one. *Sargassum* and other drifting material accumulates, in the convergent vortex, and plankton will be brought up from the layers beneath the surface in the divergence area. Since the deviations hold the animals for a longer period in the divergence area, they will be better protected there from collisions and find more suitable feeding conditions (Fig. 82). If these observations and this interpretation are confirmed it would mean that

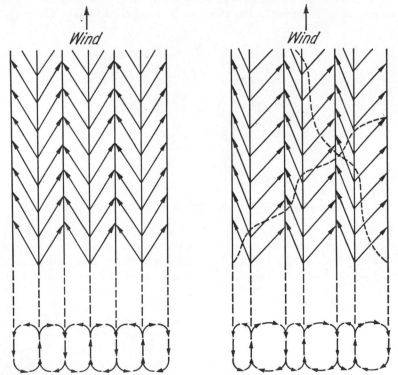

Figure 82.   The effect of the left drift of sailing *Physalia* and *Velella* on the asymmetry of the wind-induced surface currents (after Woodcock 1944)

in the northern hemisphere the right 'deviators' (in the south the left 'deviators') would be destroyed in large numbers in the convergence area. This can scarcely involve two geographically separated, genetically different types because there is insufficient isolation betweeen the populations north and south of the equator. Furthermore, according to Woltereck (1905) development takes place at great depths so that there is a considerable chance of exchange across the equator. Presumably, therefore, the different populations consist of selected phenotypes.

Mackie (1962) has recently reported that the angle at which *Velella* sails to the wind is dependent upon the strength of the wind but also fluctuates individually; in winds from Beaufort scale 4 the animals rotate so that their travel is slowed down.

Physalia can also collapse its float. It is still not known whether these animals are capable of active steering by lengthening or shortening the tentacles.

There are forms which construct a supporting float by a process of secretion; the classical examples are in the genus *Ianthina* (Fig. 83).

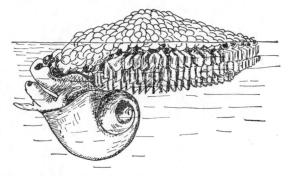

Figure 83. *Ianthina* with bubble float (after Fraenkel 1927)

There are five species in the tropical and subtropical regions of the three oceans (Laursen 1953) and these are only capable of a limited amount of active movement when hanging from the float. The lack of eyes and statocysts in the fully-grown animal must be correlated with this. The largest float found by Laursen was 12·8 cm long and it belonged to a specimen of *I. prolongata* with a shell height of 31 mm. The construction of the float has been described by Fraenkel (1927) in the following terms: the propodium creeps up along the edge of the float, bends away almost at right angles from the float when it reaches the surface so that the undersurface of the foot is directed upwards. The foot stretches out, secretes mucus at the surface and forms itself into a hollow in the shape of a spoon with rolled edges, enclosing some air. The edges close over the cavity, the foot is then withdrawn beneath the surface leaving the air in the cavity enclosed in a mucous envelope. The little air bubble is then attached to the edge of the float. The whole process takes 30 to 40 seconds and evidently represents a chain reaction which is released when the sole of the foot touches any surface, even under water. The process has been recorded photographically by Bayer (1963).

According to Fraenkel the veliger larvae are equipped with eye spot, statocyst and operculum and have a pelagic stage. According to Simroth (1895) the earliest observed stage of a float is a long rod of mucus, with a spherical swelling at the end which contains a few air bubbles. The method by which this first float is built is not known.

The float supports not only the animal itself but also on its underside the egg capsule. The older parts of the float break away continuously so that the animal always has to keep remaking it. Presumably the building process continues until the laying capacity of the animal is exhausted.

The barnacle *Lepas fascicularis* which usually lives solitarily can also be regarded as over-compensated; it has been investigated by Ankel (1950). At the basal end of the stalk there is an inflated balloon-like mass of secretion (Fig. 77) by means of which the animal is originally attached to drifting objects, mainly seaweeds, but which becomes so large that it is eventually able to support the whole animal. The mass of secretion is capable of supporting the animal by reason of its foamy structure, which is made up of concentrically arranged layers of secretion enclosing numerous gas bubbles. The origin of the gas bubbles is still not clear; the end of the stalk is conical and not disc-shaped as in other cirripedes; the cement gland opens at the surface of the stalk by numerous pores.

The Minyadidae (Actinia) have an organ comparable to the flotation apparatus of *Lepas fascicularis*. Here the basal disc is concave and a pneumatic apparatus of chitinous lamellae and honeycomb spaces is formed from the ectodermal-lined cavity thus produced. Unfortunately we still lack more detailed accounts of this organ and it is not yet clear whether the Minyadidae should be regarded as holopelagic animals or as larvae of benthonic forms.

Organisms that drift at the surface, such as *Velella, Porpita, Ianthina, Lepas* and others, are known collectively as the pleuston; the hemipteran genus *Halobates* which lives at the surface should also be included here (see p. 142). The pleustal represents only a narrow boundary zone of the pelagial, since the animals in it rely completely on the uppermost pelagial for their food; furthermore these organisms develop in the pelagial or sometimes even in the bathypelagial. A close relationship to the true pelagial is also apparent in the association of the fish *Nomeus gronovii* which lives in among the capture tentacles of *Physalia*. The morphological peculiarities of pleustonic animals and their biological relationships with one another do, however, justify our regarding them as an ecological unit.

It is noteworthy that the pleuston occurs largely if not exclusively in the warmer parts of the oceans. It is not clear whether this is correlated with the strength and frequency of the sea's movements or whether there are other causal connections. We also do not know with certainty the reasons for the occasional mass outbreaks, which have been observed in the chondrophores mentioned above and also in *Lepas fascicularis*.

Most members of the pleustal community are derived from bottom-living animals; examples are the chondrophores *Velella* and *Porpita*, actinozoans, gastropods, cirripedes and other crustaceans. The gerrid genus *Halobates*, originating in fresh water, has a corresponding way of life; *Physalia* has reached the pleustal from the pelagial. The pleustal also contains some nudibranch opisthobranchs which like *Fiona pinnata* feed on the tissues of *Velella* and climb from one prey animal to another without swimming (Bayer 1963).

5. Feeding methods

In the pelagial the methods of obtaining feeding vary considerably from the viewpoint of both form and function. The following types of feeding can be distinguished.

(*a*) Tentacle feeders: in which processes protruded from the body collect food particles which are then transported into the body. This method of feeding is found, for example, in radiolarians, foraminiferans, medusae, siphonophores and some ctenophores. The tentacles or similar processes enlarge the collecting area, sometimes to several times the body size, without significantly increasing the activity of the animal. The animal is dependent for food upon what comes within its sphere of influence. In this category we may perhaps include *Pelagothuria* in which the upward directed oral disc is somewhat analogous to the mouth and tentacle ring of the sessile actinians: it is possible that this form and the bathypelagic cephalopod *Retroteuthis pacifica* should be termed sediment-collectors. In *Retroteuthis* the arms, which are connected by a membrane, are turned dorsally through an angle of 90°, are widely spread out and together with the rear arms are fused with the dorsal body-wall (Joubin 1929) to give the impression of a sediment-collecting surface surrounded by the arm tips.

According to Joubin (1928) the cephalopod *Chiroteuthis* should also be included among the tentacle-feeders. The very delicate, semi-transparent, iridescent blue body is scarcely longer than 12 cm, but the tentacles reach a length of more than a metre. The suckers of the tentacles are modified in a characteristic way: the integument covering them forms bundles of sticky, branched filaments to which small planktonic organisms adhere. From time to time the tentacles are moved to the mouth and licked clean by the modified lips.

(b) Whirlers: These produce a water current along the body surface from which particles are trapped in mucus. Examples are the larvae of echinoderms, molluscs and worms, most ciliates and thecosomatous

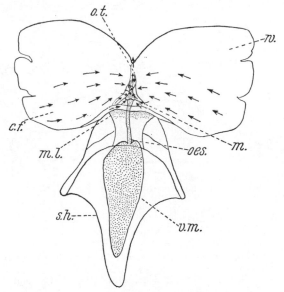

Figure 84. Ciliary currents on the wings in *Cavolinia inflexa* (after Yonge 1926) *cf* = ciliated field; *m* = mouth; *ml* = middle lobe of foot; *oes* = oesophagus; *ot* = outgoing ciliated tracts; *sh* = shell; *v.m.* = visceral mass; *w* = wing

pteropods (Fig. 84). The salps also have a modification of this method of feeding, for they obtain food particles from the water used in respiration, trapping them in mucus. It is characteristic of the whirlers that the food particles are small in relation to the size of the body. The same applies to:

(c) Filterers, which have developed various types of filtration apparatus, through which water is driven. It most cases the water movement is associated with some form of locomotion. The filtration apparatus may consist of combs of setae arranged on the extremities or on gill arches or of special filter chambers. Typical filterers are the Clado-cera, some Copepoda, Euphausiacea, Nebaliacea, Copelata (Fig. 85), some fishes such as the whale shark and basking shark, *Polyodon*

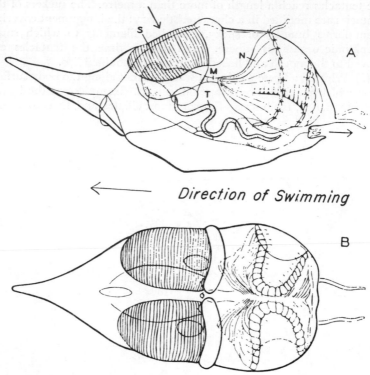

Direction of Swimming

Figure 85.   Filtration apparatus of *Oikopleura*: (A) animal in house seen from the side; (B) house from above.
M = mouth, N = net?, S = filter, T = tail (after Hesse & Dolflein)

spathula, Clupeidae, mackerel and *Brevoortia tyrannus* and the baleen whales.

The gill filtration apparatus of fishes is formed by processes which are arranged in a variety of ways on the gill arches and when well developed these effectively separate the buccal cavity from the gills. Hendricks (1908) showed that in the large sharks *Cetorhinus maximus* (basking shark) and *Rhincodon typicus* (whale shark) the long flexible gill-rakers consist of dentine with an enamel-like covering and are to be regarded as modified placoid scales. In a basking shark with a length of approximately 8 metres the gill-rakers were over 12 cm long (Fig. 86). Schnakenbeck (1955) doubted whether the gill-rakers played a part in collecting food, at any rate in the case of young animals (2·80–3·90 metres long) in which he did not find them. The papillae on the gums and at the base of the throat appeared to him to

Figure 86.   Gill-rakers of the whale shark (photo C. H. Brandes)

be more important; these become larger and more numerous towards the back, are branched at the end and up to 10 cm long behind the 5th gill slit and they encircle the throat. This view raises the possibility that the gill-rakers may only take part in food capture in combination with the throat papillae.

From various investigations on teleosts there appears to be a close relationship between the development of the filtration apparatus and the composition of the food. Thus in the Mugilidae the buccal cavity is closed off very completely from the gills; *Mugil auratus* has been shown to feed on diatoms, calanids, *Evadne* and very small amphipods. The clupeids have a very fine filter and their food consists of small pelagic organisms. Delsman (1939) found the following relations between the structure of the gill-rakers and the type of the plankton in three species of *Clupea* from the Java Sea.

Species	Occurrence	No. of gill-rakers on the lower half of 1st gill arch	Length of gill-rakers in relation to diameter of eye	Type of plankton
Clupea kanagurta	neritic	88	× 1·33	diatoms
Clupea fimbriata	intermediate	50	× 0·60–0·75	small copepods
Clupea leiogaster	pelagic	30	× 0·50	large copepods

There is a comparable situation in two species of mackerel from the same area: in the markedly neritic species *Scomber neglectus* the gill-rakers are longer, finer and narrower than in *S. kanagurta* which lives in more oceanic areas (Fig. 87).

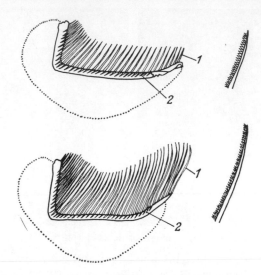

Figure 87.   Gill sieves: (a) *Scomber kanagurta;* (b) *S. neglecta.*
1 = gill-rakers of the first gill split; 2 = of the second gill split. To the right: one gill-raker (after Delsman 1939).

The clupeids *Alosa alosa* and *Alosa fallax* have 60–120 and 40–60 gill-rakers respectively; unfortunately I have found no satisfactory comparative data on the main food of these two species.

These results show that there is probably a relationship between the gill filtration apparatus and the type of food taken, and so the deep-sea fishes merit special attention. If these fishes show well-developed filtration organs one could, on a basis of their frequency and type of development draw conclusions on the size and abundance of the plankton which serves them as food. Zander (1907) gives a total of 133 gill-rakers on each side in *Cyclothone obscura,* but only 89 in *C. microdon.* Chapman (1939) records long slender gill-rakers, without giving their number, in *Myctophum oculeum, Melamphaes rugosus* and *Lestidium parri.* In contrast to this, several other bathypelagic species have short, sharp or pointed gill-rakers, which suggests that they may serve some other function. The comparative material is still too sparse to allow any definite conclusions, but it is noteworthy that the forms with long gill-rakers only have a relatively poor dentition.

Attention may, however, be drawn to the following points: gill filtration apparatus occurs on the one hand in the largest inhabitants of the pelagial (baleen whales, basking and whale sharks) and on the other hand in fishes that live in large shoals (clupeids, mackerel, *Cyclothone*). Some of these organisms are able to feed directly on the phytoplankton, or at least on its primary consumers; so they are closely associated with the early stages of the food chain. Here there is a possible relationship between shoaling and the method of feeding, such as can also be observed in invertebrates, e.g. copepods, etc., although this should not be regarded as a general ru e. It seems to me desirable that more attention should be paid to this kind of problem, including a detailed investigation of the 'gum organs' and 'throat sacks' of various fishes.

The different kinds of whirlers and filterers are connected by various intermediates. In both groups there are cases where the organism is capable of exercising a certain degree of selection; among the filterers this will be partly controlled mechanically by the mesh size of the filter, but there may also be active selection. Thus, Chang (1941) found that the stomachs of herring contained *Calanus* and *Temora* either in the same or in significantly greater proportions than in the plankton, and in *Anomalocera* also in greater proportions, but *Acartia*, *Oithona*, cladocerans and lamellibranch larvae were always present in lesser proportions. *Limacina* and *Sagitta* evidently served as a substitute for *Calanus*, because when there was a great scarcity of copepods the relative numbers of *Limacina* and *Sagitta* corresponded with those in the plankton, whereas they were otherwise only present in small numbers in the herring stomachs. Some observations indicate that *Calanus finmarchicus* filters without selection but also takes individual food particles selectively (cf. Marshall & Orr 1955). We do not know the conditions under which one type of food collection or the other is used, although this is presumably dictated mainly by the nature and amount of the available food. In feeding experiments on mysids, Lucas (1936) found that the amount of food taken varies with the density of the diatom culture offered, but that when finely suspended food is lacking these crustaceans will even seize large food particles, e.g. copepods, amphipods and mysids, and chew them. At the same time Lucas observed that in both mysids and the copepod *Eurytemora* the rate of feeding was lower in large population densities than in individual animals. The Copelata on the other hand feed exclusively by filtration and so one can distinguish between obligatory and facultative filterers. We do, however, require much more observational material in order to be able to assess the different cases (cf. also the summary by Jorgensen 1955).

(*d*) 'Seizers' take their food directly in the form of fragments, using jaws, a widely extensible mouth, chelae or even elongated arms with suckers. Fishes take their food with jaws which vary considerably in form according to the type of food, whereas in most toothed whales the dentition is strikingly uniform. The heteropods also use jaws to seize prey. Pincers or jaws are used by, for example, the chaetognaths, some polychaetes with protrusible proboscides, some copepods and

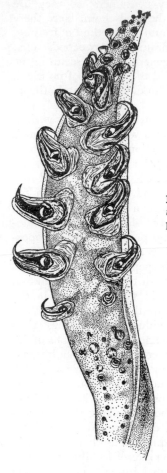

Figure 88. Arm of a juvenile cephalopod *Gali-teuthis* in which the suckers are modified into grip-ping hooks (after Chun 1910)

amphipods, and according to Chun certain cephalopods with modified suckers (Fig. 88). Cephalopods and gymnosomatous pteropods show striking parallelism in the development of capture arms or tentacles with suckers. Animals with very extensible mouths are for example *Beroë* among the ctenophores, some medusae and among the fishes the saccopharyngid gulper-eels.

There are various different ways in which the seizers take their prey. Active fast-moving hunters chase their prey which also usually moves fast; this is done for example by the toothed whales, marine turtles, fishes such as tunny, swordfish, etc., heteropods and among the medusae possibly the Cubomedusae which swim rapidly and have highly developed eyes. Among the birds the penguins are the primary exponent of this method of obtaining food. In contrast the skulkers lie in wait for the approach of prey and then rush out and seize it. They have well-developed organs of smell, taste and so on for the perception of the prey; deep-sea forms may also have luminescent organs to attract prey, but their eyes do not need to be so efficient as in the true hunters. There are still only a few live observations on this method of feeding but structure suggests that it occurs in some deep-sea fishes, e.g. Pediculati. The term collectors can be used for those forms which seize slow-moving prey without any special expenditure of energy on their own part. Thus the sunfish (*Mola mola*), which is well known to be a sluggish swimmer, is said to live mainly on medusae, for the capture which fast swimming is not required. 'Pipetters' which take their prey by sudden suction occur quite frequently among the fishes; here the mouth is usually protruded to form a long tube. The larvae of the cephalopod *Rhynchoteuthion* are also said to feed by pipetting. Among the sea-birds a special type of feeding is shown by the diving forms such as terns, frigate-birds, gannets, pelicans and others which feed almost exclusively on pelagic animals.

It has not yet been confirmed whether, for example, chaetognaths and ctenophores like *Beroë* catch their prey by purposeful pursuit or by a series of more or less random movements (chaetognaths) or perhaps by pumping water into the gastric cavity (*Beroë*, medusae?).

A special method of food capture is known in two molluscs. The cephalopod *Tremoctopus* holds the torn-off tentacles of medusae in its suckers and the nematocycts paralyse its prey (von Buddenbrook 1934). Evidently *Phyllirhoe* uses the medusa *Mnestra* in much the same way. According to Ankel (1952) this gastropod exploits the stinging action by holding the medusa with its foot which is shaped like a sucker (see p. 396). It has been reported that in the nudibranch family Fimbriidae which occurs at least occasionally in the pelagial, use is made of the extended velum as a capture net, with which crustaceans and even small fish can be entrapped (Odhner 1936).

Animals which feed by boring and sucking out their prey may be regarded as specialized hunters. Among the copepods this happens in *Sapphirina* and other semi-parasitic forms, and among the polychaetes in the family Typhloscolecidae. Species of *Trichocerca* (rotifers) bore through the body-wall of small planktonic organisms and suck out the contents. This method of feeding does not, however, appear to be very common in the pelagial.

Some of the animals which seize their prey, such as cephalopods, gymnosomatous pteropods, heteropods and perhaps crustaceans break up the prey before it reaches the gut. Lowndes (1935) reported that the copepod *Euchaeta norvegica* seizes it prey, particularly *Calanus finmarchicus*, with the maxillipeds and bites off pieces with the mandibles. In contrast to such animals there are those which swallow their prey whole, as for example the toothed whales, most fishes, ctenophores and chaetognaths. There is usually a certain size relationship between the swallower and its prey. The size of fish which sea-birds are capable of swallowing is very often quite astonishing. Scarcely credible facts have been given on the swallowing capacity of the killer whale (*Orca*): in the first chamber of the stomach of a specimen 7·5 metres long there were 'no less than 13 complete porpoises and 14 seals; a 15th seal was found in the animal's throat' (Slijper 1962, p. 274).

The principle of relative food size breaks down in deep-sea fishes which live in an environment poor in food. Many of these animals are capable of seizing and swallowing prey that is larger than themselves (Fig. 89). Stomachs that are grotesquely overfilled have been described

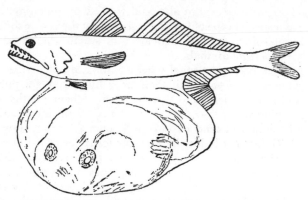

Figure 89. *Chiasmodus:* this specimen had swallowed a larger specimen of the same species (after Murray & Hjort 1912)

in several different groups of fishes, thus providing a common phenotypic character. Prey of this order of size can only be caught by specially adapted jaws, which are also to be regarded as a common characteristic of some of these fishes. In some the neurocranium and anterior vertebrae can be pulled upwards allowing the jaws to open very wide. This enables the buccal cavity to be enlarged so that oversize food can be swallowed; in association with this the front end of the vertebral column is cartilaginous and there are powerful neck muscles which pull the skull upwards and backwards. Günther has analysed various extreme types of development in some detail. In the Stomiatidae, of which the typical representative *Chauliodus sloani* is widely distributed, the jaws are equipped with very long teeth which ensure that large prey is held fast, but they can only function if the mouth can be

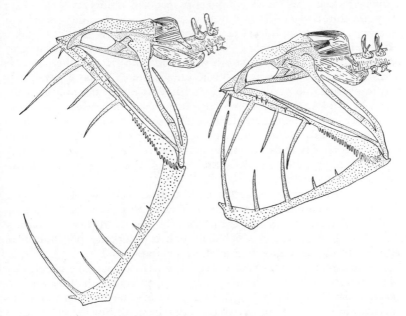

Figure 90. Skull and jaw skeleton of *Chauliodus sloani* showing mouth open and closed (after Günther 1950)

opened correspondingly wide (Fig. 90). A more detailed functional analysis of the jaw apparatus of *Malacosteus niger* shows that the jaws are shot forwards like a capture mask (Fig. 91). In this form the whole of the bottom of the mouth is lacking, so that the throat and neck region lie free, and when the jaws strike the resistance of the water is reduced and movement is facilitated. By contast, in the saccopharyngid

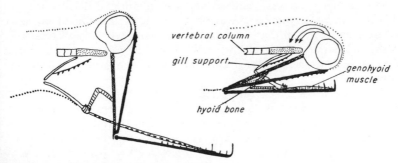

Figure 91. Diagram showing the structure, attachment and opening of the jaw in *Malacosteus niger* (after Günther 1950)

G——M.B.

gulper-eels the jaws have completely lost the ability to seize prey; they are thin and needle-like, very elongated and can be opened out sideways, so that the buccal cavity becomes enlarged to an incredible extent. These fish are poor swimmers which lie in wait for their prey and presumably open the mouth wide and place it over the prey. 'It is difficult to imagine how the fish can then close its jaws sufficiently fast to prevent the prey escaping, and how it moves the prey into the stomach; presumably there is a trigger mechanism between the upper and lower jaws, which keeps the mouth closed without further demands on the muscles' (Günther l.c., p. 65). *Monognathus toningi*, a relative of the saccopharyngids has no upper jaw, so that the prey can only reach the mouth and throat by the action of the lower jaw which can be pushed forwards.

The filterers provide a further exception to the principle of relative food size; an example of this is the size relationship between the basking shark or baleen whales and their food. To my mind there is scarcely any other living habitat which has such evolutionary possibilities in relation to the available food as the marine pelagial.

It is noteworthy that Lohmann (1914) described a striking peculiarity of the intestinal tract in the deep-water appendicularian *Bathochordaeus chuni*, which has a body length of 2·5 cm and far exceeds any of its relatives in size. The gill cavity is relatively small but the oesophagus and stomach are enlarged into sacs. Lohmann came to the conclusion that the food which probably consists of detritus and small planktonic organisms reaches the oesophageal sac. Some of the organisms will die there, but bacteria, small diatoms, *Gymnodinium* and zooflagellates, for example, may survive and convert the detrital matter which is difficult to utilize into more easily digestible food of much higher nutritional value. 'Such an improved method of exploitation would be of particular importance in the deep sea which is so poor in available food' (p. 167). This is certainly a possible interpretation, but to me it seems simpler to suggest that the nannoplankton and detritus used as food by the appendicularians are not uniformly distributed but only occur very locally so that the appendicularians, which have relatively restricted powers of movement, may from time to time suffer from a lack of food; the oesophageal sac could then be regarded as a storage container.

This interpretation is supported by Rowett (1943; according to Cannon 1946): the crustacean *Nebaliopsis* is a specialized egg-eater which has an enormous sac-like appendage to the midgut. This only has a few septa and so does not provide a large surface for resorption and it probably serves as a storage organ for the food which is only present at certain periods. The sac-like enlargements of the intestinal tract could therefore serve for the receipt and storage of large amounts of food when it is available and thus help the animal to survive during periods when food is scarce.

6. Reproduction and development

The pelagic region is characterized by its enormous three-dimensional extent and by its lack of clearly defined subdivisions. One might therefore expect that pelagic animals would show certain special features in their reproduction and development to help protect their populations. The amount of observational material available is relatively small and so one can only discuss a few examples; a more detailed assessment would require a much greater body of information.

Brood protection is known to occur in quite a number of pelagic animals, and in some phyla there are cases of viviparity. A few examples will be given:

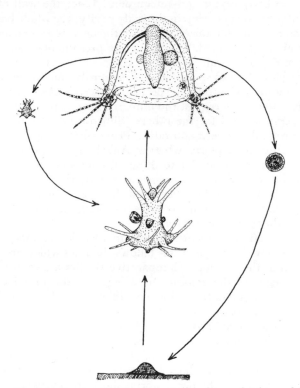

Figure 92. Diagram showing the developmental and reproductive cycle in *Margelopsis haeckeli* (after Werner 1954, somewhat modified): unique benthonic stage: overwintering eggs

Several medusae, particularly in the deep sea, are viviparous; but brood protection also occurs in neritic forms, because the polyp generation may develop on the medusa (cf. for example, Werner 1954 a on the anthomedusan *Margelopsis haeckeli*, Fig. 92). Among the cephalopods brood protection is found in *Argonauta argo*; the thin, unchambered external shell serves as a brood chamber. In the female of *Tremoctopus violaceus* the four relatively long dorsal arms are joined by a membrane and are carried curled up to form a brood chamber in which the eggs are attached to the proximal suckers.

Among the thecosomatous pteropods, Tesch (1946) mentions viviparity in *Limacina helicoides*, *Euclio chaptali* and *E. campylura* which all occur at great depths. In the gymnosomate *Halopsyche gaudichaudi*, Meisenheimer (1905) described the following conditions: a brood sac appears probably by modification of the accessory glands of the hermaphrodite apparatus; it becomes much enlarged and fills the space between the body cavity and integument. Later the wall of this sac bursts and the young develop in the body cavity; the whole body cavity thus becomes so much reduced that only the nervous system and organs of locomotion are retained. The young are presumably released by the bursting of the body wall.

The species of *Ianthina* which live solitarily and have a mucous float are mostly oviparous (4 species), but *I. ianthina* is viviparous. Evidently some of the eggs in the numerous egg capsules do not develop and serve as food for the others (Simroth 1895, Laursen 1953). In these protandrous hermaphrodites Pruvot-Fol (1925) assumed that self-fertilization took place, whereas Ankel (1926) noticed that the spermiozeugmae were similar to those of *Scala* and thought that they were capable of independent movement and that cross-fertilization took place; Laursen on the other hand thought it improbable that the spermiozeugmae could remain alive for long in open water. This problem is therefore still not solved.

Brood protection frequently occurs in pelagic crustaceans and in the Cladocera and Copepoda this follows the lines which are generally well known in these groups. Comparative studies are required on the number of eggs in the egg-sacs or of eggs in the brood chambers. *Evadne nordmanni* shows the following differences: Baltic Sea up to 5, usually only 3 embryos, North Sea 7–8, Cape Town up to 10, off Nagasaki up to 20 embryos. It is still not at all clear whether this is due to variations in season or environmental conditions.

In a comprehensive summary of the observations on euphausians made by several other authors Einarsson (1945) stated that in both cold and warm waters the neritic forms show greater variability in the number of developmental and moulting stages than the oceanic forms. The greater range of environmental conditions in coastal waters might be expected to favour the development of different phenotypes to a more marked extent than would be the case under the more uniform conditions of the oceanic region.

Methods of ensuring fertilization by means of special organs are found particularly in those forms which occur in relatively small numbers. Brinkmann (1913), for example, showed that in pelagic nemertines the gonads of the males usually lie massed together at the front end and that in some species 'head cirri' functioning as clasping organs occur at the time of sexual maturity; *Bathynectes murrayi* has elongated penis-like genital canals. In general it seems that when fertilization is internal only a few relatively large eggs are formed (Fig. 93).

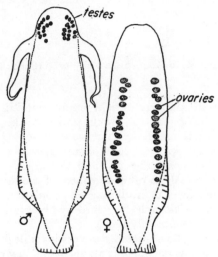

Figure 93. The pelagic nemertine *Nectonemertes mirabilis*, male with attachment organs and testes on the ventral side at the front end, female with ovaries arranged in pairs (after Coe 1943)

Copulatory organs and receptaculae are developed in the polychaete family Alciopidae.

In this context the best-known examples are the deep-sea anglerfishes (Ceratioidea) which occur mainly between 1,500 and 2,000 metres. The female reaches a length of more than a metre but the male remains small and lives attached to a female (Fig. 94). After attachment a placenta-like connection develops between the skin of the female and that surrounding the jaws of the male; this still allows the male to take in water for respiration. The male is, however, nourished by its blood circulation which is in communication with that of the female. There is no doubt that this parasitic attachment represents a method of ensuring fertilization.

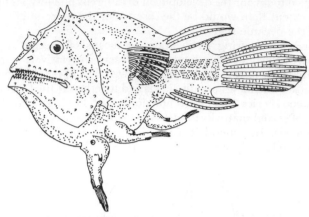

Figure 94. Female of *Edriolychnus schmidti* with three attached dwarf males (after Regan 1932).

The species-specific patterns of light organs found in many deep-sea pelagic fishes, cephalopods, etc., would probably also be regarded as having a similar function although this may not be their sole biological significance. In some species, light organs may play a part in the maintenance of the shoal (Fig. 95) and in solitary predatory forms they may serve as signals to other members of the same species. This would also be a means of maintaining the population. This, however, is a problem which would only be solved by observations made in the wild, which are still completely lacking.

In some cases it may be expected that very juvenile forms are already sexually mature, so that paedogenesis or neoteny may occur. Pelseneer (1888) regarded the pteropod *Clio aurantiaca*, which becomes sexually mature at a length of 2–2·5 mm, as a juvenile paedogenetic form of *Clione flavescens*, and in the sexually mature form described as *Paedoclione* n. gen. *doliiformis* n. sp. Danforth found larval characters, in the form of three ciliated bands, conspicuous caudal lobes and the small number of radula teeth.

In ctenophores, Chun (1892) discovered the process known as dissogony, in which very young animals become sexually mature, and after a period of reproduction the gonads become reduced. This is followed by metamorphosis to the adult animal which then has a further reproductive period. We still do not know the extent to which this phenomenon, which has been observed for example in *Bolina hydatina* and *Eucharis multicornis*, is more widely spread in the Ctenophora.

Carlgren (1917) and Gravier (1919, 1920) recorded dissogony in

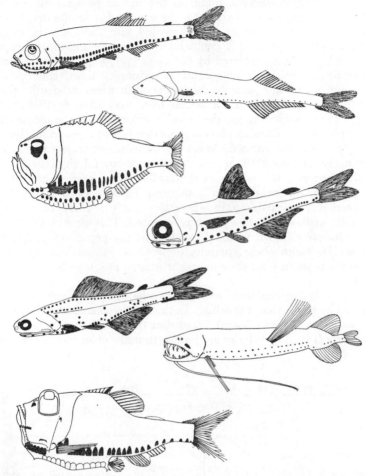

Figure 95. Arrangement of light organs in various deep-sea fishes (after various authors, some modified, from Marshall 1957): *Vinciguerra attenuata, Cyclothone microdon, Argyropelecus gigas, Myctophum punctatum, Lampanyctus elongatus, Bathophilus longipinnis, Argyropelecus affinis*

actiniarians: they found pelagic larvae with ovaries or embryos, but no males.

The precocious occurrence of sexual maturity ought to favour the survival chances of the species in the pelagial; it would be interesting to have data on the correlation between the number of eggs produced in this type of reproduction and the mean rate at which they are destroyed.

The fish genus *Schindleria*, found in the neritic pelagial off Tahiti and Samoa, is transparent except for the eyes and swimbladder, has a maximal length of 2 cm and is elongated and larva-like in form. Schindler considered that it could be the sexually mature larva of *Hemirhamphus*, a view contested by Giltay (1934); Bruun (1940) could make no final decision on the systematic position of these forms.

In many forms the juvenile stages occur in areas or depths that differ from those in which the adults live. According to Woltereck (1905), in the Mediterranean the larval development of the pleustonic chondrophoran *Velella* takes place at great depths and these animals only gradually rise to the surface. Ankel (1962) assumed the same for the Atlantic oceanic populations and ascribed a special significance to this phenomenon: the adults at the surface were carried eastwards by currents and became sexually mature. Then in deep water they were carried westwards again where the young animals once more rose to the surface and the cycle was repeated. This view could really only be substantiated if we could show that the populations were in fact associated with these currents. *Glaucus* and *Ianthina* may also behave in the same way for they too occur in association with *Velella* and *Porpita*.

In many fish species from medium depths the young are found in higher water layers than the adults. In the Norway haddock or bergylt, *Sebastes marinus*, Tåning (1949) found that the larvae were tied to the 8° isotherm (Fig. 96). Whales are known to make extensive migrations

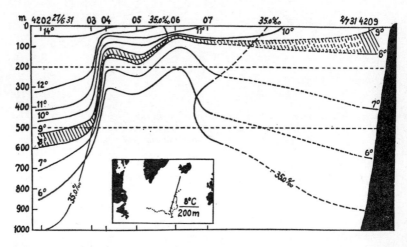

Figure 96. Correlation of the larvae of *Sebastes marinus* with a water temperature of 8–9°C in the North Atlantic (after Tåning 1949)

from antarctic waters into subtropical and tropical latitudes where the young are born. Fish migrations lead to mass aggregations in the spawning grounds (e.g. in herring) or to movements between the sea and fresh water. Anadromous fishes spawn in fresh water, where they start to grow. They then move into the sea where they usually live for several years until sexually mature when they again migrate into fresh water. Some fishes, particularly salmon, are known to return with remarkable fidelity to their native waters. Hasler has traced this back to an early and lasting imprinting of the young by the smell characteristic of the waters where they hatched. The eel is the classic example of a catadromous fish, which goes to the sea to spawn; according to Schmidt (1925) the European eel spawns in medium depths in the Sargasso Sea, whence the larvae rise to the surface and are carried by currents to the coast of Europe (Fig. 97).

Tucker (1959) has put forward a new theory on this which is still the subject of controversy. Eels from other areas of fresh water also have their spawning grounds in places far from the coasts; thus the

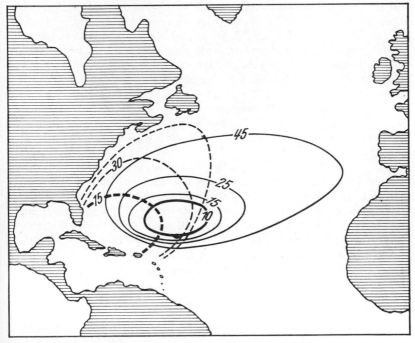

Figure 97. Distribution of the larvae of the European (solid lines) and the American eel (broken lines) plotted according to the growth of the larvae (figures show larval size in mm) (after Schmidt 1925)

South African species of *Anguilla* spawn in the Indian Ocean east of Madagascar between 10°S and 20°S and 60°E and 65°E in the region of the South Equatorial Current (Jubb 1961). The big spawning migrations of tunny in the Mediterranean Sea provide the occasion for a special fishery, after which the fish disperse again.

7. The origin of the pelagic fauna

Elements of very diverse origin can be recognized in the fauna of the pelagial: some are immigrants from fresh water or from the land, many are forms which have reached the pelagial from the benthal with which they still show clear relationships. while there are some forms which have obviously lived in the pelagial for a very long period of time and now show only a distant relationship with the benthal.

The freshwater elements in the sea have been discussed by Remane (1950); in the pelagial these include some protozoans, rotifers and phyllopods but their numbers are small. The protozoans and rotifers are to a large extent members of genera which occur mainly in fresh water, and they are often conspecific with those found in fresh water. The physiological potentiality of these species is astonishingly wide (cf. also Focke 1961). The phyllopod genera *Podon*, *Evadne* and *Penilia* are so far removed from their freshwater origins that they are now classified in genera that are purely marine.

The hemipteran genus *Halobates* contains forms which move about on the surface of the sea like the freshwater gerrids. Five species of this genus have evolved in the warmer seas, although the distribution of the individual forms is not yet known in detail. These are the only insects which have penetrated into the oceanic region (see Chapter VII, 2).

Many authors are of the opinion that fishes also migrated into the pelagial from fresh water. Thus Allen (1922) starts from the assumption that the typical fish form and the powerful swimming movements could only have evolved in an environment with strong currents. Spawning on the bottom is understandable as a protection against the eggs being washed away and this would have been retained after the move into the sea, as for example in the herring. However, this is a problem which needs further investigation.

Immigrants from the land into the pelagial are primarily the whales, which show marked specialization and two different structural types, the toothed and the baleen whales. The toothed whales have evidently undergone explosive speciation, because the eighty-five known species (including the dolphins) are classified in thirty-three genera of which no fewer than fifteen are monotypic.

The immigration of these mammals into the pelagial and their specialization are doubtless correlated with lack of competition for food and the presence of relatively few enemies. This view is supported by the evolution of the two methods of feeding, namely predation and filtration. The same point of view can be adopted in regard to the presence of the sirenians in the sea and to the evolution of the ichthyosaurs during the Jurassic and Cretaceous.

The penguins and marine turtles may be regarded as immigrants from the land which have not completely given up their connection with a terrestrial environment. Here again their evolution can be most easily explained as correlated with the lack of competition for food.

According to the view of Chun (1886, p. 56) the pelagic fauna is phylogenetically older than that of the other regions: 'we still have a very incomplete knowledge of the pelagic fauna, from which the littoral forms, the deep-sea and freshwater animals and ultimately the land animals have evolved. . . . ' Stiasny (1913, p. 14) expressed a similar view: 'the more primitive character of plankton organisms due to the favourable living conditions . . . . '. There are certainly some archaic forms in the pelagial (see below), but on the whole one has the impression that the holopelagic animals have certain structural characters derived from bottom-living forms.

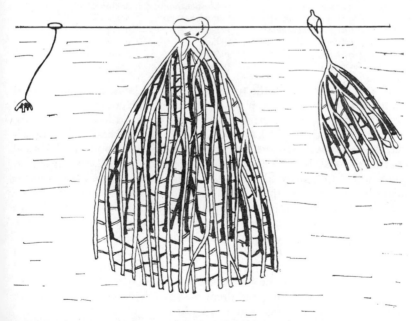

Figure 98. Pleustonic graptolite *Dictyonema flabelliforme* (from Hundt 1940)

Fossil remains of pelagic animals are only known in small numbers; this is largely due to the lack of hard parts suitable for fossilization. There appear to have been morphological types in the Cambrian which are reminiscent of recent forms, e.g. medusae (review in Mayer 1910), chaetognaths from the Middle Cambrian of British Columbia, which are very similar to the recent benthal form *Spadella* and a fossil which Lohmann (1922) regarded as an appendicularian. Other forms, such as the graptolites of Silurian seas, have died out (Fig. 98). According to Maas (1911), medusae from the Jurassic and Cretaceous are strikingly similar to recent deep-sea Rhizostomeae, e.g. *Paraphyllites* from the Solenhofer beds to *Paraphyllina*, *Atollites* from the Cretaceous to *Atolla*. On the other hand, according to Meisenheimer (1905), the Palaeozoic and Mesozoic 'pteropods' have nothing to do with the true pteropods, a characteristic group of the pelagial which probably evolved mainly in the Tertiary. Abel (1916) thought it probable that some of the dibranchiate cephalopods of the Jurassic and Cretaceous lived pelagically: he also considered that some forms lived pelagically when young but moved to the bottom when they became adult:

| | Habits | |
Species	juvenile	adult
Cuspiteuthis acuarius	nekto-benthonic	nektonic
Mucroteuthis giganteus	nekto-benthonic	nektonic
Belemnites semisulcatus	pelagic	pelagic
Hibolites	pelagic	pelagic
Acroteuthis	pelagic	nekto-benthonic
Aulacoteuthis	pelagic	nekto-benthonic

The method by which the transition from the benthal to the pelagial could have taken place can be worked out with a certain degree of probability. Originally perhaps, organisms attached to drifting objects (timber, seaweed) might have been forced to live free-floating, provided that they possessed the necessary morphological and physiological characters, that is, that there was a predisposition. *Lepas fascicularis* seems to me to be a good example of this. In other cases sessile forms might be torn up from the benthal by currents and transported into the pelagial, as is shown apparently by the hydroid stolons of *Clytia pelagica* at the present time. It is also necessary to invoke predisposition for the view that *Velella* and *Porpita* have evolved from torn-off polyp heads of forms perhaps closely related to the genera *Tubularia*, *Corymorpha* and *Branchiocerianthus* (Ankel 1962). If Ankel's theory is correct one would have to assume that at the same time as the polyp heads were released, the opisthobranchs living on them also reached the pelagial where they became endemic (*Glaucus* and *Fiona*).

Pelagic forms could also have evolved from larval stages which were able to complete their metamorphosis in the open water. In this way, for example, appendicularians, cranchid cephalopods, polychaetes such as the Tomopteridae, the Typhloscolecidae and Phyllodocidae of the *Lopadorhynchus* type could have been evolved. True neotenous forms, however, would only appear to have evolved to a limited extent, if one agrees for instance with Steuer's view (1910 p. 450) that the cranchid-like cephalopods and the appendicularians show larval characteristics. Dales (1955) regards neoteny as possible in the aberrant polychaete family Typhloscolecidae. It has been said that some gymnosomatous pteropods become sexually mature as young larvae. But since development goes further one can only speak of a precocious and thus lengthened period of reproduction. Goldschmidt (1906) regarded *Amphioxides* as a 'neotenously developed branchiostomid larva', which 'in its structure represents the most primitive chordate type so far known' (p. 447).

Many motile forms normally rise from the benthal into the pelagial, particularly at night time, and here food supply is evidently the principal factor involved. It is quite conceivable that if such animals were carried along into deeper water they could no longer return to the benthal. If they had sufficient swimming powers this would not be absolutely necessary, because by descending into deeper water during the day they would be carried away from the light. In this way for instance many crustaceans and worms could have been recruited to the pelagial. In various polychaete families, e.g. Syllidae, Nereidae, Eunicidae and Phyllodocidae, free-swimming epitokous forms appear during the time of sexual maturity; these show specific characters for pelagic life in the form of enlarged eyes and the elongation and broadening of the body appendages. If this type of structure were retained it could lead to a lasting occupation of this habitat. Finally, it is conceivable that flight reactions have led animals into the pelagial, if these originally had a fluttering type of swimming and a predisposition to this type of evolution. Flight could be stimulated by animal enemies but it could also be due to a prolonged period of unfavourable changes in the chemistry of the benthal and of the water near the bottom (e.g. oxygen lack, H_2S formation).

As an example of immigration into the pelagial that is possibly occurring at the present time one may mention: '*Chioraera* (known also as *Melibe*) *leonina*, a large nudibranch of eelgrass areas, seems to be in the process of becoming a pelagic animal. . . . When the water is smooth, and especially during cloudy weather, this queer creature puffs out its hood with air, thus floating to the surface where its tentacles rake in small crustacea for food' (Hedgpeth in Ricketts & Calvin 1962, p. 251). Unfortunately, nothing is said about the origin of the air which enables the animal to float.

The active and passive invasion of the pelagial as a habitat seems to have been primarily associated with the chance of finding food, with the exploitation of the phytoplankton production, and it was only subsequently that other types of feeding developed. In this process it is evident that an increasing number of ecological niches become available to the immigrants, as demonstrated for example by the presence and speciation of the whales, the hemipteran genus *Halobates* and the Cladocera.

Within the individual major groups immigration has usually taken place along a number of different lines. Thus according to Kerr (1931) the Eugastropoda have had four lines leading to the pelagial: from the Pectinibranchiata to the Heteropoda, from the bulloid Tectibranchiata to the thecosomatous Pteropoda, from the aplysioid Tectibranchiata to the gymnosomatous Pteropoda and from the Nudibranchiata to the Phyllirhoidae. It is noteworthy that the four lines also represent four different methods of feeding. The pelagic Polychaeta also show four main lines of immigration, namely the Phyllodocidae, Alciopidae, Tomopteridae and Typhloscolecidae, to which may be added a few others, e.g. the Aphroditidae. Among the tunicates there are three different lines, the appendicularians, the *Pyrosoma* group and the Salpidae with the Doliolidae. A comparable analysis of the Crustacea would certainly show a large number of such immigration lines within the different orders. In the ostracods, for example, only the family Halocypridae and a few members of the Cypridinidae are pelagic, and here evidently there are no structural characters which allow these forms to adopt pelagic habits. Presumably the pelagic nemertines are also polyphyletic and not phylogenetically uniform, since forms with separated muscle layers in the rhynchocoel wall occur alongside those with interwoven musculature; similarly there are forms with simple lateral nerve cords alongside those with cords composed of more than one type of fibre. These relationships are reminiscent of those in land nemertines, in which Friedrich (1955 a) has traced various different origins.

On the other hand, certain phyla or orders have only reached the pelagial by a single route. Thus in the actinians the only pelagic representatives are the Minyadidae, in the echinoderms only the Pelagothuridae. The Minyadidae are possibly only larvae, for no sexually mature forms have yet been found.

The chances of evolving a holopelagic way of life must also be strongly influenced by the reproductive biology. Thus the cephalopods normally lay eggs on the bottom, but if any of them were to evolve into pelagic animals they would have to lay pelagic eggs capable of developing in the open sea, that is, unless they evolved methods of brood protection. In the alciopid polychaetes copulation has been confirmed with the development of receptacula and spermatophores.

In the Peracarida the Lophogastridae, mysids and amphipods have numerous representatives in the pelagial, whereas the isopods have only a few and the Cumacea and Tanaidacea none at all. A subdivision of the Peracarida made by Siewing (1951) places the Lophogastridae, mysids and amphipods in one group and the Cumacea, Tanaidacea and isopods in the other. This grouping, based on morphological characters, corresponds to a great extent with the development of pelagic habits, and so it is interesting to consider whether the structural characters of the first group show a predisposition to pelagic life. The common possession of a pyloric funnel and pyloric bristles, which are associated with feeding, appears in fact to show a predisposition for the collection of finely particulate food. Furthermore mysids and amphipods retain the marsupium between two brood periods and their young hatch in a fully formed condition; there are evidently still no data available for the Lophogastridae. In the members of the second group, on the other hand, the oostegite is lacking between two brood periods and the young lack the last pair of thoracic appendages when hatched. The lengthened period of brood protection in the first group appears to provide a further predisposition for continuous settlement of the pelagial.

The latter viewpoint is especially emphasized by Russell & Rees (1960) in their description of the viviparous scyphomedusan *Stygiomedusa fabulosa* from deep water in the Atlantic: 'Many deep-water organisms are thought to have been evolved by migration from shallow or surface water, and *Stygiomedusa* is probably no exception. Encystment of the larvae, such as frequently occurs in *Chrysaora* and *Cyanea* was probably a major factor in enabling the life cycle to take place in the parent medusa. In this way the jellyfish became independent of a substratum for its polypoid phase (l.c., p. 316).

It is evident, therefore, that there are cases in which the mechanisms of feeding and reproduction are predisposed in such a way as to enable the pelagial to be invaded. Here it is self-evident that the ecological role of the sea bottom and of the open water has been changed in a basic way.

Many of the inhabitants of the pelagial must certainly be regarded as relatively young members of its fauna, whereas others are probably phylogenetically very old. Thus, for example, the ctenophores and chaetognaths are phylogenetically very isolated, having no close relationships to benthonic forms, because the few chaetognaths in the genus *Spadella* and the ctenophores in the genera *Coeloplana* and *Tjalfiella* must be regarded as secondary members of the benthal, which have evolved from pelagic forms. Pickford (1946, 1949) regarded the bathypelagic dibranchiate cephalopod *Vampyroteuthis infernalis* Chun as archaic; this form is classified in its own order, the *Vampyromorpha*, and appears to be closely related to the fossil Mesoteuthidae. In the same way *Nautilus* is an archaic form represented by a single species without close relatives, and the same can be said of *Spirula spirula*.

The maintenance of these archaic types and the isolated position of many pelagic groups indicate that these are very old and in my opinion characterize the pelagial as a habitat in which the conditions are extreme in comparison with those of the benthal and phytal. This view is supported by the fact that certain phyla, classes and orders are more or less unrepresented in the pelagial, and that evidently only a few forms have found their way back into the benthal, as for example the genera *Lucernaria* and *Haliclystus* among the Scyphomedusae, the hydromedusan *Eleutheria*, the ctenophores *Coeloplana* and *Tjalfiella*, and the chaetognath *Spadella*; an analysis of the fishes would doubtless yield a few examples.

It seems to me that there are certain parallels to this relatively limited return migration: only a few organisms have moved from fresh waters into the sea, whereas the reverse route has been well trodden; we know of no emigrants from subterranean habitats, but there are several different immigration routes into these. Kosswig (1950) has described the recent immigration of Red Sea fishes into the Mediterranean and has shown that the reverse route from the Mediterranean through the Suez Canal into the Red Sea has scarcely been used.

A further example is provided by the colonization of the deep sea by fishes; on the one hand it has been shown by Bauer (1908) that the forms involved have been geologically relatively young and, on the other hand, they have become specially adapted to the paucity of food by the development of certain structural characters (Günther 1950). There is no evidence for any return migration from the deep sea.

There is still no generally valid explanation for this predominantly unidirectional migration. It is perhaps possible that under the pressure of population and competition in areas with optimal conditions certain forms are able to move into neighbouring areas where the conditions are more favourable. Here under the selective influence of the new conditions they undergo unilateral evolution which prevents them returning to the area from which they originally came.

From an analysis of the results of several expeditions Friedrich (1954 a) has shown that the number of species in the pelagial is quite small in comparison with that in the benthal, and that they are fairly well known; even such rich collections as those made by the Thor, Dana and Discovery Expeditions have only increased the number of known species by a very small percentage. This limited number of species also confirms the extreme character of the pelagial.

The systematic status of any given group ought to provide some measure of the length of time during which it has been represented in the pelagial. On this view, those pelagic species whose closest relatives are benthonic and which are classified in the same genus ought only to have invaded the pelagial in relatively recent times. On the other hand higher systematic groups (orders, classes) occurring exclusively in the pelagial would suggest a very long period in this environment. It is also possible that the size of a systematic group would provide some useful measure in assessing this problem. In this context Friedrich (1955 b) has pointed out that within several families of various classes there may be individual genera with a large number of species, whereas a surprisingly large number of genera only contain one or very few species. Since the settlement of a new habitat has often been observed to result in explosive speciation into numerous species, it is possible that the genera rich in species have only immigrated into the pelagial relatively recently, whereas the monotypic genera are older inhabitants of this habitat. This argument is certainly justified in a few cases, for example in the genus *Conchoecia* among the ostracods, but one should not generalize because, to be consistent, it must lead to the acceptance of polyphyletism in certain groups. In the appendicularians, for example, there are the genera *Oikopleura* and *Fritillaria*, each with several species, alongside no fewer than seven monotypic or oligotypic genera, and yet no one would wish to argue that the two polyspecific genera could have moved into the pelagial more recently than the others. It is, however, quite conceivable that explosive speciation in certain genera indicates a more marked biocoenotic subdivision of the pelagial as a whole than would at first appear to exist. Thus, for example, the invasion of the oceanic region by certain neritic forms could indicate such a change in habitat with corresponding evolutionary change.

This raises the question of the chances of speciation in the marine pelagial, for it is peculiar that the evolution of new species takes place in an apparently continuous environment with relatively stable external conditions and little chance of isolation. Data on this have been summarized by Friedrich (1955 b). Brinton (1962) has also recently written on this subject in connection with his work on Pacific euphausiids.

As a supplement to this we may quote the comments of Hubbs (1951) on the circumtropical fishes *Naucrates ductor* and *Remora remora*: 'The specific integrity of such circumtropical pelagic fishes poses a problem in speciation, since widespread littoral types of the tropics, as well as most pantemperate types, in contrast, are ordinarily differentiated into allopatric subspecies or species, though their period of isolation has ordinarily been no longer, often briefer, than that of the land-separated populations of the tropical pelagic forms. The explanation seems to lie both in the uniformity of the tropical pelagic environment and in

the relationship between the population structure and the rate of speciation. Large, widespread populations that are not disrupted into effective reproductive units are now generally held, on the basis of Wright's
analysis, to be little subject to differentiation.'

8. The pseudopelagic fauna

Animals attached to plants or other animals or to other floating objects
may be termed pseudopelagic. Here we may differentiate between
those forms which reach the pelagial accidentally together with their
substrate and may even occur frequently in the pelagial, but whose
true home is in the benthal (facultative pseudopelagial), and those
species which occur exclusively in this way, their whole life cycle being
spent in the pelagial (obligatory pseudopelagial).

The best-known example is the fauna of Sargasso weed, which has
given its name to the Sargasso Sea. The following list gives the species
which Hentschel (1921) found on *Sargassum*, *Macrocystis* and floating
volcanic slag from the Brazil and Falkland Currents. To the best of
my knowledge there is no comprehensive account of the animal
populations found on floating timber. In addition to Hentschel's
data the following list also contains supplementary information given
by Winge (1928) and Timmermann (1932). This fauna has also been
discussed by Prat (1940).

Flora: *Sargassum natans* L., *S. fluitans* Borg., *S. chamissonis* Kutz;
              Dictyota cervicornis, *Iania capillacea* Harv.; *Cladophora*.
Pisces: *Antennarius marmoratus* Günther, *Syngnathus pelagicus*
              Osbeck, *Phyllopteryx eques* Günther.
Tunicata: *Diplosoma gelatinosum* Edw.
Mollusca: *Modiolarca trapezina* Lam., *Scyllaea pelagica* L., *Litiopa
              melanostoma* Rang, *Spurilla sargassicola* Krøyer, *Corambella*
              sp.
Decapoda: *Planes (Nautilograpsus) minutus* (L.), *Neptunus sayi* (Gibbes),
              Virbius acuminatus (Dana), *Latreutes ensiferus* (M.-Edw.),
              Leander tenuicornis (Say).
Isopoda: *Ianira minuta* Richardson, *Probopyrus latreuticola* Gissler;
              also isopods which belong to the Central American
              coastal fauna, which have not been identified (Timmer-
              mann l.c.).
Amphipoda: Unidentified gammarids, caprellids of the *Aegina* group.
Cirripedia: *Lepas anatifera* L., *L. pectinata* Spgl., *L. anserifera* L.,
              L. australis Darw.
Pantopoda: *Anoplodactylus petiolatus* (Krøyer), *Chilophoxus spinosus*
              (Mont).

Polychaeta: *Nereis* (*Platynereis*) *dumerilii* Aud. Edw., *Spirorbis* sp., probably *S. corrugatus* Mont. and *S. formosus* Bush; Timmermann also mentions five free-living species which belong to the West Indian benthal.

Turbellaria: The polyclads *Planocera grubei* Graff., *Stylochoplana sargassicola* (Mertens).

Bryozoa: *Membranipora tehuelca* d'Orb., *M. bellula* Heks., *M. reticulatum* (L.), *M. tuberculata* Busk., *M. savarti* Aud. *Scrupocellaria scrupea* Bush, *Aetea azorica* Jul. Calv. *Bowerbankia* sp., *Thalamoporella falcifera* Hincks.

Actinia: Found repeatedly, not identified (Timmermann l.c.).

Hydrozoa: *Gemmaria implexa* Ald., *G. costata* (Gegenb.), *Clytia johnstoni* Ald. (?), *C. simplex* Congd., *C. noliformis* (Mc Crady), *C. cylindrica* Agassiz, *Obelia hyalina* Clarke, *O. geniculata* L., *Halecium nanum* Ald., *Hebella calcarata* (Ag.), *Sertularia mayeri* Nutt., *S. versluysi* Nutt. *S. rathbuni* Nutt., *S.* (*Pasythea*) *quadridentata* (Ellis & Solander). *Plumularia sargassi* Vanh., *P. obliqua* Sound, *P. catharina* Johnston, *Monotheca margaritta* Nutt., *Aglaophenia latecarinata* Allm., *Antennella secundaria* (Gm.),

Within the groups the percentage frequency fluctuates between 5% and 90%, the dominant forms being *Membranipora tehuelca* with 90%, *Spirorbis* sp. with 85% and *Clytia* with 65%. Among the decapods the most abundant is *Planes minutus*.

According to Brandt (1892) this fauna can be subdivided into: sessile forms (hydroid polyps, actinians, bryozoans, *Spirorbis*, cirripedes and tunicates), haptic animals which cling to the weed (*Antennarius* (Fig. 99), the decapods *Planes*, *Latreutes*, *Leander* and *Virbius*, the gastropods and *Nereis*) and those forms which are mainly free-swimming but which also hide among the weed (*Syngnathus* and *Neptunus*). These forms are quite close to those which occur in the phytal of coastal regions. From analyses of the gut contents of the filterers and whirlers it appears that they feed exclusively on seston; the organisms found were coccolithophores, peridinians, diatoms, tintinnids, globigerinans, radiolarians, *Physalia* nematocysts, eggs and cypris larvae and also the remains of crustaceans. It is not known with certainty whether the *Sargassum* weed is eaten directly; in any case the vegetarians do not play such an important role as they do in the phytal. Timmermann emphasizes small size as a special characteristic of the Sargasso fauna; he speaks of a dwarf fauna and relates this to the poor supply of seston in the Sargasso Sea. But it still remains an open question whether the small size of these organisms is merely determined phenotypically by the paucity of food or whether it is determined genetically. In the

Figure 99.   Sargassum fish, *Antennarius marmoratus* (from Ekman 1935)

pantopods, Timmermann speaks of 'local dwarf races with approximately constant body sizes', and in the isopod *Ianira minuta* of a Sargasso form with specific identity. His view appears to me to be premature until we have evidence that the small size is actually independent of the genetic constitution.

The average number of species on a thallus length of 10 cm is five, including the epiphytes. It is interesting that in the hydroids one species sometimes predominates over the others, although there has been no analysis of the actual interrelationships involved. A biocoenotic analysis of these growths would be desirable in order to establish how competition for space between the different species influences the composition of the fauna in a given place. Hentschel mentions that the hydroid colonies leave the thallus tips and the young shoots free, and he concludes that the growth of the hydroids follows that of the plants. *Membranipora* and *Spirorbis*, in particular, settle on the older parts of the plants. In this connection it is worth mentioning the oft-repeated observation that the broad-leaved forms of *Sargassum* are more densely settled than the narrow-leaved and are obviously preferred by some hydroids, e.g. *Aglaophenia latecarinata*.

Hentschel has considered the development of the *Sargassum* fauna from the viewpoint that the distribution of the *Sargassum* itself is influenced by a great circular current, in the centre of which the animal populations are reduced on account of the increasing age of the weed.

Timmermann points out that the subtropical convergence, which separates the Gulf Stream water from the North Equatorial Current, runs through the Sargasso Sea. He correlates with this his finding that there is a greater number of species and denser population to the north of the convergence, and a poorer population to the south. I consider that these investigations are not yet sufficiently extensive to draw further conclusions. The *Sargassum* weed grows and reproduces vegetatively and therefore largely maintains its stock by these means, but it is doubtless supplemented by material torn off from coastal areas.

The fauna is also to a large extent independent of the littoral regions. Prat (l.c., p. 273) writes: 'Like the Sargasso weed itself, the animal inhabitants today belong to species different from those of the littoral and have evolved in their new habitat, even if their original stock has disappeared from the shores since their isolation', and he goes on: 'Here we have perhaps some fragments of the Tertiary benthos which have escaped the vicissitudes of the littoral regions, notably those resulting from glaciations, and which have evolved autonomously in a very constant environment' (p. 274).

Timmermann also considers the fauna of *Sargassum* weed to be autonomous and not subject to any significant coastal influence. His argument, however, seems somewhat drastic when he says, for example: 'at Stations 14 and 17 *Nereis dumerilii* and a few other species could in the nature of things be better regarded as forms that have been misplaced from the open ocean, than that they had the centre of their distribution in the coastal waters of Central America' (p. 306). But *N. dumerilii* is also a well-known inhabitant of the littoral benthal in European seas. So it would surely be more correct to follow the more usual view that the *Sargassum* fauna represents an impoverished coastal fauna whose population is partly recruited from coast areas and partly supplemented by the settlement of pelagic larvae.

The view that the *Sargassum* fauna is very old agrees with the observation that some flying-fishes lay their eggs in matted clumps of *Sargassum*. According to Winge (1928) these clumps consist rather uniformly of two species, which are characterized by very small air-bladders which are about the same size as that of the attached eggs. So here we have a specific behaviour pattern in the species of flying-fish concerned, which seems to indicate that the whole complex is quite old.

From various observations (Woodcock 1950) it appears that in certain wind velocities the *Sargassum* collects into bands along lines of convergence. If the water at these convergence lines sinks at a speed of 4 metres sec^{-1} it carries the weed down with it in spite of the latter's buoyancy, and in strong currents to below the limits of visibility. With increasing pressure the contents of the air-bladders will be compressed and the plants will lose buoyancy and it is conceivable that at certain depths they break up. Nothing is known of the fate of the epifauna when this submersion takes place.

There are two ways in which this phenomenon could be significant. First, it is possible that the *Sargassum* clumps floating in deep water reflect the pulses of echo-sounders and so provide evidence of intermediate reflecting layers, and secondly, when the plants decompose they will produce detritus which could serve as food for small planktonic organisms in deeper water.

The cosmopolitan cirripede genera *Conchoderma* and *Coronula* are to be regarded as belonging to the pseudopelagic fauna; the species of *Coronula* live on whales, but those of *Conchoderma* have been found on fishes, sea-snakes, turtles and decapods as well as on whales and also in the growths on ships' bottoms.

Mention may also be made of the aquarium observations made by Wagner (1939) on the biology of *Antennarius marmoratus*. These fish climb about on corals with the help of the arm-like pectoral fins and probably move in the same way among the clumps of *Sargassum*; the distal part of the pectoral fin clasps any projections rather like a hand. Wagner writes: 'This gives the impression of the climbing, which is supplemented by the undulating movements of the dorsal fin. A fish up to 11·5 cm in length feeds on fishes and prawns about 3 cm long. When the mouth is opened wide it produces a suction which draws in prey even at distances of 1·5–2·0 cm from the tip of the snout. The bait attached to the first free dorsal fin-ray plays no part in the capture of prey but it remains to be shown whether the behaviour observed in captivity corresponds with that in the wild. The short length of this fin-ray, however, makes it rather improbable that if functions as a 'fishing-rod'. The observed fish spawned twenty-three times in the course of four months. The spawn, which has been described by previous authors, is in the form of a jelly-like band about 15–20 cm long and 4 cm wide, in which 5,000–7,000 eggs are embedded. The ends of the spawn band are spirally twisted so that it can become entwined around projections.' These observations are given in some detail here because live observations on aberrant marine fishes are scarce.

There are few other observations on the ethology of the Sargassum fauna. From an observation made by Simroth (1914) it appears that the snail *Litiopa* which lives on *Sargassum* produces a long string of mucus with an air-bladder at the free end and when it becomes detached from the weed this keeps it afloat until it touches another clump of *Sargassum*. This is possibly a starting-point for understanding the mucus float of *Ianthina*.

CHAPTER V

PLANTS AND ANIMALS OF THE BENTHAL

1. Introduction

The term benthal denotes the sea floor and the marine organisms that live there. From the medium depths it extends upwards to the shore, showing a certain degree of continuity with terrestrial and freshwater habitats, and downwards into the greatest depths of about 11,000 metres in the deep-sea trenches. It is obvious that such a very extensive area presents a wealth of interesting phenomena, only a few of which can be discussed here:

(*a*) First, there is the problem of recording the general distribution of the different species. Progress in this field has been very variable both as regards the recognition of valid taxonomic units and also in the comparison of different areas; it will be a long time before we have a more detailed picture. In general, floristic and faunistic stocktaking has not been pursued for its own sake but rather in relation to other phenomena; progress in this field is largely dependent upon advances in the systematics of the different groups of organisms.

(*b*) Since the physical and chemical factors on which the organisms depend are arranged vertically there are also bathymetric differences in the distribution of benthonic organisms. This has long been known for some time and the underlying causes have been the subject of much investigation. This subject occupies a prominent place in marine biology.

(*c*) The ecological factors vary horizontally as well as vertically, thus giving a geographical dimension to the subject, which may affect not only the presence or absence of a species or group of species but also their biology, including methods of development and various physiological processes. Apart from the developmental stages we are still far from having a good understanding of the general principles.

(*d*) Knowledge of the autecology of most benthonic organisms is still only fragmentary. Consequently we know very little about the relationships between forms living in the same areas. Further investigation of the types of life form will provide more extensive information on biological interrelationships.

(e) It was shown long ago that areas of the sea floor with a similar structure had the same kind of organic populations, and this has led to much research on benthonic communities. The analysis of the factual material and the formulation of principles that are generally applicable are still the subject of much discussion.

Vertical divisions: the accessibility of certain parts of the benthal to the coasts and observations on the striking zonation found in many places led to a number of investigations on this subject. Among the early works were those of Audouin & Milne-Edwards (1832), Sars (1835), Agardh (1836); a little later came the investigations of Vaillant (1870, 1871) and in more recent times the extensive work of Stephenson (1949–1960) and Knox (1960). Local variations in the factors influencing the vertical distribution of organisms (e.g. differences in tidal range, exposure to waves), different limits to the zones owing to differing interpretations and the use of differing terminology have all contributed to give a somewhat complex picture, from which one can only formulate a very generalized outline. Neither the Colloquium of the Benthos Committee of the Commission internationale pour l'exploration scientifique de la mer méditerranée (in Genoa 1957) nor the *Treatise on marine ecology and paleoecology* (Memoir 67 of the Geological Society of America, 1957) have in fact reached a recognized standard. The vast amount of material available for discussion on this subject forbids a very detailed presentation in the present volume. The Colloquium in Genoa recommended the use of the term étages instead of zones; étages has been defined by Drach (Pérès & Molinier 1957, p. 13, footnote) 'as the vertical area of the benthonic region, in which the ecological conditions, independent of the substrate, are constant as a function of depth or vary in regular fashion between two levels, which correspond to the limiting conditions of the characteristic population'. It is easy to see that at similar depths, vertical areas on low-lying or steep coasts, on exposed or sheltered coasts, in areas with upwelling deep water or with horizontal currents of varying strengths will all be very different in character and consequently will be difficult to classify in a unified scheme of biological divisions. Any such scheme must therefore be based on an idealized section through the benthal, which extends from areas where the influences of sea water are just traceable down to the greatest depths. An attempt must also be made to find out which ecological factors or combinations of factors show vertical gradients and thus bring about changes in the distribution of organisms.

1. In the upper region of the benthal the water level fluctuates, so that the organisms are only subjected to the influence of water at

certain times; at the highest level only spray is involved, in the re-mainder of the region the covering of water fluctuates owing to tides and wind. Below the level of extreme low water the sea floor is always covered with water.

2. In areas subject to tidal fluctuations the periodic and non-periodic fluctuations in atmospheric temperature, affecting irradiation and evaporation, are much more marked than in regions with a per-manent covering of water; with increasing depth the non-periodic fluctuations in temperature and the range of the periodic fluctuations decrease rapidly so that below certain depths the temperature is almost constant.

3. With a fluctuating depth of water the amount of illumination fluctuates within wide limits. Below the level of low water the amount of light decreases and there is a change in its spectral composition which continues with depth until finally there is no light at all. The depth at which the incident light energy is no longer sufficient for plant assi-milation is of great biological importance; this is the compensation depth (see pp. 45, 54). The gradient in illumination varies considerably according to the region (see p. 51).

Turbulence and non-periodic currents are more in evidence near the surface than at depths of a few hundred metres; at greater depths they are probably scarcely important as ecological factors, although knowledge on this is extremely scanty (cf. pp. 153, 278). Constant or periodic currents do occur at great depths.

5. The water pressure increases with depth. Close to the sea sur-face where the water level fluctuates any changes in pressure are undoubtedly more important than they are at greater depths, where any fluctuation will be only a tiny percentage of the total reigning pres-sure. There is still no support for the view that pressure is a factor in-volved in the production of the zones.

6. In the region of fluctuating water level there are periodic and non-periodic fluctuations in salinity; below the low-water line the salinity is constant from the biological point of view, except in areas influenced by the inflow of fresh water, or in those, like the Baltic Sea, which are barred by a sill (see p. 40).

It follows that there are two definable horizons (sometimes deter-mined by the combination of several factors), namely the levels of mean low and mean high tide (p. 207). It is obvious that these horizons cannot be regarded as sharp lines but must show a vertical spread, since neither wind nor tide have a constant force or amplitude. In the same way all other horizons are not lines either, but are more in the nature of bands and thus represent transitions from one zone to another.

The cessation of plant production by means of photosynthesis represents a significant change in the total biological picture and it allows us to differentiate in the benthal between an upper or littoral system and a lower or bathyal system. The depth boundary between the two is often given as approximately 200 metres, but it should be emphasized that this depth fluctuates considerably according to geographical position and local factors. In a general diagrammatic representation, it is therefore preferable to avoid the use of metre readings.

2. The littoral system

The uppermost level is the supralittoral which lies above mean high-water mark and is characterized by extreme factors: evaporation which may lead to high salinity or conversely downpours of rain which may cause flooding with fresh water; extreme temperatures occur quite frequently depending upon exposure and weather conditions. Periodicity is only therefore present to a limited extent. This zone is only accessible to a few organisms: on rocky coasts there may be a few lichens and some land insects, and a few *Littorina* species and isopods (Ligiidae) may reach it from the sea.

The faunal elements of supralittoral sandy coasts have been analysed by Gerlach (1954) in the area of the North and Baltic Seas; he speaks of Cyanophyceae-sand in which a layer of pale, dry sand overlies a layer permeated by Cyanophyceae, beneath which there is pale sand again. The abundant microfauna of this area comprises a mixture of species from various neighbouring habitats, including brackish-water species.

brackish-water species	14
euryhaline marine species	11
insects	7
terrestrial species	17
casual visitors from coastal shallow waters	13
casual visitors from the marine region	34
Total	96

The region subject to the tides is known as the littoral (*sensu stricto*), eulittoral or mesolittoral. Here too there are considerable fluctuations in environmental factors but in general these show a stricter rhythm, although there are irregularities at the upper and lower limits which are associated with spring and neap tides and also with wind conditions. On account of these great irregularities the upper and lower limits of the eulittoral have been specially termed the supralittoral fringe and infralittoral fringe by Stephenson & Stephenson (1949) (cf. also the

System	Zone	Level of water	Turbulence and non-periodic currents	Temperature fluctuation	Influence of light	Salinity	Water effects
Littoral system	Supralittoral	Mean high water level		maximal	maximal	fluctuating owing to evaporation and precipitation, depending on water level	Spray — changing according to tide and wind
Littoral system	Eulittoral	mean low water level	strong	greater than the extremes of the water temperature	fluctuating considerably, depending on water level and locality		
Littoral system	Sublittoral		moderate	decrease in the amplitude and frequency of the fluctuations	decrease in intensity and a change in spectral composition / average / compensation horizon	± constant	continuous
Bathyal system			minimal	minimal	Nil		

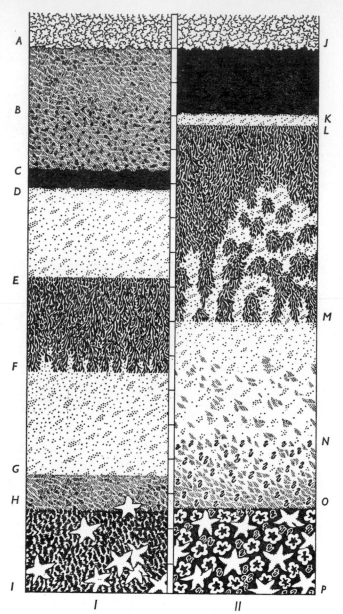

Figure 100. Zonation of the eulittoral at two positions on an island (after Stephenson & Stephenson 1948): 1, the sloping, sunny side facing south; 2, the shady, cliffed side facing north

A = lower limit of macroscopic maritime land-lichens, B = upper limit of the majority of Littorinae, C = upper limit of the densest blackening, D = upper limit of main barnacles, E = upper limit of main growth of *Fucus*, F = lower limit of main growth of *Fucus*. G = lower limit of main barnacles, H = upper limit of infralittoral fringe, I = level of extreme low water of spring tides in July 1947, J as A, K as D, L as E, M as F, N = approximate upper limit of most *Serpula*, O as H, P as I

comprehensive treatment of the littoral in the Southern Hemisphere by Knox 1960, 1961). On rocky coasts the eulittoral is characterized by algae such as *Enteromorpha*, *Fucus*, *Ascophyllum* and *Porphyra*, and coatings of diatoms also play an important role. The animals present include cirripedes, gastropods of the genus *Patella* and its relatives and also various species of *Littorina*; these are forms which are active when covered with water, but may at other times show phases of inactivity.

Stephenson & Stephenson (1948) have shown how little one can generalize from locally observed subdivisions within the eulittoral. They have given an impressive example (Fig. 100) in which the two positions compared were separated by a distance of less than 100 metres and were subjected to approximately the same amount of swell. Position I, however, was on the sloping south side of a small island and exposed to the sun, whereas Position II was on the steep, shady northern side. Here light was evidently the most important factor responsible for the differences observed, since the vegetation is largely determined by it. As a further example there is the analysis by Starmühlner (1956) of the fauna and flora of an area running from the

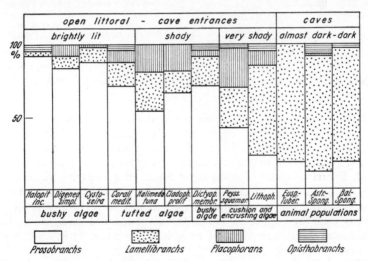

Figure 101. Percentage distribution of the prosobranchs, lamellibranchs, opisthobranchs and placophorans living among plant growths in caves in the Mediterranean (after Starmühlner 1956)

open littoral into the inside of a submarine cave (Fig. 101). Figure 102 shows in diagrammatic form some of the possible ways in which the eulittoral may be differentiated.

Along more or less muddy flat coastlines the tides sometimes form extensive shallow areas, into whose upper boundary areas a number of

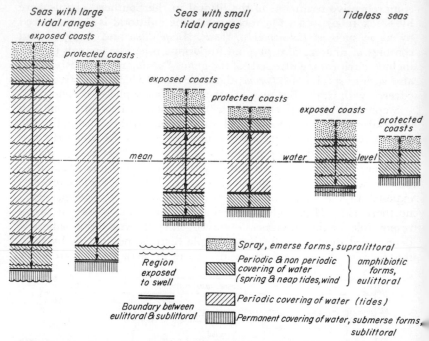

Figure 102. Interaction of different factors in the eulittoral (from Ricketts & Calvin 1962, somewhat modified)

halophytes penetrate. The degree of penetration depends upon the period of time that the different species can tolerate being covered with water; glasswort (*Salicornia herbacea*) tolerates a relatively long period under water, and therefore lives fairly far below mean high-water level. Dwarf eel-grass (*Zostera nana*) tolerates a shorter period of submersion than *Zostera marina*, and therefore flourishes on the shallow flats, whereas the larger species lives below lower water level. On muddy flats the vegetation is mainly in the form of dense, slimy mats of diatoms which are of great nutritional importance in the ecology of these areas.

Animal life in the muddy eulittoral is at first not very obvious, because it consists principally of an infauna which lives in the substrate. This can be detected partly from tracks and partly by the openings to the burrows or by faeces casts, as for instance in the case of the polychaetes *Arenicola* and *Heteromastus*, the lamellibranchs *Mya arenaria*, *Cardium edule*, *Scrobicularia plana* and others. Zonation may depend upon the period of submersion, but may sometimes be due to differences in the composition of the sediment.

Sandy flats usually have no vegetation. Schulz (1937) and Hofmann (1942), 1949) however reported the interesting phenomenon, termed 'Kniepsand', on a sandy flat on the island of Amrum. Here the surface consists of a thin layer of pure white sand which dries out quickly, beneath which lies a layer coloured greenish by a dense growth of blue-green algae; beneath this is a violet-red layer of purple bacteria, which overlies a sandy layer that is blackened by iron sulphide. Schulz used the term 'colour stripe sandy flat' (Farbstreifensandwatt) for this development of plant organisms in horizontal layers. Hofmann (l.c.) measured the translucency of the thin layer of sand and found that when dry and about 5 mm thick it had the effect of a red filter, so that the algae could utilize the red light, in particular, for carbon dioxide assimilation. The development of colour stripe flats is of course dependent upon several factors, including the level of the ground water. This colour stripe flat scarcely provides a good autochthonous source of food; the growths of diatoms in other places also appear to be less important than they are on the muddy flats.

Areas of flat coastline with gravel beaches formed and continuously reshaped by wave action are of particular biological interest; their characteristics caused Remane to speak of an *Otoplana* zone (named after the characteristic turbellarians). The ecological differentiation of sandy flats is largely determined by the particle size of the sediments and by the amount of admixed mud (see pp. 215, 280), but the period of submersion, i.e. the height above mean low water, also plays a role. Penna (1942), for example, showed a relatively narrowly restricted zonal distribution of sand-dwelling copepods in the eulittoral at Woods Hole, which was disturbed neither by the tides nor by moderate wave action. He found that the animals near low water were living in the upper 4 cm of the sediment, whereas those near high water were between 12 and 16 cm deep; this emphasizes their dependence on interstitual ground water.

Nicholls (1935) had previously investigated the distribution of the copepods on a sandy beach at Millport, and had found both a specific vertical distribution within the sediment and also a clear zonation. In particular, the species living in the upper 2-3 cm layer must, the higher their position on the beach, be able to tolerate greater fluctuations in salinity and temperature than the species living in the deeper sand layers and at the lower horizons. Different ecological potentialities ought therefore to account for the spatial distribution of closely related species.

From a comparative analysis of the macroscopic crustacean fauna of European and South American sandy beaches, Dahl (1952) came to the conclusion that here too one can recognize divisions similar to those on rocky coasts. Accordingly the following scheme can be put forward:

	Polar regions	*Temperate latitudes*	*Tropics*
Supralittoral (subterrestrial)	—	Amphipods (Talitridae)	Brachyura (Ocypodidae)
Eulittoral	—	Isopods (Cirolanidae)	Isopods (Cirolamidae)
Sublittoral	Amphipods (Lysianassidae)	Amphipods (Haustoriidae, Oedocerotidae, Phoxocephaliidae)	Brachyura (Hippidae)

This analysis also draws attention to geographical differences in the faunistic composition of the zones.

The sublittoral, and with it the true marine region, begins below the low-water line. Here atmospheric and terrestrial factors can only have an indirect effect; the term infralittoral has been used by French authors (e.g. Molinier & Picard 1954, Pérès 1957, 1961). The lower limit is in many cases clearly defined as the limit of plant growth with the exception of diatoms and heterotrophs; on sandy and soft bottoms, however, the algae will have a very patchy distribution on embedded boulders and lamellibranch shells, so that it is difficult to define an exact limit. It is also difficult to give a uniform zoological definition of the lower limit. This zone is characterized by environ-

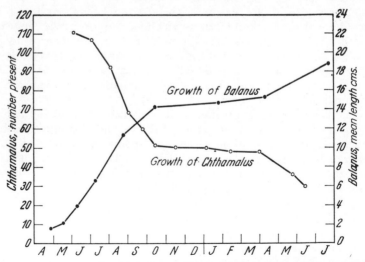

Figure 103. A comparison of the seasonal changes in the growth of *Balanus balanoides* and in the survival of *Chthalamus stellatus* being crowded by *Balanus* (after Connell 1961)

mental factors such as the marked periodicity in the illumination, by temperature fluctuations and by marked water movements.

The adjoining 'elittoral' was defined by Gislen, following his findings in Norwegian fjords, as having a lower limit where the diatom flora ceased; it therefore represents the deepest part with autotrophic plants. It has not yet been defined in zoological terms.

Before discussing the bathyal, there are a few ecological points in connection with the formation of zones in the littoral. Zones are not necessarily the direct expression of the dependence of the organisms on abiotic factors. Thus in many places it has been observed that the cirripede *Chthalamus* settles above the *Balanus* zone, but is largely absent at the level of the *Balanus* zone. In places which are free of *Balanus*,

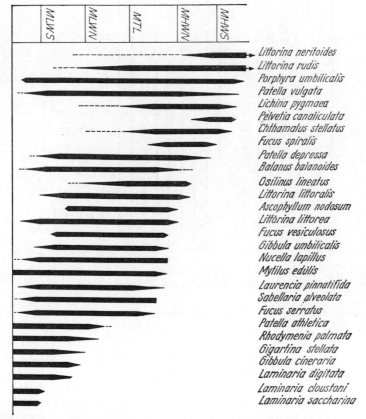

Figure 104. Vertical zonation in Cardigan Bay (after Evans 1947): M = mean water level

H—M.B.

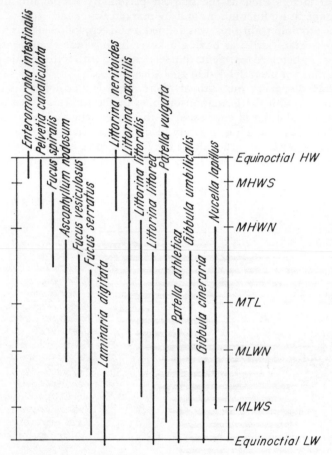

Figure 105. Vertical zonation of plants and animals in the eulittoral at Port Erin; extreme tidal range 5·5 m (after Evans from Eales 1952)

however, thriving settlements of *Chthalamus* may occur at lower horizons. According to Connell (1961 b), *Balanus balanoides* settles more densely and grows faster so that it is superior to the others when competing for food and space; *Chthalamus* has a greater resistance to heat and to desiccation but owing to competition it may settle in a higher zone although its true optimal area is lower down on the shore (Fig. 103). On the other hand the lower limit of *Balanus* may be restricted by other sessile forms, e.g. by the serpulid *Galeolaria* in Australian waters, according to Endean *et al.* (1956).

By defining the eulittoral as the zone lying between high and low water it follows that the period of submersion and desiccation of the different horizons will vary very considerably according to their height on the shore. Consequently the ecological factors also show considerable gradations. How closely organisms react to these differences will be apparent from the examples given in Figs. 104 and 105. The upper limits of distribution of the species show a clear stepwise gradation which is related to different mean water levels; these can be regarded as critical horizons above which conditions are no longer suitable for the species concerned. The upper and lower limits of distribution of a given species may lie at different levels in places where the combinations of local factors differ (Fig. 106). Hence metric data based, for instance, on the mean water level must be regarded as of doubtful validity.

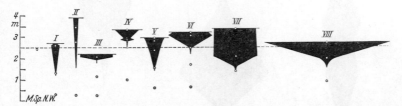

Figure 106. *Nereis diverscolor*: number per square metre in relation to tidal level at stations I–VIII on a mud-flat in the Tamar Estuary. The horizontal line at the top of a diagram marks the upper boundary of the mud-flat (after Spooner *et al.* 1940, somewhat modified)

Below the upper limit of distribution the occurrence of the individual species is by no means quantitatively uniform, as the diagrams appear to indicate. At different stations within the habitat the distribution of the individuals (Fig. 107) will to a great extent be determined by the structure of the sediment, because most species react very sharply to particle size or to the presence of fresh water (see also Fig. 136).

When comparing different areas it is also noticeable that, particularly in the eulittoral, there may be considerable fluctuations in the effective ecological factors depending upon varying climatic conditions; thus in both warm and cold climates there is a tendency for the populations of the eulittoral to be restricted to the lower horizons (Gislen 1949). The eulittoral is therefore best developed in temperate latitudes.

In plants, too, the extent to which the physiological characteristics are related to the zonation has been shown, for example, by Biebl (1938, 1939). Table 24 gives the different ranges of osmotic tolerance for three ecologically different groups of algae.

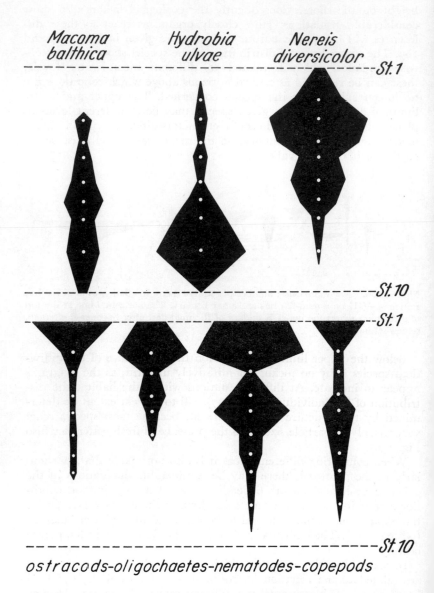

Figure 107.   Bathymetric distribution of some of the macrofauna species and micro-fauna groups in the muddy eulittoral near Cardiff (after Rees 1940)

TABLE 24

Relation between ecology and ecological resistance in algae

Ecological group	Range of osmotic tolerance
Deep-water algae	0·5–1·4 times sea water concentration
Algae at low tide and in tidal pools	ca. 0·3–2·2 ,, ,, ,, ,,
True tidal algae	ca. 0·2–3·0 ,, ,, ,, ,,

In continuously submerged algae on the west coast of Sweden, Kylin (1917) found a greater sensitivity to cold than in the amphibious forms, a result which has been confirmed by Biebl for the Mediterranean. According to this author the ecological groups mentioned above also show differences in resistance to desiccation and heat (Table 25).

TABLE 25

Relationship between ecology and resistance to desiccation and heat in algae

Ecological group	Relative air humidity tolerated	Highest temperature tolerated
Offshore	97%	+27–29°C
Low tide algae	94%	+30°C
Tidal algae	88%	+35°C

As Parker (1960) showed for *Fucus vesiculosus* on the Atlantic coasts of the U.S.A., resistance to cold is not constant but is dependent upon the time of year: an increase in resistance can be observed from October onwards. In January and February the plants can be exposed to —50°C, taking an injury rate of 50% as the standard. In the summer months this value rose to —30°C. At the same time, there was a change in the sensitivity of the parts of the thallus: in the summer months the thallus tips were more sensitive to cold than the remainder of the thallus, but the position was reversed in the winter and spring months. Furthermore, specimens from deeper water showed a greater sensitivity to cold than those from the higher, more exposed levels.

According to Kanwisher (1937) eulittoral algae lose up to 90% of their water during one tide when they are exposed to desiccation, and in frosts up to 80% of their water content may be frozen. In both cases the metabolism is much reduced. An arctic *Fucus* remained alive in spite of being frozen in the ice at temperatures down to -40°C for several months, and started to assimilate immediately after the thaw.

De Virville (1940) has given a comprehensive account of the vegetation zones in the European Atlantic littoral (see Table 26). Further

TABLE 26

Vegetation zones on the European-Atlantic littoral in relation to tidal level
(after Virville 1940)

Tidal level	Main zones								Subsidiary zones					
	Xanthoria parietina	*Caloplaca marina*	*Verrucaria maura*	*Pelvetia canaliculata*	*Fucus platycarpus*	*Fucus vesiculosus*	*Fucus serratus*	*Laminaria*	*Lichina confinis* Sheltered rocks	*Lichina pygmaea* Exposed rocks	*Rivularia bullata* Semi-exposed rocks	*Ascophyllum nodosum* Sheltered rocks	*Bifurcaria tuberculata*	*Himanthalia lorea*
Extreme H W S	→	↕	↕						↕					
M H W S		↕	↕						↕	↕				
Lowest L H W S		↕	↕						↕	↕	↕	↕		
Extreme H W N			↕								↕	↕		
M H W N				↕										
Lowest H W N				↕	↕									
M T L					↕	↕								
Highest L W N						↕								
M L W N							↕							
Extreme L W N							↕						↕	↕
Highest L W S							↕						↕	↕
M L W S								←						

information on the general problem of zone formation and also on several different examples will be found in the publications of Ricketts & Calvin (*Between Pacific tides*, 1962) and of Pérès (*Oceanographie biologique et biologie marine* I, 1961).

The narrow vertical zonation characteristic of many eulittoral species represents a linear distribution from the regional viewpoint, in contrast to the two-dimensional distribution of most land organisms and of the markedly eurybathic benthal fauna. As a biological and ecological phenomenon this type of distribution is comparable to the living populations at the periphery of flat horizontal areas of distribution, and particularly in so far as there is very close contact with a quite different environment. Bolin (1949) draws attention to this phenomenon and concludes that 'this fact greatly increases the chance for the survival of mutations and gives the organisms with a linear distribution a high evolutionary potential. It is probably an important reason for the tremendous richness of species in the intertidal zone' (l.c., p. 459).

Consideration of the fauna and flora of the upper littoral from this evolutionary aspect deserves further study, because it can throw light on both recent and past biogeography. In addition, the great fluctuations in ecological factors within a limited area (water movements, periods of desiccation and submersion, food, exposure to light, etc.) offer a wealth of ecological niches which are particularly suited to promoting the process of species formation. Examples of this are the closely related species of the genus *Littorina* and in particular of the Acmaeidae, of which 10–12 exist alongside each other on the Pacific coasts of North America (cf. Ricketts & Calvin 1962). Further information on this subject would be provided by more detailed analyses of other groups.

In *Talitrus* it has been shown that these shore amphipods have driections of flight peculiar to their own populations, which enable them to escape into the water by the shortest route (Pardi & Papi 1952). Animals bred in captivity behave according to the population to which they belong. This orientation takes place according to the position of the sun in combination with an inner rhythm which is evidently related to metabolic processes dependent upon temperature (Pardi 1957). This indicates a further complex of generally important biological problems.

From an analysis of depth distribution in the family Blenniidae Abel (1962) found a marked differentiation in the uppermost sublittoral within a few metres. This suggests that the ecological subdivisions are narrowly defined, particularly as regards the available niches, and it also emphasizes the autecological specialization of the individual species (Fig. 108).

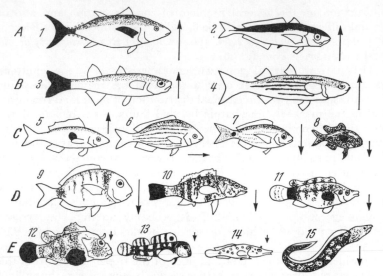

Figure 108. Relationship between form and degree of association with the bottom in fishes (after Abel 1962)

Row A: 1, *Thynnus:* 2, *Trachurus;* in the open-sea pelagial. Rows B and C: 3, *Atherina;* 4, *Mugil;* 5, *Spicara;* 6, *Boops;* 7, *Oblada;* 8, *Chromis;* in the coastal pelagial. Row D: 9, *Diplodus;* 10, *Serranus;* 11, *Crenilabrus;* near the bottom. Row E: 12, *Scopaena;* 13, *Blennius;* 14, Lepadogaster; 15, *Muraena;* living on the bottom. Note the shape of the body, tail, and pectoral fins; coloration; characteristic markings (completely black in some forms); the arrows denote direction of flight, their length the flight distance; No. 6 takes flight along the bottom.

3. The bathyal system

The upper limit of the bathyal, like the lower limit of the sublittoral, is not easy to define and it must in fact be regarded as a transition zone. On average it can be placed at about 200 metres, that is, where the Continental Shelf gives place to the Slope, because in this region there is often a marked change in the fauna; the region lying below this transition zone was originally known as the abyssal. However, it soon appeared from some of the early faunistic investigations (e.g. Perrier 1899), that some further subdivision was necessary, and this has been confirmed by the results of more recent expeditions. The bathyal region can in fact be divided into:

1. The archibenthal or bathyal *sensu stricto,* which can be defined as a broad transition zone between the lower littoral and the true deep sea.

2. The abyssal, which includes all the succeeding depths down to 6–7,000 metres.

3. The hadal, termed the ultra-abyssal by Russian authors, which is restricted to the region of the deep-sea trenches.

In the archibenthal, periodic fluctuations in the external factors still take place; seasonal changes in temperature, light and food derived from the trophic layers can still be observed, although they are not so marked as in the littoral system. It is probable that periodic fluctuations in the currents caused by changes in the wind are still effective in the archibenthal. Consequently, one may expect to find rhythms of a biological nature, although so far these have only been demonstrated in a few cases. Thus, in the macrurid fish *Trachyrhinchus trachyrhinchus* the structure of the otoliths, scales and bones show that growth takes place in the months June to November; the body weight fluctuates rhythmically and has a maximum in December, and the breeding season is around February–March. The spawning of the echinoid *Allocentrotus fragilis* off the coast of California in February is apparently triggered by upwelling water (Boolootian *et al.* 1959c).

The boundary between the archibenthal and the abyssal has been variously defined, the depths given fluctuating between 1,000 and 3,000 metres depending on the local conditions and on the groups of organisms investigated. According to Bruun (1957), the 4°C isotherm represents the upper temperature limit at which the majority of the endemic elements in the abyssal fauna can still survive. The depth at which this isotherm occurs fluctuates with the geographical area and depends upon the hydrographical conditions and this to some extent explains the different depths given for the boundary between these two zones. There would therefore be a change in the fauna at around the 4° isotherm and this would be particularly apparent in the case of cold stenothermal deep-sea animals. Among the organisms which are positively characteristic of the abyssal are the asteroids of the family Porcellanasteridae; these are only known from depths greater than 900 metres. According to Clarke (1962) the abyssal mollusc fauna is differentiated from that of shallow sea areas in the following ways: '1. Monoplacophora are presently recorded only from the abyssal zone; 2. more archaeogastropod species than mesogastropod species appear to occur in the abyssal area; 3. scaphopods comprise a larger proportion of the total deep-sea mollusc fauna, and 4. in number of abyssal species the Protobranchiata rank well above the other bivalves with the Septibranchiata and Eulamellibranchiata sharing second place' (l.c. p. 2).

There are several reasons for defining the hadal as a special zone: so far as the faunistic investigations allow any conclusion, the composition of the fauna in the deep-sea trenches differs from that of the neighbouring abyssal areas. The trenches are geographically isolated from each other (Fig. 109) and evidently have different hydrographic

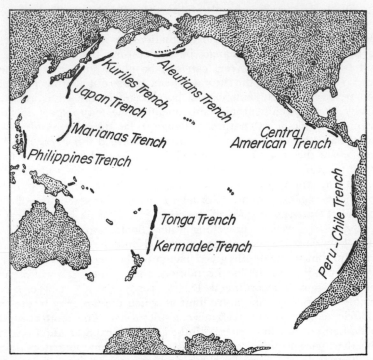

Figure 109. The most important deep-sea trenches in the Pacific Ocean (after Sheppard 1959, modified)

conditions, so that they can also be distinguished ecologically from the abyssal. A higher degree of endemism is to be expected in the trenches. As an example, the findings of Dahl (1959) on hadal amphipods may be cited: of the twenty-one species so far found only two certainly occur in several trenches, and these two species are at least facultatively pelagic. The general validity of observtions of this kind must, however, be confirmed by further investigations.

At one time the abyssal was regarded as a very uniform environment, in which a more or less panoceanic fauna was to be expected. It is indeed true that certain conditions, such as light and temperature, are very uniform when compared with conditions in the littoral. However, recent investigations have shown that the deep-sea bottom is morphologically broken up by numerous projecting sills and isolated sea-mounts (Ulrich 1964); we still do not know whether the animals living on their slopes show vertical zonation. The existence of various currents results in great differences in the structure of the sediments, producing direct and indirect ecological effects, and the water currents themselves are also an effective ecological factor. Soft sediments of planktonic origin may contain scattered areas of hard bottom, whether in the form of manganese nodules of various sizes, or as the remains of layers of volcanic ash; dredge hauls and underwater photographs have shown that these exist over wide areas of the eastern Pacific (Menard 1960). Sediment sliding down from above (p. 297) can evidently affect wide areas of the abyssal bottom and thus give rise to localized areas with differing ecological conditions. In the present state of knowledge we cannot yet draw any firm conclusions on how these differences in hydrography and sediment may affect the animal populations and their distribution. Nevertheless, the division of the abyssal into an upper and a lower zone, as advocated by some authors, appears to be justified. From an analysis of the vertical distribution of approximately 1,000 invertebrate species, Vinogradova (1958) sets the depth at which this change in the fauna occurs at about 4,500 metres; Madsen (1961) finds confirmation of this from the distribution of the Porcellanasteridae.

Table 3 (p. 47) shows the ways in which the benthal has been subdivided.

4. Benthal substrates and the relation of organisms to them

The structure of the substrate is of paramount importance for the colonization of the sea floor. On the rocky bottoms, known as hard bottoms, currents prevent the deposition of sediments or in some cases this is prevented by the slope. Hard bottoms of this kind occur not only in the vicinity of the coasts, but also on the slopes of sills and sea-mounts. They may occur continuously over largish areas or discontinuously in the form of large or small rocky outcrops in among other substrates. Remane (1940) used the term secondary hard bottoms for solid structures such as mollusc shells, small rocks and artificial constructions which are scattered in among other substrates. Conversely in areas with a continuous hard bottom other substrates may be deposited by sedimentation in crevices and on plateaus.

Dredges and bottom samplers do not function efficiently on hard bottoms and so the available material is very incomplete, particularly for the sublittoral and deeper regions; the use of diving and submarine photography will gradually yield further information.

Sandy bottoms occur in places where water movements prevent the deposition of very fine particles but allow the settlement of granular material. According to the varying strengths of the currents the sedimentary material will be separated out into different grades of sand from the very finest up to coarse grit or even gravel. It is also possible to have admixtures of clayey material and the sand may contain spaces

which allow the granules to shift about relatively easily. Conversely, the
sand may be packed so hard as to interfere seriously with the circulation of water through the substrate and the penetration of organisms.

Sand is most widely found in the vicinity of coasts; the North
Sea and Baltic Sea have very extensive areas of sand, associated with
sunken Pleistocene bottoms. In the vicinity of coral-reefs coral debris
may form sandy substrates rich in calcium; shelly bottoms are formed
by the accumulation of broken mollusc shells, which in some respects
show similarities with true sandy substrates. Sand is also found in deep-
sea areas, where it is presumably the result of sediment being washed
down from the Continental Shelf.

In the littoral and bathyal, marine soft bottoms are predominantly
terrigenous in origin and consist of mud with a large amount of de-
composed organic matter. In the oceanic region, and particularly
in the abyssal, the soft bottom sediments are derived from the plankton
and consist mainly of the remains of planktonic organisms; according
to the predominant component these can be characterized as radiolarian, foraminiferan (*Globigerina*), diatom or pteropod ooze. In the red
deep-sea clay the remains are broken down to such an extent that specific elements are no longer recognizable.

Soft bottoms are extraordinarily variable in consistency; in the
eulittoral where the substrate is not fully saturated with water it may
become quite firm like arable land, but from the sublittoral down to the
hadal the bottom is softer, sometimes almost the consistency of broth
without any clear boundary delimiting it from the supernatant cloudy
water, but some photographs of the deep-sea floor given the impression
of relatively solid deposition.

Apart from their gross structure, on which the classification is based,
the substrates may be divided, according to the distribution and habits
of the organisms inhabiting them, into soft, sandy and hard bottoms.
Some species prefer to live on the surface of the substrate, that is, in the
boundary layer between substrate and water: these make up the epifauna. Other species burrow into sand or mud or bore into hard substrates: these form the infauna of the substrate concerned. The spaces
and crevices present in hard substrates accommodate species which are
characteristic of this habitat; similarly the interstitial spaces in sandy
substrates with a certain particle size are inhabited by a typical interstitial fauna. This mesofauna is lacking in soft bottoms where there are
no true cavities. The terms used are given in detail in Table 27 (p. 225).

The inhabitants of the different structural components are characterized by certain general features, which are given in Table 27.
From this one can naturally recognize numerous life-forms, which have
a more or less characteristic distribution and form (see pp. 243 et seq.)

Species, of course, vary in the extent to which they are tied to certain

TABLE 27

Bottom-living animals

		Soft bottoms Pelos	Hard bottoms Lithion	Sandy bottoms Psammon
On the surface of the substrate	Epi-	Epipelos few sessile forms embedded in the bottom (pennatularians, hydroids, sea-lilies); creeping and striding forms, such as triclads, polychaetes, isopods, mysids, many ostracods, some asteroids	Epilithion numerous sessile forms such as sponges, hydroids, actinians, polychaetes, bryozoans, brachiopods, tentaculatans, cirripedes, ascidians; many semisessile and motile forms, such as polychaetes, crustaceans, gastropods, sea-urchins asteroids, crinoids and aspidochirotan holothurians	Epipsammon few sessile forms, motile fauna relatively poor:] some isopods, Crangonidae, brachyurans, some gastropods, the asteroid *Astropecten*, gobies and flatfishes
In cavities in the substrate	Meso-	Mesopelos no cavities in soft sediments; no true mesopelos	Mesolithion in crevices: commonly elements of sandy or soft bottom fauna, depending on the sediment	Mesopsammon rich, interstitial microfauna; animals often elongated, ciliated, adhesive; lacking in firmly packed fine sand
Lying or burrowing in the substrate	Endo-	Endopelos Motile forms, burrowers, such as polychaetes, gastropods, ophiuroids, asteroids; sessile and semisessile forms: some actinians, polychaetes, lamellibranchs; holothurians (synaptids, Dendrochirota)	Endolithion mainly forms boring into calcareous rocks and mollusc shells: boring sponge *Cliona*, polychaetes (*Polydora, Dodecaceria*), lamellibranchs (*Pholas, Zirfaea, Saxicava, Petricola*); specialized woodborers: Teredinidae and crustaceans *Limnoria* and *Chelura*	Endopsammon relatively few motile, burrowing forms, such as polychaetes *Nephthys* and *Glycera*, gastropods Natica, semisessile tube-less forms such as amphipods, lamellibranchs, holothurians, sand-eels, many tube-dwellers among the polychaetes, enteropneustans, crustaceans
Swimming above the substrate	Supra-	Suprapelos tied nutritionally	Supralithion tied nutritionally	Suprapsammon tied nutritionally

substrates or to their component parts, but apart from the relatively few euryoecous species the association is often very close, although it may change according to the different phases in the life cycle of the individual species. Many benthonic species have pelagic eggs and larvae and it is only at metamorphosis that these search for a suitable substrate. This raises the question of how larvae ready to metamorphose find a substrate that suits them (see Chapter VI, 7) and how a population living in substrates covering a small area actually survives in spite of the action of currents. Some species are known to live hidden in the substrate during the day, e.g. amphipods, and to emerge and swim around at night, while others show the reverse behaviour. Epibiotic animals often burrow into the substrate when disturbed, while other forms use it as a place to lie in wait. Table 28 gives data on the ways in which the substrate affects the habits of animals (see p. 220, Fig. 108).

TABLE 28

Ecological role of the bottom

Bottom serves as	Feeding takes place	Larval development takes place
permanent substrate for fixation of sessile epifauna	in open water (whirlers, filterers, tentacle feeders	usually in open water
permanent and exclusive living and feeding place (sessile, semisessile, liberosessile, digging, burrowing, boring endofauna, semisessile epifauna	on bottom, its surface and in water immediately above (sediment feeders, browsers, predators and carrion feeders, whirlers)	usually in open water; commonly various degrees of brood protection
mainly for living and feeding; flight routes often into open water	at the surface of the bottom and in open water	usually in open water
resting and lurking place	mainly in water near bottom, sometimes near bottom surface	usually in open water
spawning place only for pelagic animals	in open water	in open water

From measurements of the firmness of the sediment and comparison of the main animals in the Gullmar Fjord and Skagerrak, Ekman (1947) came to the conclusion that the occurrence of the pennatularian *Funiculina quadrangularis* is determined by the composition of the sediment, since both areas had the same type of sediment but differed significantly in depth, temperature and salinity.

From the investigations of Riedl (1956) on turbellarians (Fig. 110)

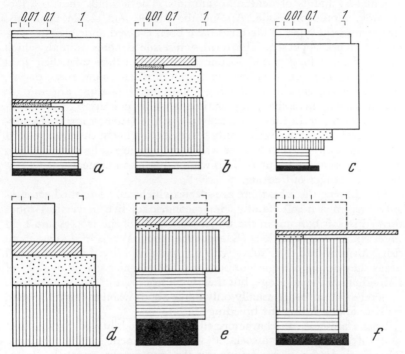

Figure 110. Histograms of the distribution of turbellarian groups in various substrates (after Riedl 1956): (a) muddy bottom in the Adriatic; (b) phytal of the Mediterranean rocky littoral; (c) pure muddy bottom in the Skagerrak; (d) muddy bottom mixed with shells, sand and detritus in the Skagerrak; (e) pure fine sand, Kiel Bay; (f) fine sand rich in detritus, Kiel Bay. The height of the blocks gives the number of species relative to 100, the surface areas denote the abundance of individuals.
Key to groups: *white;* Xenoturbellidae, Nemertodermatidae, Acoela: *oblique lines;* Microstomida: *close dots;* Polycladida: *spaced dots;* Prolecitophora: *vertical lines;* Dalyellioidea, Typloplanida: *horizontal lines;* Kalyptorhynchia: *black;* Proseriata

and of Kirsteuer (1963) on nemertines, it appears that even higher systematic groups prefer certain biotopes; thus Kirsteuer found the palaeonemertines mainly in mud, the heteronemertines in mixed sandy bottoms and the hoplonemertines predominantly in the phytal. 'The ecological behaviour of the group representatives provides a picture which corresponds to the projection of the systematic grouping (based on morphological characters) into the natural field of the environment' (l.c., p. 353), to which one can add that the morphological characters used in systematics are often not clearly related to the structure of the environment (e.g. nervous systems, nephridia).

There is naturally a close correlation between the substrate and the locomotory habits of benthonic animals. The animals may be free-moving, motile or sessile. Among the motile and sessile benthonic forms different types of life-form have been evolved and these are discussed on pp. 243 et seq. The term semisessile denotes animals which 'are tied for a long period to one place, where they take their food without having to move about in search of it' (Remane 1940, p. 50), but which are capable of moving about. Some of these are animals which attach themselves, e.g. actinians, some cnidarians, caprellids, while others live in tubes. An example of the latter group is the lug-worm *Arenicola marina*, which only leaves its burrow in the muddy sand of the eulittoral when the winter water temperature at low tide drops to 1–3°C (Werner 1954, 1956). This winter migration protects the worms from the danger of freezing.

Even among the fishes there are species which can be termed semisessile. The garden-eels (family Heterocongridae) live in vertical tubes formed in soft bottoms in the upper sublittoral of the tropics and held together by skin secretions (Klausewitz 1962). When in danger these fish withdraw into their tubes with lightning rapidity. In the occupied areas they occur in large numbers and the colonies consist mainly of individuals of the same age, but their life history is still largely unknown.

Semisessile animals usually only move about when in flight or when seeking mates during the breeding season.

Sessile animals are characterized by their inability to move about and this is frequently associated with attachment to the substrate. They include the cirripedes among the crustaceans, many hydroids, bryozoans, terebratulids, some lamellibranchs such as oysters and mussels, the slipper limpet *Crepidula* and the ascidians among the Tunicata. Animals living in tubes attached to the substrate, which they can move about in but do not leave, can also be regarded as sessile; this happens, for example in the polychaete family Serpulidae and also to a considerable extent in the Sabellidae. In contrast to these animals which can be termed 'fixosessile', there are the 'rhizosessile' forms occurring on soft bottoms. They are anchored in the substrate by basal root-like stalks and they include some sponges, the pennatularians, the sea-lily *Rhizocrinus* and the majority of the ascidians found on soft bottoms in the abyssal (Millar 1959). In addition to anchoring these animals the stalks also serve to raise them somewhat above the surface of the substrate which is a zone deficient in oxygen (Madsen 1961). It should, however, be mentioned that in spite of the stalk some forms can withdraw into the substrate; the pennatularian *Stylatula elongata*, which is sometimes abundant in the eulittoral of the American

Pacific coast, disappears completely into the mud when the tide goes out. The mechanism of withdrawal is still not clear.

Remane (1940) has applied the term 'liberosessile' to animals which live free or in the substrate without any special attachment and do not move about. Examples are the clam *Mya arenaria*, which lives in mud-flats, and the boring lamellibranchs. When there are no crevices available the echinoid *Paracentrotus lividus* bores holes in rocky substrates which it uses as a protection against wave action in the eulittoral. In deep holes this echinoid may become imprisoned as it grows so that it becomes 'liberosessile' more or less accidentally. The same has been reported for *Strongylocentrotus purpuratus* in the north-east Pacific. The behaviour of these echinoids is doubtless correlated with the need for protection in littoral areas exposed to powerful wave action.

The benthal also contains those animals which use the bottom as a place for feeding or refuge without moving on or in it. This applies primarily to those fishes which, although free-swimming, are completely dependent upon the substrate for their food and are therefore only rarely found at any distance from it. These include, for example, some of the Gadidae. The rich invertebrate fauna and the complex structure of coral-reefs provide numerous opportunities for the evolution of several different types of fish with very varied behaviour patterns. Modern methods of observation by divers or by the use of cameras have provided valuable information in this field, although only a few points can be referred to here.

Fishes living in coral reefs show a number of different shapes (Fig. 111). These include the eel form of the Muraenidae, which are closely tied to the bottom, the normal perciform types which move around among the corals, the tall laterally compressed forms such as *Psettobalistes* and the orbiculate types with plump bodies, round or polygonal in cross section, and large heads, such as *Scorpaena*, box-fishes and pufferfishes.

Among the fishes with highly specialized feeding habits are those which browse on algae, particularly on the encrusting algae on reefs; these include members of the families Chaetodontidae, Acanthuridae, Pomacentridae, Blenniidae and Balistidae. Hiatt & Strasburg (1960) make the interesting observation that 'the most primitive fishes in the phylogenetic scale are exclusively carnivores' and that they occupy a high position in the food pyramid, that is, they represent the end links in the chain, whereas 'the more advanced and specialized fishes appear to be heading morphologically and physiologically toward the lower trophic levels where the stored energy in terms of food supply is in greater quantity' (l.c., p. 122) see (p. 327). The coral-eaters are also highly specialized.

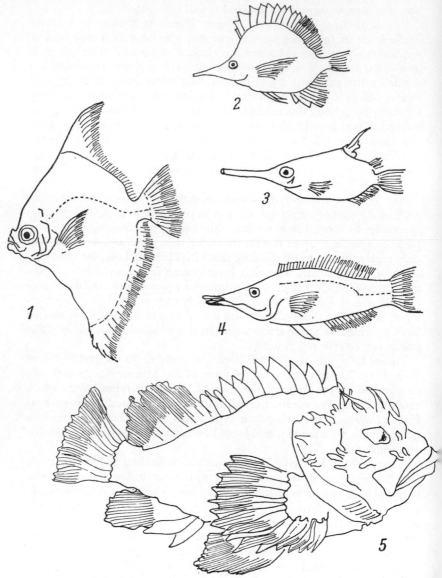

Figure 111. Examples of different types of coral-reef fishes (after Le Danois 1959). 1, *Psettus sebae*, 2, *Forcipiger longirostris*, 3, *Macrorhamphosus gracilis*, 4, *Gomphosus coeruleus*, 5, *Scorpaena porcus*

Figures 112–119. Coral reefs, 1–4 metres depth, Maldive Archipelago (photographs S. Gerlach, *Xarifa* Expedition 1958). Note the different growth forms of the numerous species.

Figure 12

Figure 113. Stagshorn coral

Figure 114.   A characteristic vertical growth form of *Millepora*

Figure 115.   *Fungia*

Figure 116

Figure 117. Brain coral

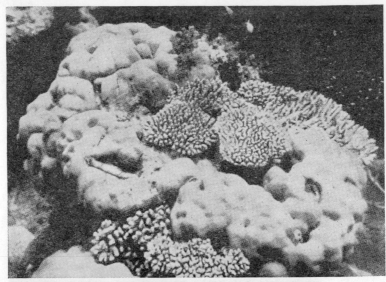

Figure 118

Figure 119

The medium-sized sharks, such as the grey shark, which swarm around the reefs occupy an intermediate position between the purely bottom fishes (e.g. ground shark) and the pelagic predators (e.g. blue shark) (Klausewitz 1962 and Fig. 122).

Klausewitz (1961) has analysed the striking coloration of many coral-fishes and has concluded: (*a*) the percentage of fishes that are not strikingly pigmented is often 70–80% and thus significantly greater than that of the brilliantly coloured species: (*b*) owing to the absorption of the red and yellow components these colours are much less conspicuous in the water then when the fish are seen outside the water; (*c*) the spaces in among the corals are broken up optically by reflexions and shadows, so that when they are moving even the strikingly coloured fishes are lit up and disappear irregularly; (*d*) the value of the striking coloration as a signal to predators is relatively small, because their ability to detect colour is less than that of the human eye, and is sometimes completely lacking; (*e*) the danger of having visually conspicuous characters is partly compensated for by long flight distances and close proximity to the reef and its hiding-places. The swarming behaviour of many species can be regarded as a protective mechanism, because a closely packed swarm gives the visual impression of a single body and also because an attacking enemy may be distracted by the multiplicity of moving objects. Figures 112–119 give

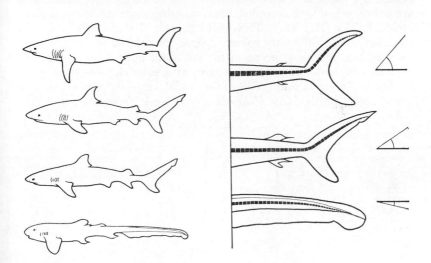

Figure 120. Shapes in the sharks
Left side, from above: blue shark (*Isurus glaucus*); grey sharks *Carcharhinus menisorrah* and *C. melanopterus*; ground shark *Stegostoma fasciatum*. Right side: angle between body axis and upper tail lobe (after Klausewitz 1962)

an idea of the spatial differentiation on a coral reef and explains why the most diverse types of animal can live in close proximity.

There are, therefore, animals which live at a distance from the bottom, rather than directly on it and yet are dependent upon the substrate for their food. This fauna must be regarded as an integral part of the benthal, and according to the nature of the substrate it may be described as suprapsammon, suprapelos or supralithion.

5. Plants of the benthal

Excluding heterotrophs and chemotrophs, benthal plants, like those of the pelagial, are tied to the photic zone; in other words the phototropic plants are characteristic of the littoral. As already mentioned, they react in different ways to external factors and have therefore been used in the ecological classification of the littoral.

The most important phototrophs in the marine region are the green, red and brown algae which are mostly multicellular organisms, the blue-green algae, the diatoms and a few phanerogams.

They owe their different colours to the formation of special pigments in addition to chlorophyll; these are phycocyanin in red and blue-green algae, phycoxanthin in brown algae and phycoerythrin in red algae. According to Strickland (1958) these additional pigments enable the plant to utilize light from regions of the spectrum which are not absorbed by chlorophyll, thus allowing more efficient use of the incident light (Fig. 121). This might suggest that there is a vertical zonation of the differently coloured algae, but this is not in fact true, since certain red algae occur in the eulittoral and in the upper sublittoral, whereas some green algae live at greater depths. Thus, from studies at Eniwetok Atoll, Gilmartin (1960) emphasized the fact that green algae, particularly *Halimeda* species and *Tydemania expeditionis*,

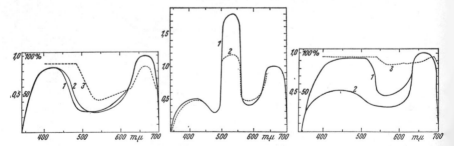

Figure 121. Photosynthesis absorption spectra: (a) green algae *Ulva lactuca* (1), *U. linza* (2), absorption curve of *U. lactuca* (3); (b) red algae *Ceramium pedicellatum* from 20 m. depth (1), *C. rubrum* from 0–0·5m. depth (2); (c) brown algae *Fucus serratus* (1) *F. vesiculosus* (2), absorption curve for *F. serratus* (3) (from Levring 1960).

are dominant at depths (20–65 m) receiving 2–4% of the surface light, whereas red algae are never abundant in this deep-water association. On the other hand, Thorson (1933, p. 58), writing of the arctic regions, remarked: 'The red algae epifauna is, as a rule, the epifauna that occurs at the greatest depth.' Accordingly, vertical distribution dependent upon geographical latitude would need to be proved. Also in some forms pigmentation has been observed to change; thus in red algae there may be a reduction of the phycoerythrin in bright light, so that the plants then appear green.

Jones (1959) has given interesting evidence of the importance of light in the zonation of certain species. He found that the spores of the red alga *Gracilaria verrucosa* will not develop if they have once been exposed to a period of 4–5 hours of full sunlight. This species can therefore only settle in the eulittoral at a lower horizon where the period of exposure is correspondingly shorter on account of the covering of water. Biebl (1939) showed that shade-loving species are killed by exposure to light for $1\frac{1}{2}$ hours, wheres typical eulittoral species such as *Ulva lobata*, *Cladophora trichotoma* and *Porphyra perforata* can easily tolerate five hours of light.

A plant, is, on the one hand, adapted to certain wavelengths of light by having different pigments and, on the other hand, the physiological differences between species in relation to the optimal light intensity lead to the development in each algal group of species that prefer light or shade. Both types of adaptation are expressed in the vertical distribution; we still lack more detailed studies on the various mechanisms and the part they play in vertical distribution.

Attention has already been drawn to the differences in resistance to pressure, heat and desiccation that occur in algae from different localities, but it should be remembered that general principles have been little investigated.

The algae offer a wealth of different types of external form from simple filaments, straps or cords to thalli that are uniformly flattened or lobed, and there are also forms that are more or less finely divided and of a delicate or a tough leathery consistency. There is some comparative information on the distribution of morphological types in fast, slowly moving or still water; there are also several records of the development of local modifications. Many forms are very small whereas others, such as *Macrocystis pyrifera*, reach an average length of about 30 metres or even more. The growth rate may be very fast: according to Sheldon (1916, cited by Moore 1958) *Nereocystis luetkeana* grows 25 cm per day from March to June and 2·5 cm from June to July.

The reasons for this gigantism are still not very clear. In some cases it may be an example of the widespread development of large size in cold climates (although *Macrocystis pyrifera* also occurs in large amounts off the coast of south California). It does, however, require an abundant supply of mineral nutrients. So far there has been little work on the nutrient supply for benthonic algae, in contrast to the position in the pelagic algae. According to Fries (1959, cited in Pérés 1961) the red alga *Gonotrichum elegans* is depdent upon a supply of vitamin B_{12}; it is presumed that the B_{12} is produced by epiphytic bacteria and taken up by the alga. This provides the conditions of growth and nutrition for algae that live epiphytically or parasitically (cf. Tanaka & Nozawa 1961) often in large numbers.

Feldmann (1937, cited in Pérès 1961) has given the following classification of algae based on the duration of life:

Annual:
Ephemerophyceae: present throughout the year, several generations in the year; spores and ova develop immediately (*Cladophora, Enteromorpha*, some *Polysiphonia*).
Eclipsiophyceae: only observed during part of the year, at other times existing as microscopic vegetative forms (*Sporochnus, Nereia, Asperococcus*).
Hypnophyceae: as the preceding group, but living in a resting stage (spores, ova, protonema) during unfavourable periods (*Vaucheria, Rivularia bullata, Ulothrix pseudoflacca, Porphyra* and others).

Perennial:
Phanerophyceae: whole plant perennial, erect (*Codium, Fucus vesiculosus* and others, *Phyllophora nervosa*).
Chamaephyceae: as the preceding, but encrusting (*Peyssonelia, Lithophyllum* and others).
Hemiphanerophyceae: only part of the erect plant is perennial (*Cystoseira, Sargassum, Laminaria hyperborea*).
Hemicryptophyceae: only the part serving for attachment survives the winter (*Rissoella, Gymnogongrus, Cladostephus, Griffithsia, Udotea, Acetabularia*).

The gametophytes of the laminarians are usually very small and have generally been regarded as short-lived; on the other hand, Scagel (1959) was able to keep the gametophytes living and fruiting in the laboratory for $2\frac{1}{2}$ years. This author emphasized the need for research on cultures of benthonic algae in order to clarify taxonomic, physiological and ecological problems.

Many algae produce dense growths which often cover extensive areas; with the simultaneous increase in size of the individuals a considerable biomass may be produced; various data have shown that in *Laminaria saccharina* this may be about 4 kg/m. This species is triennial, parts of the thallus breaking off and being renewed each year, so the annual production can be reckoned as about the same as the bio-mass. The average dry weight is 20% so the littoral algal vegetation is of considerable nutritional importance, but this is probably only so in the littoral zone itself, because the dead parts quickly decompose and fall to pieces when moved about for any length of time. The extent to which the assimilates of the living plants are released into the water and thus become important for nutrition is still an open question; the same can be said of specific stored substances which have been found in several forms (e.g. iodine and alginic acid in brown algae, anticoagulants in red algae such as *Delesseria*, *Furcellaria* and others, various polysaccharides such as mannitol in *Laminaria saccharina* and others).

The calcareous algae, which were surveyed by Lemoine (1940), are of general interest in marine biology. Here it is the red algae that are particularly important, *Melobesia* being especially abundant, widely distributed and showing a wealth of different forms. In contrast, the corallines and certain others are less conspicuous, and the green algae with calcium encrustations are of very little importance.

The main geographical centre of distribution of green and red calcareous algae lies in the tropics and subtropics; Prat (1940) considered that the greater deposition of calcium in these regions was due at least in part to the great incidence of UV light. This cannot, however, be the only factor involved because the red alga genera *Litho-thamnium*, *Lithophyllum*, *Melobesia*, *Corallina* and *Iania* also occur in higher latitudes.

Green algae with calcium deposits evidently prefer fairly still water, always remain below low-tide level and are most abundant in the upper part of the sublittoral. Red algae, on the other hand, also thrive on coasts exposed to wave action and, depending upon the illumination, occur in deeper water (down to 100 metres or more) and also on parts of the shore that dry out periodically.

The species of *Melobesia*, in particular, are often associated with animals having calcareous shells, such as foraminiferans, bryozoans, serpulids, vermetids and barnacles and provide a special habitat for numerous invertebrates. In the Mediterranean they produce the structures known as 'trottoirs' (Pérès & Picard 1958), which are ledge-like formations formed at mid-tide level on rocky coasts (Fig. 122).

In some areas calcareous algae also contribute considerably to the formation of a sediment of organic origin which may thus become a habitat for benthonic animals.

Among the phanerogams only a few of the Potamogetonaceae have become adapted to the sea. They occur mainly in warm-water areas, and mostly in the upper part of the sublittoral, although in clear water they may reach depths of 40 or even 70–90 metres. They are anchored

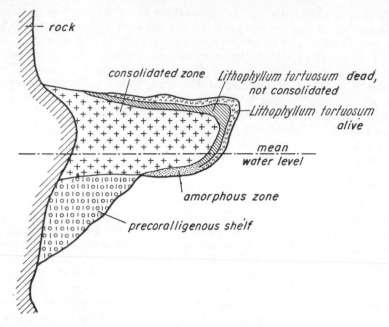

Figure 122. Diagrammatic section through a 'trottoir' formed by calcareous algae (after Pérès & Picard 1958)

in the substrate by rhizomes; as regards the type of substrate the different species vary in their requirements from pure sand to mud rich in nutrients. According to Molinier & Picard (1952), in *Cymodocea nodosa* the relation between root formation and leaf formation is dependent upon the type of substrate: on bottoms poor in nutrient the leaves are more strongly developed, but on rich bottoms the rhizome and roots. As far as I know there have been no investigations of this kind on other species.

The phanerogams are of general ecological importance mainly because they form very dense and extensive growths which are known as eel-grass meadows. The annual production of organic matter is very large (see Chapter VI, 1); as a result the phanerogams contribute considerably to the formation of detritus.

The reduction of currents within such growths encourages sedimentation; in the Mediterranean the meadows of *Posidonia* may grow upwards and in some places form a 'barrier reef'. It is a peculiarity of *Posidonia* that the rhizomes are positioned more or less vertically in the substrate, which enables them to grow vertically (Molinier & Picard 1952). Data on the growths of *Thalassia testudinea* and their fauna on the coasts of the Caribbean Sea will be found in Voss & Voss (1955) and Rodriguez (1959).

These growths provide a substrate for the settlement of epiphytes and often contain a very rich epifauna; a varied mesofauna is usually also present. Any reduction in the growths of eel-grass, such as has occurred in the North and Baltic Seas as a result of disease, will therefore have far-reaching effects on the fauna of the area.

On muddy tropical coasts there are extensive areas of mangrove swamp; these are stands of trees of different taxonomic status (including Myrtaceae, Verbenaceae, Myrsinaceae) showing striking convergence in certain structures. These enable them to tolerate periodic and extensive flooding in the tidal zone, but also more particularly to thrive in substrates that are saturated with water but deficient in oxygen. The primary adaptation for this is the possession of aerial roots which grow up from the substrate and carry oxygen to the root system.

Some mangrove trees show a form of viviparity; the seeds germinate without a resting phase and, while still on the parent plant, forming a hypocotyl that is up to 50–60 cm in length. When released from the parent tree this falls into the mud and soon produces roots which anchor the young plant.

Mangrove swamps are best developed in the Indo-Pacific region; the American and West African mangrove swamps have different species and fewer of them. These trees usually show a form of zonation, the species of *Rhizophora* and *Sonneratia* forming the outer girdle within the protection of which *Avicennia* and other forms grow. Mangrove swamps have a rich fauna, consisting partly of numerous animals, especially crustaceans, which live in the mud and partly of the various sessile, semisessile and motile forms which use the trunks and roots of the trees as a firm substrate. Fiddler-crabs occur in most mangrove swamps and show a marked zonation in their distribution within the swamps. Other typical mangrove animals are the crab *Tachypleus gigas* in the Indo-Malayan archipelago and the mudskipper *Periophthalmus koelreuteri*.

There have been several recent investigations on the mangrove swamps of the west Atlantic and east Africa (Inhaca Island). Gerlach (1958) gave the first comprehensive account of the ecology of a mangrove swamp and produced a sketch showing the distribution of some of the commoner large forms (Fig. 123). Further comparative work on this interesting habitat would be very valuable.

There is no doubt that diatoms play an important role as autotrophic producers in the eulittoral and sublittoral, because they have often been recorded in large numbers on hard and soft bottoms; on sandy substrates they also thrive in the interstitial system and in addition they occur as epiphytes on other algae. In tropical shallow waters one cannot overlook the importance of the symbiosis between unicel-

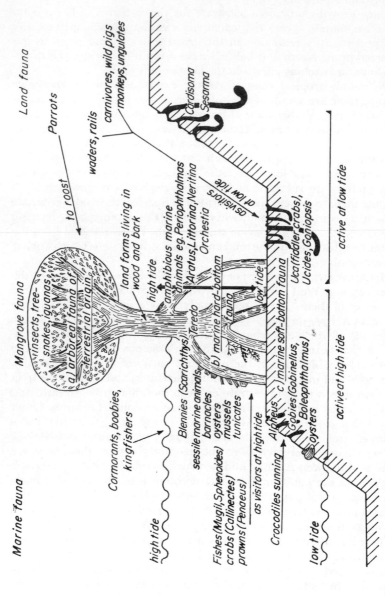

Figure 123. Diagram showing the macrofauna of a mangrove swamp and its ecological relationships (after Gerlach 1958)

lular algae and several different animals (madreporarians, actiniarians, gorgonians, lamellibranchs such as *Tridacna*, foraminiferans and others; cf. Odum & Odum 1955). On reefs, for example, there is a massive concentration of animal consumers and a relative paucity of free-living producers, whereas associations with unicellular algae are common. Yonge (1944) speaks of the symbionts as 'imprisoned plankton'; he considers it possible that there is a correlation between the presence of these numerous symbionts and the poverty of the phytoplankton, because the animals with associated algae will release fewer metabolic products and therefore less plant nutrient into the water. This would mean that only relatively small amounts of these minimal substances would be available to the phytoplankton.

Bacteria have long been known to occur in the benthal, but it was Zobell, in particular, who showed during the Galathea Expedition that they also existed in the greatest depths of the sea, and drew attention to certain physiological peculiarities. Höhnk also found lower and higher fungi in oceanic substrates down to depths of over 4,000 metres and there is scarcely any doubt that organisms of this type are also generally distributed in even greater depths. Saprophytes may be involved primarily in the breakdown of substances such as chitin and keratin which are not attacked by animal enzymes; they may also be able to utilize dissolved organic matter and to produce vitamins. Further experimental investigations are required in this field (see Chapter VII, 1).

6. Types of benthonic animals

The types of life form can be approached from two points of view. It is possible to follow the modifications of structure and function within each of the larger taxonomic groups and to relate them to habits, that is, to the peculiarities of the ecological niche of each species. Russell's (1962) treatment of the decapod crustaceans provides an example of this. This method provides an insight into the evolutionary potential of the systematic unit concerned, but it makes it more difficult to survey the analogies that exist between phyla, classes and orders within the same habitat. Here I have chosen to use the second method, and to point out analogies between members of different groups living within the same environment.

We have already seen the different ways in which benthonic animals are associated with the substrate. There is also considerable diversity in other functions and in their morphological manifestations; here it is not possible to discuss this subject in any great detail, but the main points will be indicated. It is scarcely necessary to mention that we know relatively little about archibenthal and abyssal organisms.

(a) *Reproduction and development*

In most benthonic invertebrates and also in many fishes, fertilization and early development take place in the pelagial. Larval plankton, which is often very abundant, represents an important nutritional factor in the total plankton and for this and other reasons has often attracted attention. The problems associated with it, and their relation to benthonic ecology have been discussed in great detail by Thorson (1946, 1950, 1952).

Fertilization is ensured in various different ways; true copulation takes place in some polychaetes (e.g. *Capitella*), in most crustaceans (even in the sessile cirripedes), and in higher prosobranchs, opisthobranchs and cephalopods. In certain polychaetes and nemertines the release of eggs and sperms in a mass of mucus in which several animals participate ensures efficient fertilization, and this method is probably more widespread than is at present thought. Hermaphroditism with self-fertilization occurs in some sessile forms, e.g. some serpulid polychaetes.

The free release of the gametes into the water often takes the form of an 'epidemic spawning', that is, whole populations spawn more or less simultaneously, so that the chances of fertilization taking place are very high owing to the enormous number of sexual cells present. Sexual maturity and readiness to spawn are associated with seasonally induced internal rhythms. During spawning the male gametes are released first; stimulating substances are evidently released at the same time and these stimulate the release of the eggs into the water which is already full of sperms. In several different families of polychaetes, particularly in the Nereidae and Eunicidae, the atokous benthonic forms undergo a metamorphosis to produce epitokous forms which swarm to the surface where the eggs and sperms are shed; in most cases lunar periodicity has been confirmed and hormones are involved (Durchon 1952, 1953 and Hauenschild 1955, 1956) In the well-known palolo worm, *Palola viridis*, the part of the body containing the sexual cells breaks loose and swims to the surface, while the front part of the body remains in the worm's sublittoral habitat and regenerates the lost segments. It is noteworthy that in some species of *Nereis* there is a certain degree of plasticity because different populations do not always behave in the same way, so swarming does not appear to be obligatory. Furthermore, in quite a number of polychaete families differences in egg size and mode of development have been observed in populations, and even in the same population in different years. This ability to adapt the reproductive habits to external conditions, which is widely distributed among polychaetes and has also been found in some nemertines, but is absent in other invertebrates, is possibly correlated with the great regional and vertical distribution of many polychaetes.

Thorson (l.c.) has drawn up a very valuable comparative list of the types of eggs and larvae of marine invertebrates, the main points of which are given in Table 29. Forms developing in the benthal often

TABLE 29

Types of eggs and larvae of benthonic invertebrates

(After Thorson)

Development takes place	Nutrition during embryonic and larval development	Duration of free larval life	Main distribution
in benthal	much yolk, eggs laid or viviparous		Polar regions, deep sea
	small eggs with abortive eggs	short, or completely lacking	Polar and cold temperate regions, deep sea
in pelagial	small planktotrophic eggs, poor in nutriment	long (2->4 weeks)	boreal seas, about 55–65%, tropical coastal waters 80–85% of all species
		short period	about 5% of all species in all seas
	large lecithotrophic eggs, rich in nutriment	usually long	apparently, lacking in arctic, about 10% of all species in temperate and warm seas

have large eggs rich in yolk; these are often laid in clutches but in viviparous species they may develop within the mother until the birth of the young; various forms of brood protection are also known. In contrast, there are species with small eggs with little yolk which are laid in large egg capsules and are nourished by neighbouring abortive eggs, so that the young hatch in an advanced state. The period of free larval life is therefore very abbreviated or may be completely absent. This mode of development must have special selection value in places where the feeding conditions for free-living larvae are very unfavourable; this applies particularly in the polar regions and in the deep sea where the supply of phytoplankton for feeding is poor and irregular or entirely lacking.

Pelagic larvae are either planktotrophic or lecithotrophic. The former hatch from small eggs and develop locomotory and feeding organs in the form of ciliated bands or tufts. Lecithotrophic larvae hatch from large eggs which are so well provided with yolk that they

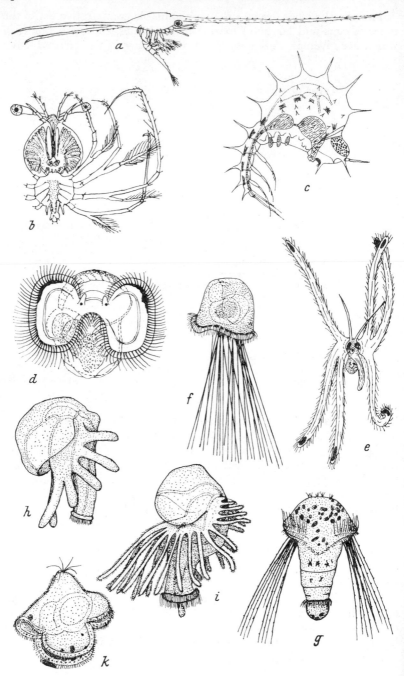

are practically independent of feeding in the plankton, even though they usually spend a long time (up to four weeks or more) in the pelagial (Fig. 124).

There is a close relationship between egg size and the number of eggs per mother animal and the period of reproduction, and this is shown in the diagram (Fig. 125) produced by Thorson (1950). In a

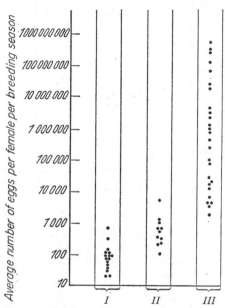

Figure 125. The number of eggs produced per female per breeding season in marine bottom invertebrates (echinoderms, nemerteans, polychaetes, prosobranchs, tecti-branchs, and decapods) with different types of development (after Thorson 1950) I, species with viviparity or a high degree of brood protection. Non-pelagic develop-ment. II, species with large eggs or a primitive type of brood protection. Non-pelagic development. III, species with small eggs and a long pelagic planktotrophic larval life

further diagram Thorson (1952) has related the size of the eggs and the corresponding larval type (Fig. 126). It would be useful to have this scheme further extended in order to discover the mode of development of forms in which direct observation is not feasible; this would be particularly valuable in the case of deep-sea animals. Thus, in the

---

Figure 124. Some pelagic larval forms (a, b, e, g after Trégouboff & Rose, c after Woltereck, d, f, h-k after Thorson)
(a) metazoea of *Porcellana platycheles*, (b) phyllosoma larva of *Palinurus vulgaris*, (c) *Thaumatops* (amphipod), (d) *Littorina littorea* (prosobranch), (e) *Carinaria mediterranea* (heteropod), (f) *Myriochele danielsseni* (polychaete), (g) *Sabellaria alveolata* (polychaete), (h) and (i) *Phoronis muelleri* – old and young stage, (k) pilidium (nemertine)

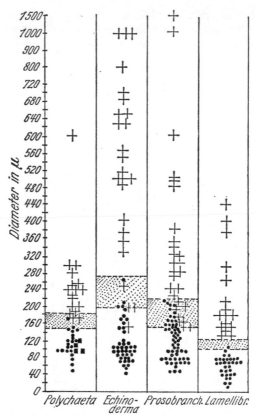

Figure 126. Relationship between egg size and type of development in different groups of benthonic animals (after Thorson 1952): *dots* = species with a long pelagic planktotrophic life; *crosses* = species with direct, non-pelagic development; *squares* = brood-protecting syllids with non-pelagic development; *stippled zone* = transition from pelagic to non-pelagic development

Porcellanasteridae for example, Madsen (1961) stated: 'The ripe eggs measure 0·5–0·6 mm in diameter which confirms an expected non-pelagic development' (p. 181).

The importance of the larval ecology of benthonic animals for marine biology in general is apparent from the geographical distribution of the developmental types, as shown for example by Thorson's (1950) analysis of the prosobranchs (Fig. 127). This shows that the percentage of species with non-pelagic larvae decreases from the arctic to the tropics, while the percentage with pelagic larvae increases. One might expect a similar result from an analysis of the vertical distribution, particularly in warm-water areas. Non-pelagic development would be selectively advantageous because the larvae would not be dependent for food on phytoplankton populations that are restricted in time, and furthermore with benthonic larvae the wastage would be lower so that smaller numbers of offspring would suffice for the maintenance of the population.

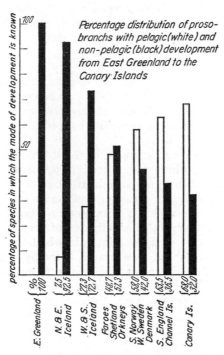

Figure 127. Percentage distribution of prosobranchs with pelagic (white) and non-pelagic (black) development from East Greenland to the Canary Islands

Forms with pelagic larvae have a greater chance of being distributed, because the larvae are easily transported over great distances; on the other hand they may be carried to completely unsuitable areas, e.g. littoral forms may arrive in oceanic regions, where there will be an increase in the numbers destroyed and furthermore the chances of finding a substrate suitable for the transition to benthonic life will be decreased. The problems of metamorphosis involved will be discussed in more detail on pages 401 et seq. It is a fact, however, that animals with non-pelagic larvae may also be distributed over wide areas, as is shown by the large number of cosmopolitan species in the abyssal fauna (p. 300). The microfauna of the mesopsammon is characterized by a very high percentage of species with direct development or with non-pelagic larvae and yet some of its species are known to have a wide distribution; we do not yet know the mechanism of distribution.

If epidemic spawning accompanied by favourable feeding conditions and a relatively low wastage occur simultaneously this may result in the production of large swarms of larvae and a big spatfall of metamorphosed animals over a wide area. This will produce fluctuations in the population of the species which in some cases may be noticeable for a few years, but which is usually soon balanced by increased predation (pp. 351 et seq.).

In some forms, pelagic breeding stages are not only present in the larvae; here one can include the well-known alternation of generations in hydroids with polyps and medusae; in some hydroids pelagic vegetative reproductive stages are released (*Halecium pusillum*) according to Huvé 1955), or the benthonic phase is much reduced relative to the pelagic.

Only brief mention can be made of the different methods of egg-laying; in many cases there are protective egg-cases but we still have no general analysis of the ecological distribution of types of egg-laying. Brooding behaviour is also common and occurs in the most diverse groups; the spawn is guarded by the mother animal or the eggs are fanned with fresh water by the activities of the mother. The most effective method occurs in those animals in which the eggs are carried on the body of the mother as, for example, in the Syllidae and species of *Harmothoe* among the polychaetes and in a great number of crustaceans.

Numerous species in several different groups practice viviparity, which is the most advanced type of brood protection; this occurs in actinians, polychaetes, frequently in asteroids and ophiuroids, occasionally in nerertines, in the Molgulidae among the ascidians and in fishes such as *Zoarces viviparus*.

Ecologically it is noteworthy that viviparity occurs very commonly in polar and deep-sea forms, although such forms are not lacking in warm seas or in the littoral. It is probable that the development of viviparity has been associated with an irregular food supply for the larvae; this view is confirmed by the existence of forms which are viviparous or oviparous according to the ecological conditions.

Attention has already been drawn to the high percentage of mesopsammic forms with direct development. It may be added that in this fauna the number of eggs produced at any one time is generally very small, that a relatively large number of forms practise a more thorough type of brood protection than comparable forms in other habitats and that many species are to a greater or lesser extent neotenous (*Halam-*

mohydra, *Psammodrilus*, Mystacocaridae, opisthobranchs; see Delamare Deboutteville 1960). Small body size, method of development and the structure of the interstitial environment are evidently interdependent so that here too the adaptive character of these developmental peculiarities is unmistakable.

The seasonal periodicity of reproduction in various species may also show adaptation to special ecological niches. A particularly important part is played by spawning and hatching at a time when the larvae can obtain the necessary food. This phenomenon can be recorded by quantitative observations but the underlying mechanism requires experimental investigation. In *Balanus balanoides*, Barnes (1957) attributes this synchronization to a substance present in the diatom diet of the female parent, which accumulates during the diatom outburst in late winter and eventually stimulates the hatching of the nauplii. The participation of other factors should not be excluded. Thus, in the Murmansk area Pzepishevsky (1962) found that mass spawning was stimulated by the uptake of small naked flagellates, which developed before the diatoms and nauplii. The metabolic products of the flagellates were not observed to have any stimulatory effect.

In discussing the work of Bolin (1949), Gohar has pointed out a special form of adaptation in the eulittoral. The larvae of the gastropod *Nerita forskali* cannot hatch when continuously submerged; but when the eggs, which are shaped like an hour-glass, are in the tidal zone the egg-shell becomes indented when desiccated during periods of low tide and swells again by the uptake of water when the tide is in. The repetition of this process leads to the bursting of the egg-capsule so that the larvae can escape.

(b) *Types of movement*

Several different types of movement have been evolved among the diverse types of animal living in the benthal. Here we can only present a few general examples and cannot cover the field exhaustively. It should be remembered, of course, that there is a very close connection between method of movement, general structure and the ecology of the animals.

Animals that run about on sandy or soft substrates are particularly well represented among the crustaceans. Here there is a striking difference between inhabitants of the deep-sea floor and of the bottom in the littoral; the former mostly have long extremities that are carried spread out and are often equipped with long bristles, whereas in littoral forms the body is relatively longer than the legs. Examples of such deep-sea forms are isopods such as *Munopsis*, amphipods of the genera *Lepechinella* and *Rachotropis*, crabs like *Platymaia*, *Scyramathia* and *Homolochunta*; some deep-sea prawns may also be mentioned here (Fig. 154).

Among the animals which walk about on stilt-like legs are the pycnogonids; this has been confirmed on several occasions from photographs of the bathyal and abyssal floor. At the same time abyssal species often have long powerful spines and 'hairs' on the body and extremities, but their function, like that of the dorsal spines in amphipods, is still not clear (Stock 1963). The development of single fin-rays from the pelvic and caudal fins in fishes of the family Benthypteroidae is particularly striking; from observations made from the bathyscaphe these structures serve not only as organs of taste but are also used as stilts to walk along the sea floor (Fig. 128).

The adaptive nature of this type of character to soft substrates is very well demonstrated in crabs such as *Platymaia*; this crustacean does not walk on the tips of the distal joints of the legs but lays these flat on the bottom; this evidently prevents it from sinking into soft deep-sea sediments.

However, by no means all deep-sea crabs show this characteristic; some such as *Geryon* scarcely differ from the littoral crabs of which

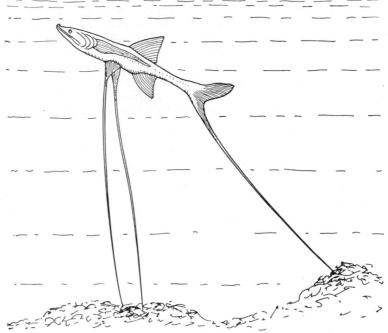

Figure 128. *Benthosaurus.* The elongated ventral fins and the caudal fin appendage serves as 'stilts', enabling the animal to hop along on the bottom. (Redrawn from a photograph by G. S. Houot, *UNESCO Courier,* July-August 1960, p. 23 and an illustration in Murray & Hjort 1912)

the best-known example is *Carcinus maenas*. This is a further indication of the diversity of habitats in the abyssal and benthal, comparable to the diversity in the littoral as expressed in such different types as *Cancer pagurus* and *Pachygrapsus*.

Movement by crawling is much commoner than walking and occurs in several different phyla because it does not require jointed extremities. The type of crawling differs considerably according to the structure of the animal: turbellarians and nemertines crawl on mucus bands or in mucus tubes by the movement of the cilia which cover the body; many polychaetes crawl by the undulatory movements of the body, supplemented by the paddling of the parapodia; the ophiuroids crawl along the substrate by the flexion of the narrow slender arms, while the asteroids, echinoids and holothurians move by means of the tube-feet, which extend, attach and then contract.

Certain modifications may occur which are correlated with the nature of the substrate, as for example the development of a creeping sole in deep-sea holothurians (Elasipoda) and some forms show more than one type of movement. Thus, after stimulation many nemertines move like earthworms by alternate contraction and extension of succeeding body segments; nereids and other polychaetes as well as the Cerebratulidae among the nemertines can swim free of the bottom, a method which provides a good flight route. When touched by the starfish *Asterias rubens* the cockle *Cardium echinatum* can jump a distance of 15–20 cm, using its long foot; this response is lacking in the related *Cardium edule*, which lives in shallower and often brackish water, where the starfish is rare. Even some lamellibranchs are able to swim: the scallops (*Pecten*, *Chlamys*) lie on the substrate; by the rapid closure of the shells which expels a jet of water they are lifted from the bottom and swim off with shells opening and closing to produce further jets. The direction of movement depends upon the exact place along the perimeter of the shells at which the water is expelled. MacGinitie (1949) has reported analogous behaviour in the razor-shell *Solen sicarius* from the east Pacific. The principle of jet propulsion is well known among the cephalopods and highly developed in many of the benthonic forms. The long-tailed decapod crustaceans can move backwards in jerks by powerful flapping of the tail against the belly. The caprellids, which can be regarded as semisessile forms, can spend short periods in the open water by convulsive bending and stretching of the body. Amphipods, isopods and decapods such as the prawns can swim by the paddling movements of the limbs; for them the open water often serves for flight, and sometimes also for feeding, but they have no structures specially developed for swimming.

In the swimming crabs (Portunidae and Neptunidae), on the other hand, the fifth pair of thoracic limbs is flattened and broadened and

thus adapted to function as swimming paddles; other structures associated with swimming have been described by Schäfer (1954). For
these crabs the open water serves as a place to feed because they can
take off from the bottom and chase passing prey, in so far as their
habits are known in any detail.

There are several opisthobranch gastropods which can rise in the
water by swimming, which is effected mainly by the undulatory movements of the broadened foot. The ecological role of the open water may
therefore vary considerably in benthonic animals: it may serve as a
place for feeding, e.g. in *Chioraera leonina* and the swimming-crabs
(see p. 253) or as a place to escape to, e.g. in the lamellibranchs mentioned; there are also some examples from among the fishes.

The infauna, that is the animals living in the substrate, is either liberosessile (see p. 229), semisessile or more or less motile. Evidently the
motile forms, the burrowers, are the most abundant, although in most
cases they cannot of course be directly observed. In polychaetes such
as the Nephthyidae, Glyceridae, Arenicolidae and others the pharynx
is protruded into the substrate; there it swells like a balloon and an-
chors the animal, so that the body can be drawn up behind it. The
same applies in the Echiuridae and Priapulidae. Other forms, like
the Capitellidae, move in the manner of earthworms, although the
anatomical basis for this is completely different. The lamellibranchs
also used their distensible foot in order to burrow or move into the
substrate, but the degree of mobility is extremely variable.

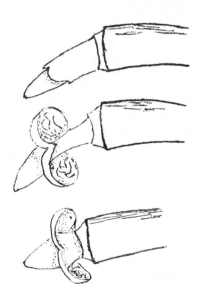

Figure 129. *Dentalium entalis;* successive stages in the burrowing of the foot
showing its extensible lobes (after Mor,
ton 1959)

Scaphopods also burrow into the substrate, particularly into sand; according to Morton (1959) the foot of *Dentalium entalis* has a transverse fold of skin which can be inflated by pumping fluid from the haemocoel, thus providing an effective anchorage in the substrate (Fig. 129).

Predatory gastropods (Naticidae) burrow a few centimetres below the surface in search of their lamellibranch prey; *Nassa* and opisthobranchs such as *Philine, Akera* and others also burrow in soft bottoms.

Several crustaceans burrow; some of them such as the crabs *Corystes, Calappa* and others only bury themselves in the substrate without then moving any further, others dig tubes to varying depths (several brachyurans, the anomuran *Callianassa*) or burrow actively through the sediment (e.g. the amphipods *Haustorius, Bathyporeia, Corophium*). The morphological characters associated with burrowing or digging are very diverse, and are not understood in every case. The structure of the substrate is changed by burrowing and the construction of tunnels; MacIntyre (1963) differentiates between those animals which have this effect—the modifiers—and those that only hide in the substrate—the non-modifiers.

Remane (1933) has demonstrated the close relationship between the structure of the substrate and the locomotory habits of the interstitial fauna. A very large percentage of this microfauna moves by creeping with the help of cilia; this happens in numerous turbellarians, rotifers, gastrotrichs, archiannelids, polychaetes such as *Psammodrilus* and *Ophryotrocha*, the medusa *Halammohydra* and others. There are considerable differences in the arrangement of the cilia; the whole surface of the body may be ciliated, only some surfaces may carry cilia or there may be of ciliated bands. In addition to this type of locomotion many forms move by undulations, particularly of the main trunk of the body, in which the body is in touch with the substrate on at least two opposite sides. This method is particularly characteristic of the nematodes, although some copepods and others show something approaching it. The haptic behaviour of numerous animals in the mesopsammon is also particularly striking; these either attach themselves by means of glandular secretions or are able to climb with the aid of their limbs, parapodia or special bristles (Fig. 130).

This behaviour can be regarded as an adaptation to living amongst sand grains that are easily shifted about relative to each other.

In the epifauna, climbing forms occur particularly on plant vegetation, sponges and corals. Characteristic types are the caprellids (Fig. 131), brachyuran crabs and pycnogonids; certain polychaetes climb about on hydroid colonies. Some crabs also climb on rocky substrates; Schäfer (1954) has shown that they have peculiar spines and teeth on the cephalothorax and limbs. In many forms the distal joint of the last pair of thoracic limbs is moveable and forms a claw or pincer, which

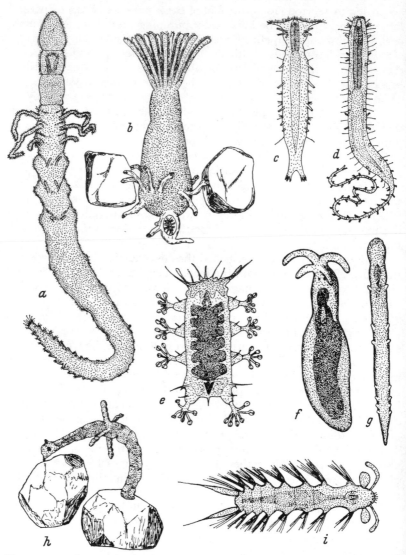

Figure 130. Some animals of the mesopsammon (redrawn after various authors): (a) *Psammodrilus balanoglossoides* Swedmark (polychaete); (b) *Monobryozoon ambulans* Remane (bryozoan); (c) *Dactylopodalia cornuta* Swedmark and (d) *Urodasys viviparus* Wilke (gastrotrichs); (e) *Batillipes mirus* (after Marcus) (tardigrade); (f) *Unela odhneri* (Delam) and (g) *Pseudovermis axi* Marcus (opisthobranchs); (h) *Psammohydra nanna* Schulz; (i) *Nerillidium gracile* Remane (archiannelid)

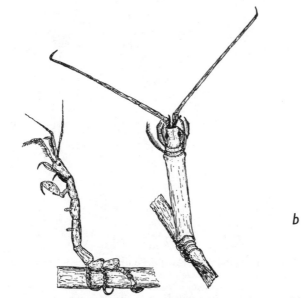

a   *b*

Figure 131. Parallel morphological types in semisessile crustaceans: (a) *Caprella* (amphipod); (b) *Astacilla* (isopod) (combined from various authors). Filter formed in *Caprella* by the 2nd antennae, in *Astacilla* by the mouthparts

allows the animal to attach itself; in the Trapezinae on the other hand the ends of the claws are covered with a dense mat of hairy bristles which prevents injury to the corals on which the crabs crawl. Many crabs camouflage themselves by attaching algal fragments to small curved chitinous hooks; they are then very difficult to discern when climbing among algae. Among the fishes the mudskipper *Periophthalmus* can crawl around on the roots of mangroves.

Various forms bore in hard substrates; boring sponges of the genus *Cliona* penetrate calcareous rocks and lamellibranch shells; the polychaete genus *Polydora* makes U-shaped burrows in mollusc shells and calcareous rocks; lamellibranchs belonging to various genera (*Saxicava, Pholas, Zirfea, Petricola* and others) bore into a variety of rocks and also attack timber and hard clay; timber is attacked by the Teredinidae (*Teredo, Bankia, Xylophaga*) and by crustaceans (the isopod *Limnoria* and the amphipod *Chelura*) and largely destroyed; the echinoid *Paracentrotus lividus* bores into rocks.

The boring is mostly carried out mechanically with the aid of bristles, shell edges, spines, but chemical action is certainly involved in the case of *Cliona* and probably in *Lithophaga*. If the populations are dense there may be considerable damage to the substrate; when the shells

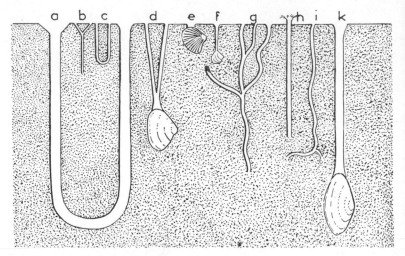

Figure 132. Diagram to show form of the burrow or tube in various in faunal species (after Remane 1940): (a) *Arenicola marina*, (b) *Pygospio elegans*, (c) *Corophium volutator*, (d) *Scrobicularia plana*, (e) *Cardium edule*, (f) *Macoma baltica*, (g) *Nereis diversicolor*, (h) *Lanice conchilega*, (i) *Heteromastus filiformis*, (k) *Mya arenaria*

of living oysters are attacked there may be widespread economic loss and such organisms also help in the destruction of coral reefs or rocky coasts and are thus of geological importance.

These boring animals are to a certain extent comparable with those which build tubes in sandy and soft substrates, of which there are many examples among the actinians, polychaetes, molluscs and crustaceans. In dense populations the structure of the substrate will be considerably influenced by the cementing of the tubes and by the diffusion into it of oxygen from the water used for respiration by the tube-dwellers. Many of the tubes are U-shaped; one arm is inhalant, the other exhalant (Figs. 132 and 133).

(c) *Feeding mechanisms*

The benthonic fauna has naturally evolved an even greater diversity of feeding mechanisms than the pelagial, and so only the main types

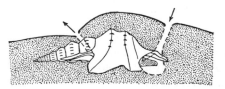

Figure 133. The gastropod *Aporrhais pes-pelicani* buried in mud (after Yonge from Ankel 1938)

can be discussed here, particular attention being paid to those of special biological interest.

Among the herbivores there are some that browse on the low-lying plant mats growing on firm substrates; examples are the chitons and the limpets and other gastropods, in which the radulae appear to be particularly well adapted for this method of feeding. The limpets are interesting because of their homing behaviour; after a feeding sortie they return to their 'home' (Fig. 134). Some echinoids also graze in the same way, using the teeth of the complicated Aristotle's lantern.

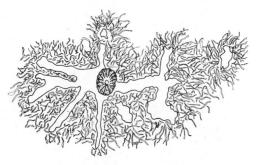

Figure 134. *Patella* in its 'home', showing the tracks of its feeding sorties into the surrounding algal growths

On the coast of Oregon, Castenholz (1961) investigated the grazing effect of *Littorina* and *Acmaea* and found that there was a rapid decrease in the diatoms as the number of gastropods per unit area increased. Even a few gastropods are sufficient 'to demonstrate the prime importance of grazers in maintaining this anomalous summer situation of high diatom reproduction rate and low standing crop' (l.c., p. 784). The coating of diatoms, bacteria and young epiphytes growing on the larger seaweeds was grazed in the same way, without the latter being eaten to any great extent, although the prosobranch *Lacuna* does penetrate deep into the thallus tissue of *Laminaria*. The main food of the echinoid *Strongylocentrotus droebachiensis* is also *Laminaria*.

The microfauna of the mesopsammon includes a number of diatom-eaters which can be recognized as such by the yellowish-green coloration of their gut contents. They graze the surfaces of the sand grains which are often densely covered with diatoms and bacteria; adsorbed fine detritus also provides a basic form of food.

In warm seas, herbivorous fishes play a considerable role on rocky coasts and coral reefs; Hiatt & Strasburg (1960) classify the alga-eating fishes of a reef in the Marshall Islands into four ecological groups (cf. Le Danois 1959) (Fig. 135):

1. Forms which live mainly on unicellular algae: Family Mugilidae

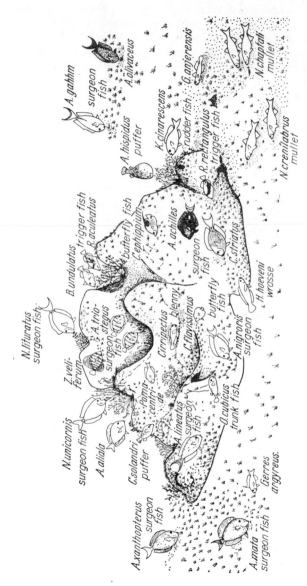

Figure 135. Herbivorous fish in characteristic feeding attitudes (after Hiatt & Strasburg 1960)

—*Neomyxus chaptali, Crenimugil crenilabrus;* Family Blenniidae—*Istiblennius paulus.*

2. Scraping and scratching forms which gnaw algae from the substrate: these include, Family Acanthuridae—*Acanthurus olivaceus, A. gahhm, A. guttatus, A. mata, Zebrasoma veliferum;* Family Pomacentridae—*Pomacentrus nigricans, P. albofasciatus, Abudefduf sordidus, A. glaucus, A. amabilis;* Family Blenniidae—*Exallias brevis, Istiblennius coronatus, Cirripectus sebae, C. variolosus;* Family Balistidae—*Balistapus undulatus, Rhinecanthus rectangulus, R. aculeatus;* Family Monacanthidae—*Amanses carolae;* Family Ostraciontidae—*Ostracion cubicus.*

3. Grazing species bite off pieces of alga above the substrate: these include, Family Chaetodontidae—*Chaetodon ephippium, C. auriga, C. reticulatus;* Family Acanthuridae—*Acanthurus triostegatus, A. chilles, A. guttatus, A. aliala, Naso unicornis, N. lituratus;* Family Pomacentridae—*Dascyllus aruanus, Pomacentrus vaiuli, P. jenkinsi, Abudefduf saxatilis, A. dicki, A. biocellatus, A. amabilis;* Family Balistidae—*Balistapus undulatus, Rhinecanthus aculeatus;* Family Tetraodontidae—*Acrothron hispidus.*

4. Forms which take algae more or less accidentally with their predominantly animal food: these include various members of the families Chaetodontidae, Pomacentridae, Labridae, Scaridae, Gobiidae, Tetraodontidae.

In the scraping and grazing forms the dentition is specially adapted with a beak-like protruding mouth, often with broad 'cutting teeth'. The Chaetodontidae, on the other hand, have a pointed snout with small protruding 'cutting teeth'. In structure these fishes resemble those which are more or less specialized coral-eaters; these include fishes which bite off individual coral polyps protruding beyond the skeleton (some Chaetodontidae) and those which consume parts of the coral colony including the skeleton (members of the Scaridae, Balistidae and Tetraodontidae).

Algal fragments are also frequently eaten by polychaetes of the family Nereidae; on account of the powerful pharyngeal jaws these worms have long been regarded as essentially predators, but in several cases they appear to be omnivores with a preference for plant material and detritus.

Among the vegetarians we can include those animals which consume timber, in particular the ships' worms (Teredinidae) and the gribbles (*Limnoria*). As in wood-eating insects, the Teredinidae have been shown to have symbionts, which enable the wood to be broken down; the gribbles appear to attack timber which has already been attacked by fungi, and it is the fungi which form the principal food of these crustaceans (Becker et al. 1957).

Substrate-feeders can thrive in sediments which have a high content of organic matter, either in the form of finely particulate detritus or as a microflora (bacteria, yeasts, fungi). Benthonic diatoms are also important to some substrate-feeders, at least in the euphotic zone.

These substrates contain very large quantities of indigestible ballast and so the animals have to consume huge amounts and therefore pass out correspondingly large amounts of faeces. McConnaughey & Fox (1950) investigated the polychaete *Thoracophelia mucronata* on the coasts of California, which lives in the sand of the eulittoral and feeds on sand; they did not observe any accumulation of detritus in the gut and assumed that the nutriment consisted of material adsorbed on the surfaces of the sand grains. This was probably diatoms and also the 'quartz fungi' found by Höhnk (1955) (Fig. 51). By feeding with coloured and uncoloured sand alternately it was shown that the gut contents changed four times during one hour; with a population density of 3,000 worms/o·1 m² they estimated that there must be an annual turnover of about 21% of the sand in the area occupied, which means that the sediment must be considerably influenced by its living inhabitants.

The excretory casts of the polychaetes *Arenicola* and *Heteromastus*, which in some places occur in large numbers, are well known because the shallow waters where they live are accessible to direct observation; in recent years *Arenicola* has been the subject of numerous ecological and physiological investigations (Fig. 136). Like these two polychaetes, the substrate-feeding members of the Flabelligeridae, Amphictenidae and Terebellidae also belong to the infauna of muddy-sand bottoms.

Substrate-feeding has evolved several times in the echinoderms, e.g. in holothurians (such as *Molpadia arenicola*), echinoids, ophiuroids and in the asteroid family Porcellanasteridae. The last-named are

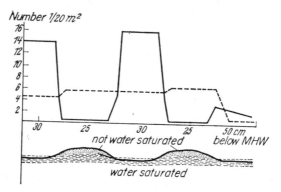

Figure 136. Density of *Arenicola marina* (solid line) and *Nereis diversicolor* (broken line) on the shore at Königshafen, Sylt (after Wohlenberg 1936)

typical inhabitants of the deep sea; Madsen (1961) considers that this method of feeding is a prerequisite for the settlement of the abyssal by some forms and for the development of high population densities; in the abyssal the amount of food available for predators, such as the majority of asteroids, is very small.

The coral-eating fishes mentioned above (members of the families Scaridae, Balistidae, Tetraodontidae) may also be regarded as substrate-feeders in so far as they bite off parts of complete coral colonies. These forms digest the tissues of the coral polyps and the numerous filamentous algae living within the corals (see p. 330). Here again the gut has to be refilled continuously on account of the large amount of ballast taken in.

Among the substrate-feeders one can differentiate between selective and non-selective forms, but there is still relatively little evidence from direct observation and one generally has to rely more on examination of the gut contents. As an example, the lamellibranch *Scrobicularia*, which lives in muddy substrates, pipettes off the surface of the sediment with its long inhalant siphon and is thus selective (Fig. 137).

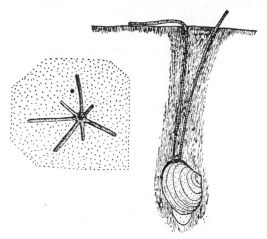

Figure 137. *Scrobicularia* in the substrate with extended siphons; on the left, the star-shaped pattern of tracks on the surface (drawn from observations and various specimens)

The fiddler-crabs of the genus *Uca* take up the surface sediment which they move towards the mouth with the pincers; the mouthparts, however, act selectively and reject the uneaten parts in the form of a small ball of mud. Altevogt (1957) has described this process and also the special structures on the mouthparts involved, the so-called 'spoon-hairs' (Fig. 138). Mortensen (1938) found that in several

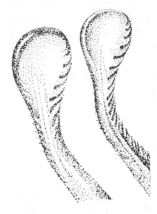

Figure 138. Spoon-shaped hairs from the edge of the 2nd maxillipeds of *Uca triangularis* (drawn from a photograph of Altevogt 1957)

offshore echinoids the gut was filled with plant material, and this was mainly from land plants which has been washed from the estuaries of rivers with mangrove and pandanus swamps; since the offshore echinoids are mostly substrate-feeders this probably indicates a facultative selection of plant material.

A large number of benthonic animals live on suspended particles, that is, on seston, which is brought to the vicinity of the bottom by water currents. Most of the animals concerned are sessile or semisessile. Several methods of catching the particles have been evolved among these suspension-feeders, and these are basically comparable with the methods used by pelagic animals.

Those that catch the particles with tentacles extend processes into the surrounding water where the particles become attached and the tentacles are then moved to the mouth. This is the characteristic method of feeding in coelenterates, in which the crown of extended tentacles is provided with nematocysts which paralyse and hold the prey. In most cases selection takes place for the nematocysts do not react to every tactile stimulus; passing detritus is usually ignored, but there are even specialized adaptations to the touch of some animals. The best known example is the relationship between the fish *Amphiprion* or *Dascyllus* and certain actinians; whereas the actinians are dangerous enemies of other fishes, *Amphiprion* is able to nestle down among the tentacles and thus enjoy protection from predators. In this case the fish must either have developed an immunity to the actinians' venom (Le Danois 1959) or the actinian must fail to react to the touch of the *Amphiprion*.

Strictly speaking, the crinoids should also be included among those animals which catch prey by means of tentacles; here prey that floats or swims past is caught by the extended arms. This method of feeding is also found in certain protozoans, e.g. Foraminifera and Suctoria.

The behaviour of the mollusc *Vermetus gigas* can also be regarded as a type of tentacular feeding. These animals are sessile, living in calcareous tubes attached to the substrate. They expel mucus threads, probably under pressure, from the foot glands; these spread out as they float in the water, without becoming detached, and after some time they are drawn in and eaten, together with the particles which have become attached to them (see pp. 266 et seq. for other instances in which mucous secretions is used for catching food). It is noteworthy that one asteroid can also feed in this way. *Patiria miniata*, an omnivorous form from the Pacific coast of North America can browse on diatoms by extruding the stomach on to the substrate on which they are growing; when suspended matter is abundant these starfish secrete mucus on the oral side and then flex the body so that the sediment collects in the mucus (MacGinitie 1949).

Tentacular feeders do not only find their food in the open water; several forms test the substrate for food particles with the help of very distensible and contractile tentacles. These therefore represent a special group among the tentacular feeders. The surface of the substrate is explored in this way by the terebellid polychaetes, while among the scaphopods the tentacles or 'captacula', search through the interstitial spaces in sand, into which the front part of the body protrudes.

The food particles become attached to mucus secretions and are taken to the mouth either by cilia (e.g. in terebellids) or by contraction of the tentacles (e.g. in scaphopods). The fact that scaphopods feed principally on foraminiferans suggests a considerable ability to select (Morton 1959).

Ciliary feeders which catch food particles from water they set in motion are very common, particularly among the polychaetes, bryozoans, brachiopods, lamellibranchs, gastropods, tunicates and others. The tentacles of the Spionidae and the filaments of the 'branchial crown' in sabellids and serpulids have grooves with ciliated epithelia permeated by mucus glands. Food particles are caught in these grooves and transported to the mouth where selection of the particles is carried out by a labial apparatus that is frequently complicated. These filamentous processes also have a respiratory function although this is subordinate to their feeding function; it should be noted that these organs have more than one function.

In microphagous asteroids, such as *Luidia* and *Pontaster* there are ciliary currents which lead to the ambulacral grooves and thence to the mouth. In lamellibranchs, particles carried into the mantle cavity with the water used in respiration collect on mucus and are carried to the mouth by ciliary currents. Infaunal forms, such as *Mya arenaria*, which live several decimetres down in the substrate, can obtain food from the substrate surface and from the water. Particles that are not eaten are wrapped in mucus and passed out as 'pseudofaeces'; such animals will therefore play an important part in the process of sedimentation as can be seen by examining a bed of mussels or by observing the clarification of turbid water by lamellibranchs.

In general, filterers also produce their own water currents but they filter out the particles present by some form of filtration apparatus. The development of this type of filter can be found in animals with bristly or hairy body appendages, in which the development of parallel elongated bristles produces an effective filter. Thus many crustaceans are filterers.

The barnacles beat their elongated thoracic limbs rhythmically and thus move their filtration apparatus through the water; some microphagous decapod crustaceans also behave in a similar fashion. Thus, the small crab *Porcellana platycheles* grasps the underside of rocks and beats the hairy third maxillipeds, a process which reminded its discoverer of the beats of barnacles (Grosse 1854, cited in Russell 1962). The same has been reported in other forms (MacGinitie 1949). The hermit-crab *Orthopagurus* lives in worm tubes and also filters microplankton with the third maxillipeds. The females of *Hapalocarcinus* perch on madreporarians, particularly in the fork between two recently formed branches; the stimulus thus produced causes the coral branches to grow parallel to each other instead of diverging and by growing thicker to envelop the crab which thus becomes enclosed in a chamber, in which it remains and is even fertilized by the free-living male. It obtains food 'by the movements of the nets of setae on the maxillipeds, resembling the sweeping action of the legs of cirripedes' (Potts 1915, cit. Russell 1962 p. 40).

The thalassinid (*Macrura reptantia*) which lives in tubes dug by itself extends its hairy first and second limbs in the form of a basket and filters off plankton and detritus from the stream of water driven into the tube; the filtrate is taken to the mouth by the third maxillipeds.

According to the observations of MacGinitie (1949), the crab *Lopholithodes foraminatus* stirs up the sediment and filters off the suspended detritus with the feathery maxillipeds. Mysids take detritus deposited on the surface of the substrate in the same way.

In a few cases nets of mucus are produced for catching food. On account of its jaws the common shallow-water polychaete *Nereis diversicolor* is regarded as predatory or as feeding on sizeable fragments. Harley (1950) found that these worms build a funnel out of mucus filaments with the help of the parapodial glands in the anterior segments; this funnel is constructed at the front end of the burrow and is closed at its hind end by the peristomium of the worm. The powerful undulations of the hind part of the body suck water into the tube, from which the mucus net filters off the suspended particles; from time to time the whole funnel is eaten and then a new one is made.

The mechanism is considerably more complicated in the sedentary polychaetes of the family Chaetopteridae, of which *Chaetopterus variopedatus* has been investigated by MacGinitie (1939). (Fig. 139). Mucus is secreted from two wing-like processes of a segment in the front of the body; the mucus is drawn backwards by cilia lining a groove, thus forming a long bag which is open at the front. The mucus bag is rolled up into a pellet in a cup-shaped organ at the end of the groove. Further back on the body there are wing-like processes from the segments which by fan-like movements suck water into the front opening of the tube; this water is filtered by the mucus bag. When the rolled-up pellet of mucus has reached a certain size, the production of mucus ceases and the ball is transported to the mouth by reversal of the ciliary beat and then eaten. Since the leathery tube is restricted at both ends it is scarcely possible for large fragments to enter; however, if this does happen the fragments are led along the body outside the mucus bag. MacGinitie (1945) found that the mucus net of *Chaetopterus* and also of the echiurid *Urechis caupo* retains proteins such as serum globulin and haemocyanin; from this it is reckoned that the mesh size is about 4nm. It is also possible that adsorption plays a role.

Comparable conditions are found in a number of prosobranch gastropods. Several different lines (not phylogenetic) leading to the sessile or semisessile forms, such as *Crepidula* and *Aporrhais*, have been studied by Yonge & Iles (1939) and by Werner (1953). The formation of a mucus net has reached its highest peak in the sessile *Crepidula* (Fig. 140), which produces an endless mucus band consisting of longitudinal and transverse fibres, which are wound up and eaten together with the attached particles. Werner (1959) has found the same behaviour in the ascidian *Clavelina lepadiformis*.

In these examples of catching food by filtration, the water movements necessary for the filtration process are produced by the animal itself, either exclusively for feeding or in conjunction with the circulation of water for respiration.

The exploitation of currents for filtration is relatively rare in the sea; the permanent currents are certainly too weak, while in surf areas the tides often cause changes in levels which sometimes produce fluctuations in the water movements and also the strength and direction of the current changes with every incoming and receding wave. Nevertheless there are two filterers from sandy areas with breaking waves which make use of the currents produced by waves. Both examples will be described in somewhat greater detail because they also

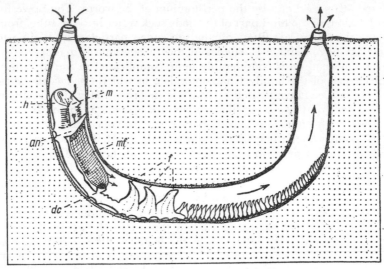

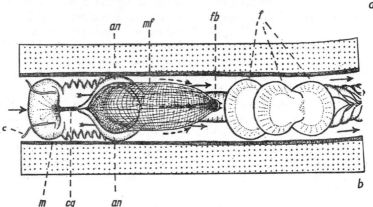

Figure 139. *Chaetopterus variopedatus:* (a) natural position in tube; (b) front part of body from above (after Werner 1959)
an = aliform notopodium, *c* = cirrus, *cg* = dorsal ciliated groove, *dc* = dorsal cupule, *f* = fans, *fb* = food ball, *h* = head, m = mouth, *mf* = mucus filter

show the relationship between the type of feeding and the settled area (Seilacher 1959, 1961).

On the coast of El Salvador the gastropod *Olivella columellaris* lives buried in the sand with the longitudinal axis of the body vertical to the shore and the front end directed towards the land. When the waves are running up the shore and while they are beginning to recede the animal remains completely hidden in the sand. During the latter

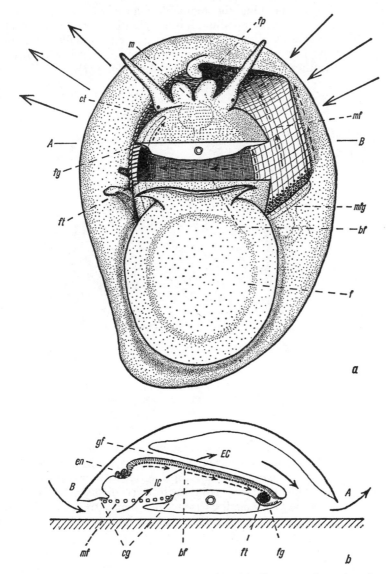

Figure 140. *Crepidula fornicata:* (a) underside, (b) diagrammatic section through front of body at A–B in a) (after Werner 1959)

bf = branchial mucus filter, cg = ciliated grooves, ct = ctenidium, EC = exhalant chamber, en = endostyle, f = foot, fg = food groove, fp = food pouch, ft = food thread, gf = gill filament, IC = inhalant chamber, m = mouth, mf = mantle mucus filter, mfg = mantle filter gland

part of the back-streaming the tentaculate propodia are extended. Between these and the metapodium a mucus net is expanded on each side, which inflates like a balloon as the water recedes and filters off fine particles. After a short time (less than a minute) the net is withdrawn and eaten together with the filtered particles (Fig. 141).

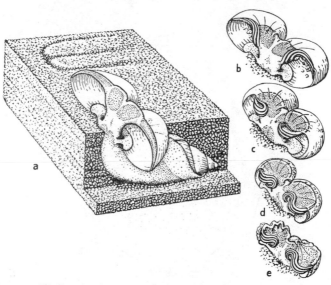

Figure 141. *Olivella columellaris:* (a) body buried in the sand and directed towards the shore; the propodium extended above the surface has expanded the mucus net which becomes distended by the tidal back-flow; (b-e) retraction of the net (after Seilacher 1959)

The animal compensates for the fluctuating water levels caused by tidal changes by coming up out of the sand and allowing itself to be transported by the waves to the appropriate level; it travels with an incoming wave when the tide is rising, with a receding wave when it is falling. The process of digging itself out, being transported and digging itself in again is so rapid that the animal is able to extend its nets in time for the next receding wave. The animal evidently chooses those horizons in which the current is no longer strong enough to transport large particles but where the fine detritus is still moving.

Analogous phenomena are found in the crustaceans of the family Hippidae, which sit buried in the sand along the coasts in all the warmer seas. Their first antennae are apposed to form a breathing tube through which fresh water from the surface is carried to the gills (Fig. 142). The long second antennae are feathery and are held out in

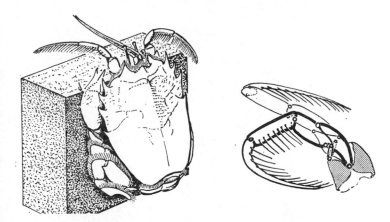

Figure 142. *Blepharipoda*: (a) in position in the sand, natural size. The respiratory tube (1st antennae) between the eye-stalks; alongside the eye-stalks are the 2nd antennae with bristles; (b) two positions of a single antenna (different strength lines) for filtration as water ebbs and flows (after Seilacher 1961)

the current so that floating particles can be filtered out of the moving water. The individual species show characteristic differences in the way they do this. According to the MacGinities (1949, *Hippa* (*Emerita*) *analoga*, for example, sits in the sand with the front end directed towards the sea and only unfolds the filtration apparatus during the passage of a receding wave. On the other hand, *Blepharipoda occidentalis* takes up a position parallel to the shore and by changing the position of the antennae uses both incoming and receding waves for filtering. This filtration apparatus shows corresponding morphological differences and the different species, even of the same genus, occupy different zones on the shore.

Magnus (1963) has made underwater observations on the behaviour of the crinoid *Heterometra savignyi* in shallow water in the Red Sea. These animals extend their arms across the current (Fig. 143) and by means of the tentacles on the pinnules each arm forms a fine-mesh net; the filtered food particles are embedded in mucus and conveyed to the mouth along the oral side in a food channel. These crinoids are active by night, while during the day they remain inactive with the arms coiled; according to the light conditions the feeding may start at twilight or not until complete darkness, so this is a good example of adaptation to the conditions of light in a given locality.

Table 30 is based on one produced by Werner (1959), but it includes these animals which exploit exogenous water currents by means of fixed nets,

TABLE 30

Types of suspension feeders

Type	Water moved by	Food caught by	Occurrence
I. Fixed net fishers	Swell Bottom currents	Mucus filter Setal filter Tentacle filter	*Olivella* Hippidae Crinoid *Heterometra*
II. Whirlers 1. Ciliary feeders	Cilia	Cilia	Ciliata, Kamptozoa, Phoronidea, Bryozoa, rotifers, polychaetes (Sabellidae, Serpulidae), thecosomate Pteropoda, Crinoidea, ciliated larvae of invertebrates
2. Setal strainers	Setose limbs	Setose limbs	Cirripedia, some Decapoda
III. Filterers 1. Sponge type	Flagellar beats	Filtration by choanocytes, etc.	Sponges
2. Mucus filterers (a) external mucus filter	Muscular action	Mucus filter	Polychaetes (*Chaetopterus, Nereis, Areniola*), Echiurids (*Urechis*)
(b) internal mucus filter	Ciliary beats; muscular action in salps	Mucus filter	Prosobranchs, Lamellibranchs, Tunicates, *Branchiostoma*
3. 'House' filterers	Muscular action	Grid filter in 'house'	Copelata
4. Setal filterers	Muscular action	Setal filter	Crustacea. Phyllopoda, Ostracoda, Copepoda, Mysidacea, Amphipoda, Euphausiacea, Porcellanidae.

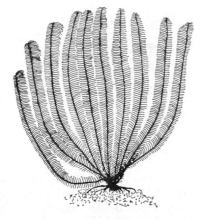

Figure 143. The crinoid *Heterometra savignyi* showing arms extended to form filtration fan (after Magnus 1963)

There are several different types of carnivore among both the invertebrates and the fishes, but a detailed treatment of these is beyond the scope of the present volume. Carrion-eaters are not uncommon among the crustaceans, gastropods, cephalopods and echinoderms. The behaviour of the gastropods in the genus *Nassa* is of some interest; these animals burrow into the substrate, usually in sandy bottoms, but protrude a tube-like elongation of the mantle, rather like a snorkel, into the water above the surface of the sediment (Fig. 144). Through the incoming respiratory water they smell prey at a considerable distance and will quickly congregate in surprisingly large numbers around any bait offered (cf. Henschel 1932). However, like many other carrion-feeders, they will also take living prey when they can get hold of it, *Crangon* and certain other crustaceans also lurk in the sediment and have a good sense of smell for carrion.

Among the typical predators are many turbellarians, nemertines, polychaetes, crustaceans, molluscs, echinoderms and fishes. Several types can be distinguished according to the method of feeding and the degree of specialization. Many asteroids have a peculiar method of feeding by everting the stomach and digesting the tissues of the prey extra-intestinally before sucking them up; they feed mainly on lamellibranchs but some species are more or less adapted for feeding on other prey animals, thus *Asterina* feeds mainly on ascidians, *Solaster papossus* on sea-urchins; *Patiria miniata* (see above) browses on diatoms and is largely omnivorous. The Luididae and Astropectinidae have no suckers on the tube-feet; they cannot therefore open lamellibranchs and digest them externally, but will swallow these, and also gastropods, whole. The hard shells of many marine animals such as lamellibranchs,

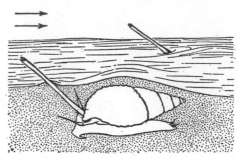

Figure 144. *Nassa reticulata.* Two animals buried in the sand; the siphons are extended above the surface facing the direction of the current (after Ankel 1938)

gastropods, asteroids, ophiuroids and echinoids do not in fact give complete protection. This armour can be breached in various ways: the animals may be swallowed whole and killed by the digestive juices (e.g. in flatfishes and in the echinoderms mentioned above) or they may be crushed (e.g. by the chelae in *Homarus* or by teeth in *Anarrhichas* (Fig. 145), or they may be opened by prolonged tension (asteroids).

The fish *Chilomycterus* will attack intact sea-urchins in spite of their spines. It is an astonishing sight to watch this fish bite into the spines and down to the calcareous test and then to consume this and its contents as well as the spines.

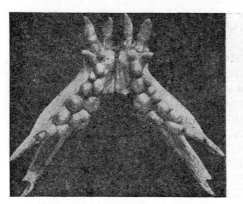
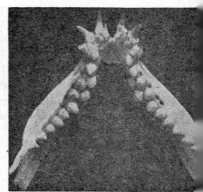

Figure 145. Lower jaws with gripping and crushing teeth of (a) *Anarrhichas lupus*, (b) *A. minor* (after Lühmann 1954)

The carnivorous prosobranchs in the family Naticidae bore holes into the shells of lamellibranchs and consume the contents (Fig. 146). The mechanism of boring has been described several times but there is some controversy about whether this is done by acid secretions, by the

Figure 146. Left shell of *Tellina baltica*, bored by a naticid (after Ankel 1938)

mechanical action of the radula, or whether both play a part; Ankel mentioned Hirsch's suggestion that a specific enzyme (a 'calcase') was involved. Turner (1953) covered shells of *Mya arenaria*, some with a layer of paraffin about 0·5 mm thick and some with a powder insoluble in dilute acids and organic solvents; in both cases starving *Polynices duplicata* bored their holes without hindrance, so one must conclude that mechanical boring takes place. It is possible that the so-called boring gland functions as a sucker, which presses the front end firmly on to the substrate and thus facilitates the action of the radula.

In *Lunatia*, Ziegelmeier (1954) was also only able to observe radular action during boring, with alternating periods of rasping and rest; the animals took about four hours to bore through a shell 0·1 mm thick. Since the boring gland has no external openings, Ziegelmeier assumed that the secretion was passed into the lumen of the proboscis sheath, where it may function as a 'lubricant' to reduce the friction of the rear pharynx wall on the inner wall of the proboscis sheath.

Among the prosobranchs, the Pyramidellidae have completed the transition to parasitism. They pierce lamellibranchs and gastropods from the mantle edge (Fig. 147), and also asteroids and polychaetes, and suck the body fluids, without however killing the host. It is indeed astonishing that a serpulid, for example, which is normally so sensitive to touch, does not withdraw its crown of tentacles into the protection of its calcareous tube when one of these gastropods pierces a tentacle. Evidently the gastropod has found a gap in the host's reflex physiology, which does not allow for any defence reaction on the part of the polychaete (cf. also Ankel 1938 on the relationship between pyramidellids and lamellibranchs).

Suction-feeders which pierce and suck their prey have been evolved among the syllid polychaetes, and *Autolytus* and other forms are specialized to feed on hydroids. The polychaete pierces the base of a polyp

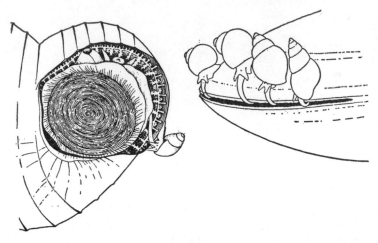

Figure 147. Left, the pyramidellid *Brachystomia ambigua* sucking at the mantle edge of *Turritella communis* (after Ankel 1959): right, *Brachystomia rissoides* on the hind end of *Mytilus* (after Ankel 1936)

with its eversible toothed pharynx and then, by the pumping movements of the very muscular gut, sucks out the body fluid (Okada 1928, Hauenschild 1945).

Some nudibranchs are also said to suck their prey, particularly in those species in which the radula is much reduced. According to Barnes & Powell (1954), *Onchidoris fusca* mechanically pushes in the operculum of *Balanus* and then sucks out the soft parts.

Other general ecological phenomena are closely associated with the development of different types of feeding. Predatory forms, being the third or higher link in the food chain, will naturally only attain a relatively low population density in comparison with the suspension- and substrate-feeders, and the order of size of the predators also plays a part. Clark (1962) gives an instructive example from the polychaete genus *Nephthys* which generally lives a predatory life. The population density of the various species is usually less than $100/m^2$, densities of $500/m^2$ being less frequent. In San Francisco Bay the small *N. cornuta* reached densities of up to $780/m^2$. Even more astonishing was the discovery by Sanders that in Long Island Sound *N. incisa* lived in densities of over $1000/m^2$, particularly as the normal density of this species is less than $100/m^2$. It turned out that these dense populations were feeding on detritus, and were therefore behaving ecologically in a different way from the other known populations. It is undecided whether the high population was made possible by a change in feeding habits or conversely whether the aberrant method of feeding was forced upon the animals by a local fluctuation in the population density (See pp. 313 et seq.).

This example is significant in so far as it shows the possibility of modifying such an important function as feeding within a single species. Genetic fixation of such aberrant habits might in certain circumstances lead to the isolation of populations and thus play a part in the process of speciation.

Some of the ascidians provide a further example of a change in feeding method; in general, the tunicates are filtering suspension-feeders with small mouths, a delicate gill-filtration apparatus and a ciliated pharynx. Some forms, however, although still sessile, have

Figure 148. *Above,* left foreground, *Heliometra glacialis*; behind it, an actinian; right, the faeces of a shark (depth 1650 m). *Below,* left, a leptomedusa; right, a swimming pycnogonid (depth 700 m) (photographed from the bathyscaphe, after Pérès 1959)

K—M.B.

come to feed on larger fragments, the remains of copepods, amphipods and others having been found. In these, the mouth has become greatly enlarged and the pharyngeal ciliation is lacking (the genera *Octacnemus* and *Hexcrobylus* according to Millar 1959). These modified ascidians are typically deep-sea forms, but they live alongside unmodified forms which are suspension-feeders, so that one cannot postulate any valid correlation between the method of feeding and a preference for deep water.

The asteroid family Porcellanasteridae has already been mentioned (p. 262) in connection with the importance of changed feeding habits in the colonization of the deep sea. In general, it seems that the relative rarity of predators in the abyssal is correlated with the lower population density of the primary links in the food chain. The same ought to apply to some parasites, which require a certain minimum density of their hosts in order that fresh infection may take place.

In those animals which catch their food with tentacles, a high population density requires sufficient seston, which must be kept moving by currents. This applies, for example, to the phytal and particularly to reefs in which dense populations of coral polyps and other tentacle-feeders seem to indicate large amounts of seston, particularly plankton (see p. 243). The epifauna on bathyal and abyssal bottoms evidently consists largely of tentacle-feeders, such as octocorals and crinoids. While diving in the bathyscaphe off Japan, Pérès (1959) observed that

Figure 149. *Ophichthus urolophus* (right), *Coelorhynchus kishinouyei* (left) (depth of 1,000 m, as Fig. 148)

at a depth of 1,650 metres there was one crinoid (*Heliometra glacialis*) per 2 m², which is a very high density. In these animals the calyx was facing upwards, and so Pérès concluded that the number of downward sinking particles was greater than that of those moved horizontally by the bottom currents; in the Mediterranean, on the other hand, *Leptometra phalangium* was positioned vertically to face the bottom current. Although individual observations of this kind do not permit any final conclusions they do at least point to the value of observations made from the bathyscaphe and of pictures taken by underwater television and photography (Figs. 148, 149). It is possible that the height to which such stalked filterers protrude above the substrate may

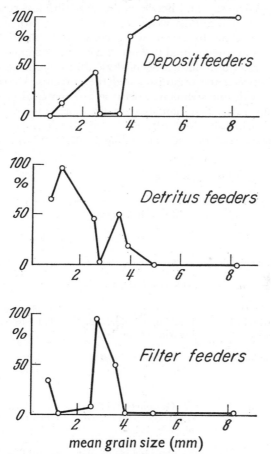

Figure 150. Percentage distribution of different feeding types in marine substrates of different particle size (after McNaulty *et al.* 1962 b)

have given some indication of the thickness of the layer of turbid bottom water, but we still lack information on this.

Attention has already been drawn (p. 109) to the biology of fiddler-crabs (Altevogt 1957). With the proviso that there may be other reasons, it can be concluded that a zonation of closely related substrate-feeders, feeding in the same way, must be due to differences in the chemistry of the substrate.

Biological observations may therefore lead to the detection of physical and chemical differences and thus stimulate a more detailed investigation. There is a further example of the relationship between the method of feeding and the substrate: filterers are dependent upon an optimal particle size and hence on currents with a certain velocity, because currents that are too weak do not hold the particles in suspension and those that are too strong will carry material that is too coarse; on the other hand, the optimal particle size is not the same everywhere, so that one cannot automatically conclude that there is an established correlation between the organisms and the currents. Detritus-feeders are commonest in fine sediments, because fine material is washed out of coarser substrates. Sediment-feeders are more numerous in coarse sediments than in fine (McNulty et al. 1962 b (Fig. 150)).

This discussion on methods of feeding leads to the question of the

TABLE 31

Varying function and form in crustacean chelae

(after Schäfer 1954)

Available food	The grasping function	The form of the chelae
Large swimming animals	(a) Grasping—holding —cutting—pulling with capture chelae	Long-limbed, cutting cone-toothed chelae
Shelled or armoured large animals with little little or no mobility	(b) Grasping—holding breaking—pulling with breaking chelae	Tubular shears Plaster-tooth chelae
Small moving animals	(c) Grasping—holding with nipping chelae	Straight long-limbed (often toothless) beak-like chelae
Sessile or semisessile epifauna and epiflora	(d) Grasping—holding —plucking with curved nipping chelae	Curved, short-limbed (often toothless) beak-like chelae
Organogenous sediments, algae	(e) Grasping—holding —scraping with scraping chelae	Shovel-like chelae (often hairy) with papilla or toothless.

structural basis and evolution of this function and also to physiological investigations which provide, for example, a closer understanding of the coefficient of exploitation and thus of the cycle of matter in the sea (Kinne 1962, 1963). These problems are, however, beyond the scope of the present study.

A number of autecological and synecological problems, as well as those of a general biological nature are involved. Thus, Schäfer (1954) has analysed the form and function of the chelae in the Brachyura and has shown that in many species this organ may have several functions related to the habitat and the method of feeding and may therefore vary considerably in form. Its basic function is to grip, and Table 31 shows how it becomes modified in form and function, and how various modifications of the basic pattern lead to great biological diversity.

The behaviour of prey animals towards their enemies is a very wide subject which has not been much investigated as regards the invertebrates; Bullock (1953) has given an interesting account of the reactions of gastropods in the presence of starfishes.

7. Animal communities

The aim of investigations on the benthonic fauna is not only to identify the individual species and find out about their geographical distribution and functional structure in relation to the environment in which they live; using oyster beds as an example, Mobius (1871, 1877) demonstrated the existence of a natural association of species with mutual relationships. In 1911–1913 Petersen published the results of his quantitative investigations on the bottom fauna in Danish waters, the aim being to estimate the available food for fishes. In doing so it struck him that large areas of the same type of bottom were characterized by the same dominant species and that some of the accompanying species were also typical of the area concerned. He used the term communities for this correlated occurrence of several species; nowadays the term associations has also been introduced. Since Petersen's publications many other investigations based on similar methods have been carried in different parts of the sea. These have been concerned almost exclusively with the macrofauna; since, however, this only comprises a part of the total marine fauna these results only provide a partial picture of the total relationships. The estimation of the microfauna presents much greater technical difficulties and so far there has scarcely been any quantitative assessment.

Associations are biological units which must be characterized and distinguished from each other; the basic principles have been discussed in detail by Remane (1940) and others; a more recent summary is given by Jones (1950). Definitions of these communities are based on the statistical evaluation of quasi-quantitative samples. Any discrepancies that may arise will depend largely upon the aim and methods of the investigation: if, like Petersen, the purpose is to define the biological characteristics of extensive areas then one must choose large areas that are not very homogeneous; if, on the other hand, one wants to characterize strictly defined areas, as Gislen (1929) and Molander (1928, 1962) attempted to do in the Scandinavian fjords, then emphasis must be laid on the greatest possible homogeneity. Further uncertainties arise because the samples necessary for statistical calculations, which must be distributed over a certain period of time, are not always available. Inaccuracies are also to be expected when the work covers only a section of the population and not the whole, and so certain conclusions on the existence of associations are somewhat uncertain.

The traditional method of representing an association (Fig. 151) gives the impression that the species and individuals found are distributed homogeneously. But this is completely erroneous because such a diagram is only a graphic representation of a statistical result which in no way corresponds to the natural arangement, in which there is considerable heterogeneity.

Uncertainties arise particularly in localities where the biotope is not clearly defined. The following are the most important characteristics of benthonic biotopes, although these have not always been sufficiently well investigated: the depth, the currents and water movements and the structure of the sediments (which is related to the water movements) which varies from coarse gravel to soft mud with every kind of intermediate, the aeration of the sediment and its content of organic matter.

It is noteworthy that bottom sediments of the same type in areas far apart are characterized by representatives of the same genera. In the North Sea, the soft substrates close to the shore have the lamellibranch *Macoma baltica*; in Greenland this is replaced by *M. calcarea* (Thorson 1933, 1934) and in the north-east Pacific by *M. nasuta* (Shelford 1935). In the same way sandy bottoms in the Kattegat have a community characterized by the lamellibranch *Venus gallina*; similar substrates in East Greenland have *V. fluctuosa*, in the Adriatic *V. verrucosa*. Since the other members of these faunas are similar Thorson (1952) speaks of 'parallel animal communities', which therefore have alternative species of the same genera occurring as dominants. In the same way alternative species are also known in the upper littoral of many areas, e.g. in the balanids and *Littorina*. Parallel communities can also be recognized in the subterranean waters of the littoral, particularly in the interstitial system of sandy beaches (cf. Deboutteville 1960).

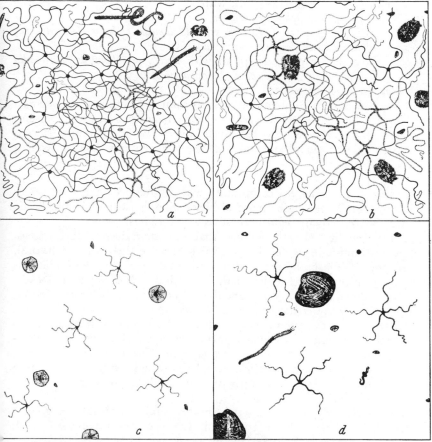

Figure 151. Diagrams of the composition of four associations in Biscayne Vay (after McNulty *et al.* 1962 a): (a) ¼ m² of *Amphioplus-Ophionephthys* community; (b) the same in a different area; (c) 2 m² of *Mellita-Tellina* community; (d) ¼ m² of *Venus* community

The composition of communities varies according to local conditions and it is often possible to distinguish different zones within a single area; as an example we may take the *Macoma calcarea* community in Franz Joseph Fjord and Scoresby Sound described by Thorson (1933, 1934). This community occurs in depths of 4–45 metres in the Arctic and the following zones can be recognized:

(a) *Astarte borealis* zone: shallow water 4–15 m with:
   Macoma calcarea, Astarte borealis, Priapulus caudatus, Nereis zonata, Harmothoe imbricata, in places some *Pectinaria granulata*. Ophiuroids completely absent.

(b) *Ophiocten* zone: 14–45 m, with:

M. *calcarea, Ophiocten sericeum*

b 1, upper part, 14–25 m, with *Philomedes globosus, Diastylis* sp., sometimes *Pectinaria granulata*

b 2, lower part, 25–45 m with *Pecten groenlandicus, Portlandia lenticulata, Neaera obesa, Macoma moesta.*

(c) *Halcampoides* zone:

M. *calcarea* frequent but less so than the actinian *Halcampoides purpurea. Ophiocten sericeum* and *Axinus flexuosus* also present. This zone is only developed in very favourable places, since a dense population of *Halcampoides* requires special feeding conditions.

The *Macoma calcarea* community with its main component species also occurs in the inner parts of Norwegian fjords and in the Baltic Sea. Spärck (1936) regards these as arctic relicts.

Petersen had already noticed that representatives of the *Macoma baltica* community which in the North Sea are tied to the eulittoral, occupy a wide bathymetric distribution in the Baltic Sea. This has been confirmed several times; thus one finds M. *baltica* and *Scoloplos armiger* down to over 100 m, *Hydrobia* to about 75 m, *Corophium volutator, Pygospio elegans, Cardium edule* and *Mya arenaria* to about 50 m. Petersen (1918) sought to explain the presence of M. *baltica* in these depths as due to the absence of its enemy *Asterias rubens*, which prevents the penetration of M. *baltica* into deeper water in the North Sea. Spärck (1936, p. 19) considered that in spite of its relatively high oxygen consumption the presence of M. *baltica* at great depths in the Baltic was due to lack of competitors.

Remane (1955) has analysed another similar example in greater detail. According to Ford (1923), sublittoral coarse sand on the English North Sea coast is characterized by a *Spatangus purpureus–Venus fasciatus* community. In the same substrate in the western Baltic, however, Remane (1933) only found parts of the *Macoma baltica* community together with other forms not represented in the *Spatangus–Venus* community and a rich microfauna and so he established the *Halammohydra* community (named after the medusa *H. octopodides*). On further investigation these elements were then found in coarse sand areas of the eulittoral in the North Sea. Here therefore there had also been a shift in depth distribution from the eulittoral into the sublittoral, a submergence. At the same time there had been a change in the composition of the communities.

The central and northern Baltic Sea provide further examples of this kind of change; according to the investigations in Finnish waters of Segerstråle (1933), the numbers of the amphipod *Pontoporeia affinis* and the isopod *Mesidothea entomon* increase with depth, whereas the chironomids are more numerous in shallow bays.

The comparison made by McNulty *et al.* (1962a) between temperate seas and the warm waters of Florida also shows evidence of changes in composition. They found:

	marine sand	marine mud	brackish mud
Temperate seas	*Tellina*	*Macoma*	*Scrobicularia*
Florida	*Codakia*	*Tellina*	*Macoma*
	Laevicardium		

Similar substrates in different areas may therefore have quite different associations. This appears to contradict the statement made above about the existence of parallel animals communities. This contradiction will be explained if we consider that changes in associations will be found in areas in which certain abiotic environmental factors have changed radically, sometimes reaching extreme value, as, for example, the reduced salinity in the Baltic Sea. The existence of associations over extensive areas having constant environmental conditions and the fact that they change when the environment changes suggests that the associations are dependent more on the similar environmental requirements of their species than on the biological interplay of these species. The association will be maintained so long as all the individual factors within the environmental complex suit the requirements of the species. If, however, one of the factors reaches a minimum even for a single species, then this species will disappear from the association, which will become impoverished or will be changed in composition owing to the entry of other species with more appropriate ecological requirements. If, in characterizing an association, one chooses species with wide ecological potentialities, then this association will be found over large areas and it will have few divergent sub-communities due to variation in external factors. Conversely, communities characterized by species with restricted ecological potentiality can only exist in restricted areas (cf. also Stickney & Stringer 1957).

Since the associations are primarily determined by the combinations of abiotic factors, Jones (1953) has given a classification for the Atlantic-boreal region, which is based on these factors and not on the names of the animals. This classification is as shown in Table 32.

This method of classification appears to be better than using the faunistic composition, particularly when making comparisons between large areas, as for instance between the Atlantic and Pacific boreal or between the boreal and antiboreal (cf. also McNulty *et al.* 1962 a).

Pérès (1957) has attempted a classification of benthonic communities

TABLE 32

Bottom associations according to Jones 1950

Terms used by Jones 1950	Terms used by other authors	Principal animals in the association	Occurrence
Boreal shallow sand association	*Tellina tenuis* community (Spärck 1935)	Lamellibranchs: *Tellina tenuis*, *Donax vittatus* Polychaetes: *Arenicola marina*, *Nephthys caeca* Amphipods: *Bathyporeia pelagica*	on relatively exposed coasts in N.W. Europe
Boreal shallow mud association	*Macoma-baltica* community (Petersen 1913 and others)	Lamellibranchs: *Macoma baltica*, *Mya arenaria*, *Cardium edule* Polychaetes: *Arenicola marina* Amphipod: *Corophium volutator*	on less exposed coasts in N.W. Europe, also in estuaries and Baltic Sea
Boreal shallow rocky association	*Mytilus* epifauna (Petersen 1913)	Lamellibranch: *Mytilus edulis* Gastropods: *Littorina* spp., *Patella vulgata* Cirripedia: *Balanus balanoides*	on rocky coasts and other suitable substrates in N.W. Europe
Boreal shallow vegetation	Phytal (Remane 1933)	Gastropods: *Littorina littoralis*, *Rissoa* spp., *Lacuna vincta* Crustaceans: *Hyale prevosti*, *Idothea* spp., *Hippolyte varians*	on firm substrates in the littoral
Boreal offshore sand association	*Venus community* (Petersen 1913) *Echinocardium* cordatum-*Venus gallina* community partim (Ford 1923) *Acrocnida-prismatica*	Lamellibranchs: *Dosinia lupinus*, *Venus striatula*, *Tellina fabula* *Abra prismatica*, *Ensis ensis* Polychaetes: *Sthenelais limicola*, *Nephthys* spp. Amphipods:	Outside the coastal region wherever sand is present

Terms used by Jones 1950	Terms used by other authors	Principal animals in the association	Occurrence
	community (Molander 1928) ?*Tellina-fabula* community (Spärck 1935)	*Ampelisca brevicornis, Bathyporeia guilliamsoniana,* Sea-urchin: *Echinocardium cordatum*	
Boreal offshore muddy sand association	*Echinocardium-filiformis* community (Petersen 1913, Molander 1928) *Echinocardium-flavescens* community (Einarsson 1941) and others	Lamellibranchs: *Nucula turgida, Abra alba, Cyprina islandica, Cultellus pellucidus, Spisula subtruncata* Gastropods: *Turritella communis, Philine aperta* Ophiuroids: *Ophiura texturata, Amphiura filiformis* Sea-urchins *Echinocardium cordatum E. flavescens* numerous polychaetes	widely distributed
Boreal offshore mud association	*Brissopsis-chiajei* community (Petersen, Molander)	Lamellibranchs: *Nucula hitida Abra nitida* Ophiuroids: *Amphuria chiajei* Sea-urchins: *Brissopsis lyrifera* several polychaetes	widely distributed
Boreal offshore gravel association	*Spatangus-purpureus-Venus fasciata* community (Ford 1923), *Spatangus-montagui* community (Molander 1928), *Venus-fasciata* community (Spärck 1935), *Modiolus* epifauna	Lamellibranchs: *Nucula henleyi, Lima loscombi, Venus casina, V. fasciata, Paphia rhomboides* Polychaetes: *Glycera lapidum, Serpula vermicularis,* Archiannelida: *Polygordius lacteus* Crustaceans: *Balanus porcatus, Galathea* spp.,	distributed

Terms used by Jones 1950	Terms used by other authors	Principal animals in the association	Occurrence
		Eupagurus spp., Echinoderms: *Asterias rubens, Ophiothrix fragilis, Echinocyamus pusillus, Echinocardium flavescens, Spatangus purpureus*	
Boreal deep mud association	*Brissopsis-Sarsii* and *Ampelisca-Pecten* community (Petersen 1913)	Lamellibranchs: *Nucula tenuis, Chlamys vitrea, Thysaira flexuosa, Portlandia lucida*	widely distributed
Boreal deep mud association	*Mellina-tenuis Terebellides-flexuosa* community (Molander 1928) Foraminfera *Pecten-vitreus* community (Spärck 1935)	Ophiuroids: *Ophiura sarsi, Amphilepis norvegica* Sea-urchin: *Brissopsis lyrifera* numerous polychaetes	widely distributed
Boreal deep coral association		*Lophohelia prolifera, Paragorgia arborea, Gorgonocephalus caput-medusae*	Epifauna of deep water hard bottoms in a fringe off the European Atlantic coasts

on a world-wide basis and this was modified and supplemented in 1961. He starts from the zones in the benthal and within each zone differentiates according to the substrate, e.g.:

Mesolittoral (mediolittoral) zone

      I. Hard stationary substrates
         A. seas with small tides
           (a) upper mesolittoral
           (b) lower mesolittoral
         B. seas with large tides
     II. Movable substrates
         A. sands and slightly muddy sands
         B. muds

III. Minor mesolittoral communities on
various substrates
   A. large boulders
   B. shingle
   C. gravel

In addition to depth and substrate, Pérès makes use of other abiotic
factors (e.g. temperature, light, currents) to varying degrees and
characterizes the associations by naming the principal animals present.
In many places he had to draw attention to our lack of knowledge
when comparing different areas of sea, but he was able to provide further examples of the parallel animal communities postulated by Thorson (l.c.). Pérès pointed out that we still have only fragmentary
and localized information on the mesobathyal, infrabathyal and hadal,
but it seems to me that even for the littoral zones we still lack the basis
for a generally valid scheme of animal communities. The data assem-
bled by Pérès will be very useful for any future attempt to work out
and define the interrelationships within communities.
   Even though the associations are primarily determined by the
abiotic factors the effect of the biotic factors must not be disregarded
because, for instance, every predatory species in an association will
influence and restrict the species on which it preys. So far little is
known definitely about these relationships; Thorson (1953) showed
that the cessation of feeding by ophiuroids during the breeding period
enables other species to enter the community, from which they were
previously excluded by the voracity of the ophiuroids, so that the
composition of the community may be temporarily changed. Further
investigations will doubtless reveal similar phenomena caused by biotic
factors.
   Little is known about the mutual interactions of the species compos-
ing marine associations and in most cases we can only hypothesize.
We will not therefore speak of biocoenoses. This term goes back to
Möbius and his investigations on oyster beds; it covers those animal
communities in which the biotic relationships of the members of the
association are particularly close. Detailed analyses of biocoenoses
and their dynamic structure mostly relate to the terrestrial and freshwater regions, which are directly accessible to this kind of treatment.
Knowledge of these in the marine field is relatively sparse, as is shown,
for example, by the fact that such common species as *Nereis diversicolor*
and *N. pelagica* have been regarded as predatory species because they
possess powerful jaws, whereas they are filterers and feeders on vegetation and detritus. The position of their dense populations in the biology
of the eulittoral community is therefore quite different from what was
originally thought; so that even in this very accessible habitat the dyna-

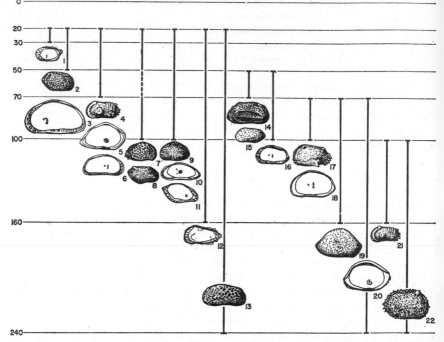

Figure 152. Bathymetric distribution of characteristic ostracod species in the eastern part of the Gulf of Mexico (after Benson & Coleman 1963)

(1) *Hemicytherura sablensis* Benson & Coleman, (2) *Loxoconcha sarasotana* Benson & Coleman, (3) *Haplocytheridea gigantea* Benson & Coleman, (4) *Puriana rugipunctata* (Ulrich & Bassler), (5) *Bairdia gerda* Benson & Coleman, (6) *Haplocytheridea proboscidiala* (Edwards), (7) *Aurila conradi* (Howe & McGuirt) *floridana* Benson & Coleman, (8) *Loxocorniculum postdorsolatum* (Puri), (9) *Aurila amygdala* (Stephenson), (10) *Paracypris? sablensis* Benson & Coleman, (11) *Pellucistoma magniventra* Edwards, (12) *Paracytheridea tschoppi* Van Den Bold, (13) *Hulingsina ashermani* (Ulrich & Bassler), (14) *Protocytheretta daniana* (Brady), (15) *Cytheretta? sahni* Puri, (16) *Campylocythere laevissima* (Edwards), (17) *Pterygocythereis* sp. aff. *P. americana* (Ulrich & Bassler), (18) *Perissocytheridea laevis* Benson & Coleman, (19) *Bairdoppilata triangulata* Edwards, (20) *Bairdia victrix* Brady, (21) *Puriana rugipunctata* (Ulrich & Bassler), (22) *Echinocythereis garretti* (Howe & McGuirt)

mic relationships between species are still scarcely known. The same applies to many other animal species and so the more general term 'association' is to be preferred. Furthermore, as already emphasized above (pp. 33, 281), we are still largely ignorant about the quantitative and qualitative share of the microflora and microfauna in the composition and dynamic processes of marine communities. Benson & Coleman (1963), working on ostracods from the Gulf of Mexico, have shown how promising the microfauna may be in characterizing zones. But there are also certain anomalies; at one station in a depth of about 10 metres these authors found offshore species living alongside those which were elsewhere only known from shallower depths, and no explanation was found for this (Fig. 152).

Further investigations will be needed to determine the extent to which the members of an association are actually interrelated, producing a self-regulating unit or biocoenosis. There are a few data (e.g. Vatova 1949) on the quantitative relationships of lamellibranchs and echinoderms, which are mainly living in the relationship of prey and predator. These results should be extended and enlarged by the inclusion, for instance, of the polychaetes and crustaceans. It should be remembered, of course, that in most associations the origin of the energy flowing through it (that is the plant food) lies outside the community; this primary source of energy will be affected by water movements and therefore be subject to more or less extensive periodic and non-periodic fluctuations. It is also possible that an inflow of pelagic larvae from distant areas may supplement a population or allow invasion by other species. In this context there are undoubtedly regional differences depending upon the development of pelagic and non-pelagic developmental stages, as shown by Thorson (see pp. 245 et seq.). In the case of many marine associations, this and any non-periodic changes in food supply will differ considerably from what happens on land and in fresh waters.

Investigations on coral reefs have revealed a multiplicity of benthonic associations. A coral reef has a rich mosaic of in-, meso- and epifauna, not only with vertical zonation, but also with considerable differences between fauna on the exposed face and that on the sheltered side (Fig. 153).

Deeply shaded caves undoubtedly deserve particular attention because there the darkness and the food supply, consisting solely of inflowing detritus and plankton, present a special set of factors. In such places the principal consumers are sponges, hydroids, sedentary polychaetes, bryozoans, certain molluscs and ascidians; there are also many visitors.

In the Tyrrhenian Sea, Riedl and his colleagues have carried out extensive ecological studies on cave populations (Riedl 1956). In addition to the two factors mentioned above these caves, which are quite close to the surface (1–5 m), are subjected to considerable water movements. A number of the forms found were only known from depths of 500–2,000 metres. In the same way Gurjanova et al. (1929, 1930), working in the Kola Fjord, had already noticed the occurrence of a number of sublittoral algae and animals in dark places in the littoral (caves, crevices). This cryptofauna therefore showed ecological relationships with the bathyal, which are presumably determined by the

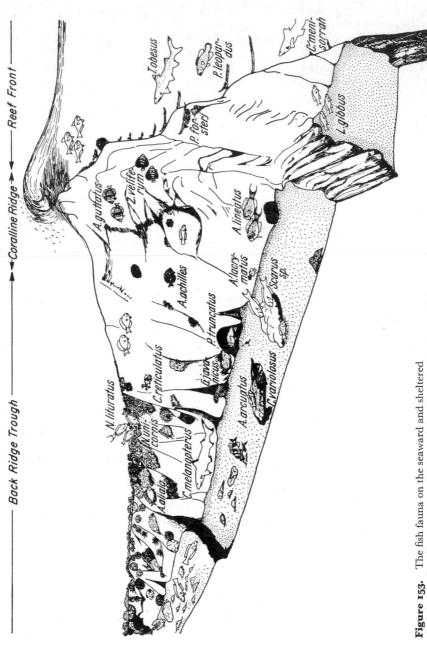

Figure 153. The fish fauna on the seaward and sheltered side of a coral reef in the Marshall Islands (after Hiatt & Strasburg 1960)

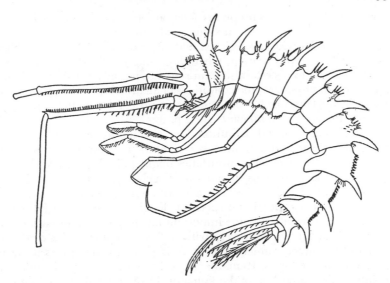

Figure 154. Deep-sea amphipod *Lepechinella wolffi* (pereiopods 3, 5 and 6 and the pleopods missing) (after Dahl 1959)

common factor of poor light. It remains to be seen whether such cave forms have risen from deeper water or whether this is in fact a fauna characteristic of shade and darkness. Some cave-dwellers develop long processes or appendages and are thus similar to certain deep-sea forms (Fig. 154); as in the deep sea these striking cave forms live alongside others which do not seem so specialized.

An indication that this cryptofauna may possibly be of more general interest is provided by the discovery of a new species of calcareous sponge belonging to the Pharetronidae (Vacelet & Levi 1958). These sponges were known as fossils from the Devonian to the Cretaceous and seven recent species have now been found in the Indo-Pacific region. While diving at depths of 10 metres off Marseilles these authors found a further species in rock crevices. This discovery is of biogeographic interest because it provides a closer link between the Mediterranean and the Indo-Pacific region, and it is also a further example of the existence of a geologically old form in a biotope with extreme conditions.

Further investigations are needed to show whether associations could be better characterized by the relative proportions of their life forms. The degree to which life forms are dependent upon the substrate would indirectly provide a clue to the nature of the bottom; for example, the percentage of suspension-feeders would indicate the impoi tance of an allochthonous food supply, and by comparison of widely separated areas the taxonomic differences due to the regional limitation of individual species would cease to exist. On this problem Marshall (1953) has written, when discussing the littoral *Balanus* zone '. . . the life forms as established in response to and contributing to the climatic and physiographic conditions of an area are of primary importance in interpreting biomes and, by contrast, taxonomic categories are not themselves significant' (l.c., p. 435).

The circumoceanic distribution of certain life forms, such as the littoral Balanidae and Arenicolidae or the brachyuran Ocypodidae, and the proven parallel communities might provide a clearer picture of synecological divisions in the sea. Biogeographical problems could be better explained if we had more information on taxonomic units and an analysis of the individual species belonging to such associations in different areas would provide important autecological information.

In the meantime, however, with a few exceptions, the available material is scarcely sufficient for carrying out a general classification according to life-forms. Remane (1940) has indicated certain relevant points in his comparison of the fauna of the phytal with that of secondary hard-bottom associations. Gerlach (1960) endorsed the opinion of Stephenson 1958, when he wrote 'that the coral reef can be regarded as part of the phytal. There are great similarities between the fauna of coral reefs and that of algal areas, whereas the forms adapted to live commensally with corals are in a minority, and it would not be justifiable to give the coral reef equal rank as a major habitat with the benthal and phytal' (l.c., p. 362).

8. The deep sea

A brief account may now be given of the deep sea, an area which is of great general as well as scientific interest. The upper limit of the deep sea cannot be easily and clearly defined; this is partly because there are regional and local differences in the occurrence and vertical distribution of the ecological factors, and partly because we do not know with any degree of certainty the biological standards which are essential for any clear definition. This account is concerned with the region that was defined as the bathyal on pp. 220 et seq.

In considering the deep sea it should be mentioned that the number of hauls, on which our knowledge depends, decreases sharply with depth and that these are very widely scattered geographically. Consequently, negative hauls, in particular, provide no evidence of the actual presence or absence of species. Furthermore, information on the relative abundance of individual species only has limited statistical value.

A single example may be given to show the gaps in our knowledge: Barnard (1963) worked on material comprising 590 mollusc specimens taken from depths between 600 and 1,800 fathoms in an area west of Cape Point, South Africa. This material contained representatives of 78 species, of which 25 (about 33%) were new and 2 of these new species represented new genera. A further 19 species, previously known from elsewhere were found in this area for the first time, which means that 44 out of the 78 species were new to the fauna of this region.

With the exception of bacteria (Zobell 1954, 1956), deep-sea organisms have not been examined in the living state. Consequently, some interpretations of biological phenomena in the deep sea are largely speculative and require further intensive research, particularly of an experimental nature. In my opinion, the application of metabolic data derived from littoral forms to species living in the eurybathyal or deep sea to explain biological processes in the bathyal (Schlieper (1963) must be approached with great caution, because the significance of longterm adaptation in autochthonous deep-sea organisms cannot yet be assessed.

The following factors are of importance in the deep sea:

1. The hydrostatic pressure increases with depth and alters the physico-chemical characteristics of the water and of any dissolved substances (see p. 78); it also probably affects membrane permeability and enzyme function. We still do not know the extent to which an ability to adapt to these special conditions is a prerequisite to settlement of the deep sea.

2. The temperature is only a little above zero and remains rather constant, so that deep-sea organisms live in a region of cold water. In analogy with shallow-water arctic forms, Thorson believes that deep-sea animals have a very slow rate of growth. According to Gislén (1930), the ratio of organic dry weight: live weight decreases with increasing depth, that is, the percentage content of water increases, and the biomass per m² appears on average to be low (Table 33), hence there would appear to be a low production of animal matter. This interpretation, however, requires to be tested because we do not, for example, know the effect of hydrostatic pressure and of the greatest constancy in temperature and other factors on the intensity of metabolism and the rate of growth.

3. The absence of sunlight prevents primary production of organic matter by photosynthesis. Deep-sea organisms are therefore dependent upon the food which reaches them by sinking or by vertical migration from the productive euphotic layers. We cannot yet assess the extent to which sediment washed down from the Continental Shelf (Fig. 155) and turbid currents contribute to the food supply for the deep sea.

TABLE 33

Frequency of benthonic animals in the abyssal of the Pacific

Animal group	*11 dredge hauls from the Costa Rica Deep, 3290–3700 m*			*3 dredge hauls from the Peru-Chile Trench, 5821–6324 m*		
	Mean numbers per dredge haul	*Estimated density per 1000 m²*	*Order of frequency*	*Mean numbers per dredge haul*	*Estimated density per 1000 m²*	*Order of frequency*
Porifera	2	1		3	0·7	
Hydroids	0·4	0·2		3·3	0·8	
Scyphozoa	9·1	4·5	VI	40	10·0	IV
Corals	0.2	0·1		0·3	<0·1	
Alcyonaria	1·3	0·6		0·6	0·1	
Actiniaria	1·5	0·7		5	1·2	
Aschelminthes	4·7	2·3	X	14	3·3	IX
Polychaeta	99·0	45·0	I	148	37·0	II
Solenogastres	7·2	3·6	VIII	6	1·5	X
Gastropoda	4·6	2·3	X	20	4·8	VII
Lamellibranchs	34·0	17·0	II	244	61·0	I
Monoplacophora	0·09	<0·1		3	0·7	
Scaphopoda	3·0	1·5		0	0·0	
Copepoda	0·5	0·3		0	0·0	
Cirripedia	0·09	<0·1		2	0·5	
Ostracoda	1·4	0·7		0·3	<0·1	
Nebaliacea	0·7	0·3		0	0·0	
Amphipoda	22·0	11·0	III	44	11·0	III
Isopoda	18·0	9·0	V	26	6·2	V
Tanaidacea	20·0	10·0	IV	6	1·4	
Cumacea	3·6	1·8		15	3·5	VIII
Decapoda	0·17	<0·1		0	0·0	
Pantopoda	0·09	<0·1		0	0·0	
Holothuroidea	2·2	1·1		0·3	<0·1	
Ophiuroidea	6·0	3·0	IX	23·3	5·5	VI
Asteroidea	0·6	0·3		2	0·5	
Crinoidea	0·1	<0·1		6	1·4	
Brachiopoda	0·09	<0·1		0	0·0	
Echinoidea	0·2	<0·1		0	0·0	
Bryozoa	8·0	4·0	VII	0	0·0	
Pogonophora	0·27	0·1		4·6	1·1	
Ascidians	1·4	0·7		3	0·7	

There is no doubt that there must also be a certain amount of secondary production of particulate organic matter in the deep sea, and that this must serve as a source of food for animals. Bacteria, yeasts and other fungi can take up dissolved organic matter and they can also utilize substances such as chitin, keratin and spongin which most animals cannot digest, and thus return these to the economy of the sea. Primary production by chemosynthesis cannot be estimated.

Figure 155. A fall of sand about 10 m high in the submarine Cape San Lucas Canyon, photographed by Conrad Limbaugh during the Vermilion Expedition of the Scripps Institution of Oceanography. (By permission of the Scripps Institution of Oceanography, University of California, 15 May 1963)

Quantitative data on the heterotrophic plants living on the deep-sea floor and on the amount of their production are only available in the case of bacteria (Zobell 1954). On the assumption that the average number of bacteria is 10^6/cc of the bottom and that each cell contains 2×10^{-13}g organic carbon, there will be 2 mg of combined carbon per m² of the bottom (to a depth of 1 cm). On the further assumption that these bacteria each divide once in the course of 20 hours there will be an annual production of 0·876 g organic carbon per m².

This seems to be a relatively high production, to which must be added the potential production by the fungi (including yeasts) and

heterotrophic flagellates which cannot yet be estimated. This to a
certain extent contradicts the assumed low production of animal matter.
To estimate the latter one would need to know more about the structure of the food pyramid, that is, the number of intermediate links in
the food chain between bacteria and fungi on the one hand and the
known members of the macrofauna on the other. The intermediate
microfaunal links in the deep-sea have hitherto been largely neglected
(thus, for example, Dahl (1943, p. 43) writes: 'The number of deep-
sea Harpacticoidea is likely to increase with improved collecting
methods') so that it is not possible to estimate the rate at which the

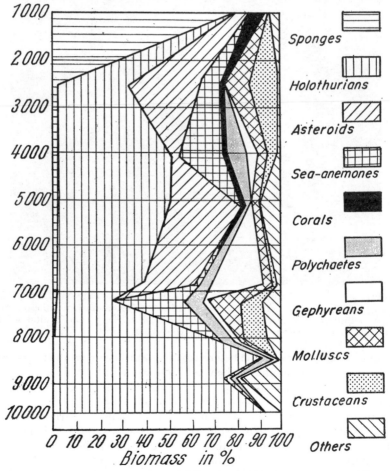

Figure 156. Relative percentages of different animal groups in the benthonic biomass at different depths (after Zenkevitch 1954)

substances produced by heterotrophic plants are reduced. The vertical distribution of food animals so far found and the relative decline in predatory forms suggests a low production of macrofauna.

In the open water of the deep sea the number of heterotrophic plants is significantly lower than on the bottom, so that here one must assume a low autochthonous food supply. The population density in the pelagial and the production of food animals in the deep-sea pelagial will therefore be seriously affected by the scarcity of food.

With these points in mind, attention can be drawn to the following points:

The number of species in the bottom fauna decreases with depth (Table 35 p. 305), so that the deep-sea becomes increasingly monotonous. The biomass also decreases (Table 33). Taken together these two facts mean that on average the individual species present do not occur in large numbers, as is frequently the case in other extreme habitats, such as shallow or brackish waters. Here food supply must be regarded as the decisive limiting factor. In some cases, however, individual catches with large numbers of individuals have been reported (e.g. Zenkevitch 1954). In view of the lack of uniformity in methods of capture, comparisons are scarcely valid, but large population densities are not impossible if more favourable feeding conditions were present locally; photographs taken underwater have shown that conditions do vary.

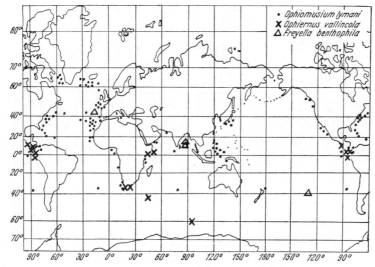

Figure 157. Records of three deep-sea echinoderms with cosmopolitan distribution (after Madsen 1954)

The number of species in the individual phyla decreases at different rates so that there is a change in the percentage contribution of the groups to the biomass present at different depths (Fig. 156). This diagram, produced by Zenkevitch from work on the Kurile-Kamtschatka Trench, should be largely confirmed by the findings of other expeditions.

In view of the uniformity of living conditions and the supposed continuity of the deep-sea habitat one might assume that benthonic animals would be widely distributed. Several species have been shown to have an almost cosmopolitan distribution (Fig. 157), whereas according to present records other species have a restricted distribution and sometimes appear to be confined to a relatively small area (cf. Fage 1954 for pycnogonids, Kirkegaard 1954 for polychaetes, Madsen 1954 for echinoderms (Fig. 158), Nybelin 1954 for the brotulids, Zenkevitch 1954 for pogonophorans and Clarke 1962 for molluscs (Fig. 159). Endemism with restricted distribution appears to occur particularly among the inhabitants of the deep-sea trenches which are isolated from one another. However, the number of hauls taken is still insufficient to allow any firm conclusion.

Distribution is dependent not only upon the extent of suitable habitats and on external distributing agents such as currents, but also on the geological age of the systematic groups and on specific methods of distribution such as mobility during the different developmental

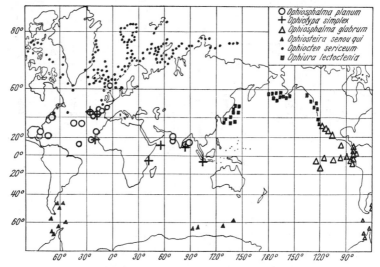

Figure 158. Restricted distribution of some deep-sea echinoderms (after Madsen 1954)

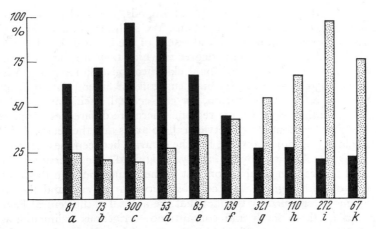

Figure 159. Percentage distribution of the abyssal molluscs from ten basins in the North Atlantic Ocean: *black* = N. America Basin species; *stippled* = Canaries Basin species

(a) Caribbean, (b) Gulf of Mexico, (c) North America, (d) Newfoundland, (e) Labrador, (f) Norway, (g) W. Europe and Iberian, (h) Mediterranean, (i) Canaries, (k) Cape Verde (after Clarke 1962 b)

stages. This indicates the need for research on breeding biology and types of locomotion. Many deep-sea forms appear to lack pelagic larvae and to practise extensive brood protection, so that one important means of distribution is lacking. However, even in forms such as the barnacles in the genus *Scalpellum*, some of which have pelagic larval stages, distribution may be restricted. Thus Dahl (l.c., p. 47) writes: 'It is interesting to note that genera like *Scalpellum* and *Verruca* which existed already in the Mesozoic have a world-wide distribution, while their species have often a very restricted range.' Future observations must confirm this restricted distribution of species before we can draw further conclusions on ecological subdivisions in the deep sea. In particular, we are still not clear about the relation between the distribution of species and the current systems.

Madsen (1961) estimated that not more than $2 \cdot 10 - 5\%$ of the sea bottom in depths greater than 3,000 metres had ever been touched by dredge hauls. It follows, therefore, that insufficient material has been collected for making biogeographical comparisons. In spite of the relatively small amount of material collected, a large number of species has been found to occur in all the three major oceans and this supports the idea that the abyssal fauna has a large cosmopolitan component; this includes, for example, several species of pennatularian, the unique abyssal scyphopolyp *Stephanoscyphus simplex*, the bryozoan *Levinsanella magna*, a number of polychaetes, pycnogonids, the cirripedes *Megalasma hamatum* and *Verruca gibbosa*, several isopods, decapods, brachipods, gephyreans, ophiuroids and ascidians. We still need to know whether there is a correlation between the extent of the bathyal distribution of the individual species and the degree to which they are eurybathyal.

Attempts at defining a regional classification (Ekman 1953, Vinogradova 1959, Madsen 1961) have suggested a differentiation into an arctic and an antarctic region, an east-central Pacific area and an Atlantic–Indian–West Pacific region. There are differences of opinion concerning the boundaries between these areas and the relation of the Atlantic to the Indian abyssal, but the contrast between the east Pacific and the west Atlantic abyssal is fully recognized. The diversity of these abyssal faunas is in striking contrast to the homogeneity in the littoral faunas of both regions. The contrast between the east and the west Pacific is determined by the fact that the abyssal of the central Pacific extends north to south and contains no terrigenous sediments; as a result of its poverty in food the large populations which would be necessary for faunal exchange cannot develop.

It was not until the more intensive investigation of the deep-sea trenches after the Second World War, particularly by the Danes (Galathea Expedition) and Russians, that it became necessary to define and name this region more accurately. Zenkevitch and other Russian authors have spoken of super-oceanic depths and of the ultra-abyssal, while Bruun (1956 a) used the term hadal for the trenches with depths exceeding about 6,000 metres. One would expect peculiarities in the

TABLE 34

Number of species of the principal animal groups from depths of over 6000 m, in the Kurile-Kamtchatka, Phillipine, Banda, New England, Kermadec, Sunda, Marianas and Tonga Trenches

(after Wolff 1960)

Porifera		8–10	Pycnogonida		2
Coelenterata		23–30	Mollusca		45–55
Hydrozoa	4		Gastropoda	15–20	
Scyphozoa	1– 2		Lamellibranchiata	25–30	
Anthozoa	18–23		Bryozoa		1
Nemertini		1	Echinoderma		43–55
Nematodes		1– 3	Asteroidea	8–10	
Polychaeta		50–55	Ophiuroidea	6– 7	
Echiuroidea		7– 9	Holothurioidea	25–35	
Sipunculoidea		4– 5	Pogonophora		5
Priapuloidea		1	Enteropneusta		1
Crustacea		58–65	Tunicata		2
Tanaidacea	3		Pisces		3– 5
Isopoda	28–30				
Amphipoda	20–25		**Total**		250–310

TABLE 35

Ratio of genera and species in recent echinoderms at different depths

(after Madsen 1954)

	Ratio total genera: species	>1,000 m	>3,000 m	>4,000 m	>5,000 m	>6,000 m	>7,000 m	>8,000 m	>9,000 m	>10,000 m
Crinoidea	1 : 3·77	1 : 2·4	1 : 1·38	1 : 1·82	1 : 1·5	1 : 1	—	—	—	—
	1 : 3·62	1 : 2·5	1 : 1·17	1 : 2·0	1 : 2·0	—	—	—	—	—
Echinoidea	1 : 3·26	1 : 2·26	1 : 1·57	1 : 1·8	1 : 1	1 : 1	1 : 1	—	—	—
Regularia	1 : 3·8	1 : 2·29	1 : 1·64	1 : 1	1 : 1·3	1 : 1	—	—	—	—
Echinothuridae	1 : 3·74	1 : 2·23	1 : 1·8	1 : 1	1 : 1	—	1 : 1	—	—	—
	1 : 5·0	1 : 2·87	1 : 1·66	1 : 1	1 : 1	—	—	—	—	—
Irregularia	1 : 3·78	1 : 2·34	1 : 1·55	1 : 1	1 : 1·5	1 : 1	1 : 1	—	—	—
Spatangoida	1 : 3·67	1 : 2·45	1 : 1·55	1 : 1·8	1 : 1·5	1 : 1	1 : 1	—	—	—
Ophiuroidea	1 : 8·63	+	1 : 2·65	1 : 2·62	1 : 1·6	1 : 1	—	—	—	—
Asteroidea	1 : 6·38	+	1 : 3·23	1 : 2·33	1 : 1·63	1 : 1·3	1 : 1	—	—	—
Porcellanasteridae	1 : 3	1 : 3·2	1 : 4·8	1 : 3	1 : 1·4	1 : 2	1 : 1	—	—	—
Pterasteridae	1 : 9·2	+	1 : 5·8	1 : 7·5	1 : 2	1 : 1	—	—	—	—
Brisingidae	1 : 5·2	+	1 : 8·3	1 : 4·5	1 : 3	1 : 1	—	—	—	—
Astropectinidae	1 : 10·8	+	1 : 3·5	—	1 : 2	—	—	—	—	—
Holothurioidea	1 : 6·47	+	1 : 3·9	1 : 3·32	1 : 1·2	1 : 1·23	1 : 1·3	1 : 1	1 : 1	1 : 1
Synallactidae	1 : 8·0	+	1 : 3·37	1 : 2·86	1 : 1	1 : 1	1 : 1	1 : 1	1 : 1	—
Elasipoda	1 : 5·01	+	1 : 5·46	1 : 4·26	1 : 2·3	1 : 1·5	1 : 1·5	1 : 1	1 : 1	—

populations of the trenches associated with their restricted extent and the fact that some are widely separated from each other.

Although the investigation of these regions is still only beginning, Wolff (1960) has produced an interesting summary of the widely scattered literature. According to this, some 250–310 species have so far been recorded from the hadal, and their systematic distribution is given in Table 34. It seems, however, that certain reservations are appropriate; thus Stschectrina (1958) identified no less than twenty-seven foraminiferan species from the Kurile-Kamtschatka Trench, without taking any special account of a possible influx from shallower areas. Nevertheless, the number is astonishingly high and will increase since little attention has yet been paid to the microfauna. The absence from this list of any decapod crustaceans is striking; even if some species are eventually found it cannot be expected that this group will have any significant share in the total fauna of the hadal.

Many of the species found in the trenches appear to be endemic in the hadal, but it is not yet possible to assess any significant differences in the endemism of different trenches. The Kermadec Trench appears to be characterized by a considerably larger number of isopods and amphipods than the other trenches. Wolff correlates this with the fact that the principal centre for the abyssal forms in these two groups lies in high latitudes.

There are still few comparable quantitative data on the populations of the deep-dea trenches, since only a small number of bottom samples has been taken. Using the 0·2 m² Petersen grab, the Galathea Expedition took five living animals in seven hauls; in the Banda Trench the best catches were from 6,580 metres (12 individuals belonging to 8 species with a total weight of 2·2 g) and from 7,270 metres (5 individuals belonging to 3 species with a total weight of 2·5 g). In the Kurile-Kamtschatka Trench the Russians found a biomass of 0·3 g/m at 8,330–9,950 metres. The number of individuals is evidently subject to very great fluctuations, and in some cases individual species may be clearly dominant; thus Wolff (l.c.) mentions the following striking dredge hauls: about 3,000 *Elpidia glacialis sundensis* at 7,160 metres in the Sunda Trench, about 1,800 *Elpidia glacialis kermadecensis* at 8,200 metres in the Kermadec Trench and in the Kurile-Kamtschatka Trench at 9,000 metres about 1,000 *Elpidia glacialis* and almost 2,000 pogonophorans belonging to the species *Zenkevitchinia longissima* and *Spirobrachia beklemischeri*.

Wolff regards pale coloration and probably also total blindness as being characteristic of the hadal and hadopelagic fauna; gigantism occurs among the Isopoda, Tanaidacea, Mysidacea and Amphipoda and occasionally in other groups. Thus, the solitary hydroid *Branchiocerianthus* (*Monocaulus*) is over 2 metres tall. The giant growth of some forms is in contrast to the dwarf growth of others. In the meantime, however, it does not seem possible to find any single cause for the coexistence of such contrasting growth; the view that the hydrostatic pressure promotes metabolism so that growth increases (Zenkevitch & Bierstein 1956, Bierstein 1957 cit. Wolff) can scarcely be generally valid. It remains to be seen whether the relative uniformity of temperature and of other factors in the abyssal together with the reduction in predatory forms, permits an increase in longevity, leading, if growth continues, to a corresponding increase in size.

The problem of food supply for the hadal fauna is of special interest; this will depend partly on transport by currents and partly, as in the deep sea as a whole, on the presence of heterotrophs as producers. So far this has only been demonstrated in the case of bacteria (Zobell 1952, 1959); since the animals are mainly sediment-feeders it is probable that heterotrophs are very important as a source of food.

There are still differences of opinion about the origin and geological age of the deep-sea fauna. Forbes (1859), Huxley (1896), Ekman (1953), Zenkevitch & Bierstein (1960) and others have stressed the presence of numerous archaic faunal elements in the deep sea and have deduced that the deep sea has been settled for a relatively long period of geological time. On the other hand, Murray (1895), Bruun (1956), Menzies & Imbrie (1959) and Clarke (1962) have regarded the Recent deep-sea fauna as relatively young geologically, although they have also emphasized the presence of some archaic forms.

The colonization of the deep sea has certainly not been a process that has happened once only; during the course of the earth's history this must have taken place more or less continuously, although sometimes in spurts; the actual organisms must have come mainly from the densely populated Shelf region. There is no good evidence for the spread of individual species into deep water, so any views put forward must be purely theoretical. Benthonic forms could be transported into deeper oceanic regions by passive means, e.g. by a downrush of sediment; such forms would be able to stay there permanently if they possessed a certain predisposition to the special living conditions of the region. The same might have happened in the case of sessile forms on floating objects which have gradually sunk. It is conceivable that dynamic population processes might have caused an extension of the original distribution of certain species from the Shelf or from the epipelagial into greater depths; Lohmann has shown that overcrowding of planktonic organisms causes them to extend their range into deeper water layers. It is possible that some species have moved in owing to competition from ecologically superior species. The primary reason may have been changes in abiotic factors; thus, cool stenothermal animals may have migrated downwards into cooler deep water to avoid incoming warm water, or some forms may have invaded deep water with small fluctuations, in e.g. temperature and light, in order to avoid an area where these factors fluctuated more widely. Finally, there is also the possibility of a spontaneous change in behaviour patterns, which has allowed the colonization of areas not previously accessible.

Beurlen (1929) developed a theory that with the regression of the extensive marine transgressions during the Jurassic, Cretaceous and Eocene periods, some at least of the richly differentiated littoral crustacean fauna migrated or sank into deep water. According to this, there have been three thrusts since the Jurassic; on this theory the presence in the deep sea of archaic forms once littoral in distribution would be explained by the course of geological events. The biological prerequisites for the process of downward migration or passive downward movement could be inferred merely by analogy with recent phenomena.

In addition to these possible methods by which the deep sea has been colonized there is also the question of the evolution of specific deepwater forms. In most cases one can reckon on the more or less extensive genetic isolation of these populations, which will be all the more effective the deeper a group of individuals has penetrated. The external conditions would probably exert a considerable selective action on new migrants into abyssal depths. Once this 'filter' has been overcome further evolution—the resultant of genetic variability and environmental factors—ought to proceed more slowly than in those environments which are subjected to greater fluctuations in external conditions. Two points emerge from this, which have already been mentioned: 1, that many abyssal endemic genera are monotypic; 2, that there are large numbers of archaic forms.

The high proportion of monotypic genera in the abyssal is emphasized, for example, by E. Dahl (1953) in the case of the isopods, amphipods and cumaceans. He finds a certain parallel in the pelagic fauna, although the causes responsible may not be the same in both cases.

Archaic forms are in no way restricted to the abyssal, but also occur in considerable numbers in the Shelf region, so their presence by itself proves nothing. However, their numerical relation to those forms which are obviously young is significant, and there is also the question whether the groups regarded as archaic have not speciated to a greater extent in deep water than in the upper horizons. It is not yet possible to carry out a detailed analysis of the material from this viewpoint, but it is very probable that the percentage of archaic species is low in the Shelf region, whereas in the deep sea such forms make up a significant proportion of the fauna. Zenkevitch & Bierstein (1960) put the ratio for the deep sea at 16%, but for the Shelf region at only 0·00005%.

The echinoderms provide an excellent example of the degree of speciation in the deep sea. Madsen (1954) has summarized in tabular form (p. 305) the bathymetric distribution of all the known echinoderms, giving both the total number of genera and species in the different classes and orders and the corresponding values for different depths. Comparison of the number of species with the number of genera gives a measure of the speciation at different depths. As might be expected, the ratio changes with depth and in every case ends with a ratio of 1:1. In most orders the values decrease continuously with depth, relative to the total ratio. The exceptions, however, are the Porcellanasteridae and Brisingidae among the asteroids and the Elasipoda among the holothurians which do not show their highest speciation figures until depths of more than 3,000 metres; these are often regarded as archaic forms. Madsen (1961, p. 191), however, states that 'the Porcellanasteridae are regarded not as archaic, but rather as paedomorphic phanerozoniate sea-stars of a comparatively young phylogenetic origin; and a similar point of view may be adopted for the Elasipoda; both groups are without fossil records.' Evidently, therefore, it is not the rule that forms regarded as archaic show greater speciation in deep water; thus, in Madsen's summary the crinoids behave just like the other echinoderms, and Gislén (1953) has emphasized that many primitive forms are present in medium depths (down to about 2,000 metres), but not in greater depths.

In addition to the archaic forms there is no doubt that the deep-sea fauna also contains younger elements, which appear to have been living for a relatively short time in the deep sea, because some belong to eurybathic species, some to eurybathic genera.

By measuring the ratio of oxygen isotopes in foraminiferan tests taken in core samples by the Albatross Expedition, Emiliani & Edwards (1953) concluded that the temperature of oceanic bottom water in the upper half of the Tertiary decreased by about 8°C. Deep-water temperatures must have been higher in the lower Tertiary than during the Pleistocene and Recent because the polar water masses, whence the bottom water originates (see p. 65), must have been relatively warm since reef-building corals were living in the polar regions at that time. Correlated with this there must have been less circulation in deep water. From the lowering of temperature, Bruun (1956) concluded that when the cooling started there was widespread mortality of the Mesozoic and lower Tertiary deep-sea animals and that the old fauna was replaced by the immigration of a new fauna.

These theoretical remarks require further analysis; on the one hand Zenkevitch & Bierstein (1960) have cast doubts on the argument put forward by Emiliani & Edwards, because the foraminiferans investigated may not have been autochthonous in the deep sea but may have been washed down, and would then only document temperature changes in the upper water layers. On the other hand, Madsen (1961) has estimated an annual decrease in temperature of $0.23 \times 10^{6}°C$, if the abyssal temperature has actually dropped a total of about 8° since the middle Oligocene. This would allow the older inhabitants of the deep sea a chance to adapt and survive.

By and large, the ecological, physiological and evolutionary problems associated with the deep sea can probably only be explained by analogy with those of the littoral; to do this, however, one would require a greater amount of comparable material and a basic knowledge of the structure and dynamics of the abyssal environment.

9. The phytal

Alongside the benthal and pelagial, Remane (1933) used the term phytal to denote the third main habitat. This habitat contains not only large areas with major vegetation but also growths of sessile animals such as hydroids, corals and bryozoans. If the term phytal is to be meaningful it is essential that the plant and animal growths contain a fauna, the composition and distribution of which are clearly different from that in the sediment and which also remains the same within a given population of a single plant species, even when the latter occurs on different bottoms. (Similarly Pérès, 1957, p. 425, regards the sponges with their sessile and motile epibionts and the endobionts as a community independent of the substrate, because the composition remains the same whether the sponges are living on rocks, gorgonians, dead corals, plants or sand. The substrates would therefore sometimes house different communities. Consequently one must reckon that representatives of the phytal fauna, or at least the same life forms, will occur in the aphytal although plants are absent there.)

The term phytal as defined by Remane, therefore, means something quite different from the 'système phytal' of Pérès (1961), which was used for the littoral system, 'since it is characterized by the presence of benthonic plants containing chlorophyll' (p. 31), in contrast to the 'système aphytal' of deep water.

Detailed analyses of the phytal fauna of different areas are only available to a limited extent, and a number of investigations have been restricted to the macrofauna that can be taken by dredge hauls (e.g. Thorson 1933, 1934); the best-known surveys are those carried out in the North and Baltic Seas. The information obtained there on the occurrence and distribution of life forms ought also to apply to other areas, but the species-specific differences will depend on geographical factors. Naturally, considerable biocoenotic differentiation can be recognized within the phytal because, on the one hand, the structure of the plants is very variable (e.g. finely divided to broadly foliaceous) so that conditions will vary considerably according to the dominant plant species; on the other hand, some plants themselves show different structural modifications (e.g. leaf-like thallus, stipe and holdfast in laminarians) and therefore offer different conditions for settlement; the giant kelps extend through several levels of illuminination. In addition, of course, the character of the phytal fauna will be affected locally by abiotic factors, such as currents, waves, temperature and so on.

The phytal fauna is characterized by the presence of numerous sessile animals, among which the small forms, such as foraminiferans, sponges, hydroids, polychaetes and bryozoans generally predominate. The plants usually do not provide a suitable substrate for the medium-sized and large forms or for species with attachment surfaces, such as actinians, although the laminarians and *Fucus* occasionally carry clumps of ascidians like *Ciona intestinalis*. Dense epiphytic growths of diatoms and other algae and also probably of bacteria and fungi (including yeasts) are frequently present. Together with the epifauna this epiflora provides a rich source of basic food for the semisessile and motile animals that are also present. Thus hydroids are eaten by opisthobranchs of many different types which are usually brightly coloured, or their tissues are sucked by polychaetes such as *Autolytus*. Plant growths are often grazed by large numbers of small gastropods, such as *Rissoa*, *Lacuna* or *Cerithium*, or by *Chiton*, copepods and ostracods. Tentacle-feeders like the terebellid polychaetes live partly on the growths and partly on the detritus that settles in large quantities on the plants, although tentacle-feeders and the filterers are largely dependent upon the fine detritus floating in the water and on small motile invertebrates.

In the fjords of Greenland, Thorson (1933, 1934) recorded large motile forms, such as *Buccinum hydrophanum*, *Ophiura robusta* and the echinoid *Strongylocentrotus droebachiensis* which is one of the few macrophytophagous animals. The phytal also has a large number of fishes, including pipefishes and the related sea-horses, the latter having characteristic prehensile tails. The arms of the ophiuroids *Asteronyx loveni* and *Gorgonocephalus caputmedusae* are also prehensile organs.

A number of morphologically and biologically divergent types live in the phytal. Among the amphipods the Caprellidae have the hind thoracic limbs modified to form clasping organs with which the animal attaches itself to algae while the front part of the body and the front limbs are spread out and act as capture organs. An analogous morphological type is found in the isopod *Astacilla*, which lives especially on the 'stems' of hydroid polyps (see Fig. 131, p. 257). The small hydromedusan *Eleutheria* which crawls with the help of its marginal

L—M.B.

tentacles also lives in the phytal, as do the semisessile scyphomedusans *Lucernaria* and *Haliclystus*. Crabs of the genera *Macropodia*, *Inachus* and *Hyas* attach pieces of alga and hydroid on to curved chitinous hooks and thus camouflage themselves in much the same way as the small echinoid *Psammechinus* which holds similar camouflage material with its aboral tubefeet. Remane has drawn attention to colour adaptation in phytal animals, particularly in the isopod *Idothea*.

CHAPTER VI

DISTRIBUTION OF MARINE ORGANISMS IN
SPACE AND TIME

1. Biomass and productivity

Attention has been drawn in the preceding chapters to productivity as the basic biological process responsible for life in the sea. This process is a subject of special interest to modern marine biology, particularly as regards the phytoplankton. Recent summaries have been given by Steemann Nielsen & Jensen (1957), Strickland (1960) and Steeman Nielsen (1960). Strickland cites almost 500 references, which is an indication of the great interest taken in this subject.

Research in this field was started by the early quantitative investigations on plankton, particularly those of Brandt, Hensen and Lohmann. They endeavoured to determine the quantity of phytoplanktonic organisms in a given volume of water by counting the number of individuals, and then by comparison of different samples to obtain a picture of the population density in different areas. They were interested not only in the number of organisms of different sizes, but also in their weight, and so introduced several different units: wet weight of plankton, dry weight, sedimentation volume, total surface area of all the cells in a given volume of water, amount of protoplasm present after subtraction of skeletal substances and vacuole volumes. Numerous measurements and calculations were carried out to enable the values for one unit to be converted into those for another (see p. 314 et seq.). The term 'standing stock' or 'standing crop' has been used to express the amount of phytoplankton or zooplankton present at a given time. The term 'biomass' usually refers to the wet weight of a plankton sample.

Several different methods have been used to obtain material. Net hauls have the disadvantages that they only catch a fraction of the planktonic organisms present corresponding to the mesh size and that the volume of water fished can only be determined approximately by installing special apparatus. Also in dense phytoplankton the meshes of the net become clogged so that the filtration rate changes; also the efficiency of net, whether in vertical or horizontal hauls, depends upon the towing speed. Hardy (1936) avoided certain disadvantages of the open net by using his continuous plankton recorder, but this too does not give absolute values, although it does provide valuable information on the distribution and relative abundance of the zooplankton and of the larger phytoplankton organisms.

The sedimentation volume of net hauls has also been used as a measure of the biomass but this gives somewhat inaccurate values because spiny forms occupy a greater volume than is justified by their actual size, and delicate organisms, such as naked flagellates, easily disintegrate and are therefore unrecognized.

A new method involving centrifugation of water samples was introduced by Lohmann (1908); this enabled him to obtain a good picture of the distribution of nannoplankton in the North Atlantic, particularly during the voyage of the *Deutschland* in 1911. In 1932–1936 Hentschel published the results he obtained on the Meteor Expedition using the same method, and thus complemented and confirmed Lohmann's interpretations. The efficiency of the centrifugation method is dependent upon the specific weight of the plankton organisms, which is generally somewhat greater than that of sea water (exceptions are some Myxophyceae (Braarud 1957), which may, however, only play an important role in brackish coastal waters) and on the speed of the centrifuge. Estimates of the validity of this method vary considerably: Gran (1932) gives losses of 30%, Allen (1919) of even 90%, whereas according to Nielsen & Brand (1934), Davis (1957) and others the results obtained are quite satisfactory.

Only small amounts of water can be used with this method nuless, a large continuous flow centrifuge is available. The use of biomass values obtained from small amounts of water to represent large volumes or areas must therefore be treated with reserve. This disadvantage is to a certain extent overcome by filtration through paper or a membrane filter, which has been much used although in ways which are by no means standardized. Here, too, one must reckon on losses of material, because ultraplankton of the size order of the pores will pass through the filter. The efficiency of the method can be increased by the addition of aluminium hydroxide.

These very time-consuming methods give the number of phytoplankton organisms present, apart from losses which cannot be estimated with any degree of certainty. They have the advantage that the species themselves mostly remain recognizable and can be identified. Nevertheless, one cannot make a direct calculation of the numbers of each species and their size because this would be affected by the following factors: 1, the variability in size of the individuals of a species

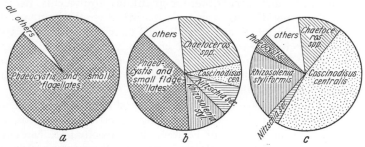

Figure 160. Relative proportion of the different algae in the total catch at a station in the European North Sea: (a) number of cells, (b) surface area of cells, (c) cell volume (after Paasche 1960)

is dependent upon environmental factors; 2, the varying proportion of cell sap in relation to formed organic matter, which is characteristic of the species but may vary within a species according to physiological conditions; 3, the variable chemical composition of the species (Fig. 160).

By fixation (usually in formol) and sedimentation, Utermöhl (1931) obtained material which was counted under the inverted microscope. This method has been largely adopted and seems to have stood the test so well that it has been specially recommended by the Plankton Symposium which discussed these problems at Bergen in 1957. In a critical study of methods, Gillbricht (1962) came to the conclusion 'that a single plankton sample ought on average to represent about 15 to 25% of the mean plankton content of the relevant body of water' (p. 217).

Since 1930 great efforts have been made to determine the biomass content in a volume of water, a net haul or a filter residue by chemical means. Kreps & Verbinskaya (1930) estimated the amount of phytoplankton present from the colour intensity of an alcohol or acetone extract of the plant pigments, particularly of chlorophyll, using standardized solutions for comparison. As an important development of this method, Krey (1939) and Riley (1939) introduced the extraction of filtration residues. Of the pigments only chlorophyll α and β-carotene can be utilized because these are the only main pigments present in the principal groups of photoplankton, whereas the other pigments are of varying importance. The numerous investigations carried out by this method in recent years have only given relative results, because it appears (Gillbricht 1952) that sometimes a considerable part of the extracted chlorophyll comes, not from living algae, but from dead plants or even from animal faeces. For this reason Kalle's (1951) use of the fluorescence of chlorophyll to determine the amount of plant matter is also only of limited value. Furthermore,

Gillbricht has found very different plankton equivalents to 1 μg chlorophyll: in peridinians, 277·4 × 10⁶ μm³ phytoplankton, in diatoms 131·1 × 10⁶ μm³; finally we still need to confirm that, even within a single species, the chlorophyll content varies, depending upon the physiological condition of the cells. There is, therefore no completely satisfactory method of determining the amount of plankton from measurements of chlorophyll values.

Marshall & Orr (1962) developed a method of carbohydrate determination, which has been used for some years in other fields, and obtained very good agreement between the carbohydrate content of filtered plankton and the number of diatoms. Observed discrepancies were correlated with the physiological condition of the cells (Fig. 161).

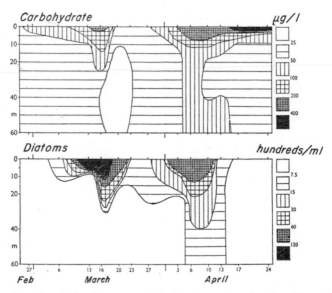

Figure 161. Correlation between carbohydrate content and spring diatom increase in the same locality (after Marshall & Orr 1962)

Data on the total content of organic matter have been obtained by direct or 'wet' incineration, without differentiating between phytoplankton, zooplankton and detritus. From what has been said it is not however possible to determine the amount of phytoplankton by subtracting a value obtained from measurement of the chlorophyll.

Krey (1952 et seq.) and his co-workers have compared the total organic matter with the true living matter, the protein, which they determined by the Biuret reaction. In other cases Folin's reagent was used. This allowed the determination of the total protein, but it was not

TABLE 36

Composition of three different plankton samples based on several methods of determination

Further details in text. (After Parsons and Strickland 1960)

Sample	Oxidizable carbon (mgm/m³)	Carbohydrate (as glucose) (mgm/m³)	Protein as egg albumin (mgm/m³)	Lipid as stearic acid (mgm/m³)	Chlorophyll a (mgm/m³)	Plant carotenoids (m SPPU/m³)
49°13′N, 123°57′W Departure Bay (Nanaimo), 21 April 1959	475 (100)	550 (115·7)	398 (83·8)	43 (9·05)	12 (2·5)	4·5 (0·94)
Culture of oceanic phytoplankton (mixed coccolithophores and small diatoms)	730 (100)	455 (63·7)	325 (44·5)	91 (12·4)	11·3 (1·5)	3·7 (0·5)
50°N, 145°W 0–50 m. Composite sample, 4 August 1959	130 (100)	72 (55·4)	150 (115·4)	22 (16·9)	0·27 (0·2)	0·08 (0·06)

possible to differentiate between zooplankton and phytoplankton, and this method does not provide reliable conversion factors for the determination of the amount of plankton.

The points made above illustrate the methodological inaccuracies involved in the determination of the biomass in a given volume of water. Further sources of error will arise owing to heterogeneity in spatial distribution when calculations based on a single or a few determinations are used to represent a large volume or area of water. Table 36 illustrates how it is almost impossible to apply the conclusions obtained from one method to the values derived from another or to the total amount of phytoplankton. This table shows the values found by Parsons & Strickland (1960) at two stations in the Pacific and in one culture (the figures in brackets) converted in such a way that in each case the value for oxidizable carbon is expressed as 100; the range of variation of the values in the other columns shows the lack of a common conversion factor of sufficient accuracy.

The equivalent values recommended by the Committee on Terms and Equivalents at the Plankton Symposium are therefore only approximate and must be checked in each individual case (Tables 37

TABLE 37

Phytoplankton equivalents

(after Cushing *et al.* 1958)

	Carbon 1 mg	Organic matter dry weight 1 mg	Oxygen equivalent 1 ml	Plankton biomass 1 mg	Plankton dry weight 1 mg
Carbon	1	0·43	0·53	0·024	0·30
Organic matter Dry weight, mg	2·3	1	1·2	0·055	0·69
Oxygen equivalent, ml	1·9	0·83	1	0·046	0·57
Plankton biomass, mg	42	18	22	1	13
Plankton dry weight, mg	3·3	1·4	1·8	0·08	1

and 38). Since one is dealing with heterogeneous living material which may often react to environmental factors in quite different ways, it would be very hazardous to employ these equivalents in any arbitrary way.

Quantitative determinations of the amount of plankton present at a given time (the 'standing stock') does not, however, provide any information about productivity, that is, about the production of new organic matter by assimilation, because this involves a time factor. Assimilation primarily involves the binding of the carbon derived from

TABLE 38

Zooplankton equivalents

(after Cushing *et al.* 1958)

	Carbon 1 mg	*Nitrogen* 1 mg	*Phosphorus* 1 mg	*Plankton biomass* 1 mg	*Plankton dry weight* 1 mg	*Displacement volume* 1 ml
Carbon	1	6·0	75	0·12	0·60	96
Nitrogen	0·17	1	13	0·020	0·10	16
Phosphorus, mg	0·013	0·078	1	0·0016	0·008	1·3
Plankton biomass, mg	8·3	50	620	1	5·0	800
Plankton dry weight, mg.	1·7	10	130	0·20	1	160
Displacement volume, ml.	0·010	0·060	0·75	0·0012	0·0060	1

Note: in zooplankton the dry weight of the organic matter is very similar to that of the dry plankton.

carbon dioxide, and so the productivity may be expressed as mg carbon per m^3 sea water or per m^2 of surface, when the total thickness of the actively assimilating layer is being considered. The production due to assimilation is known as the primary production.

Assimilation is accompanied concurrently by respiration, that is, matter is oxidized, and so the gross productivity is greater than the net productivity: the amount of assimilation measured during daylight must be reduced by the amount lost through respiration during the 24 hours. First attempts at the measurement of primary production were based on the fact that phytoplankton development is dependent upon the nutrients, so that the consumption of nutrients should be a measure of the scale of production (e.g. Atkins, 1923, 1930). But this overlooks the fact that some of the nutrient salts are returned to the water by decomposition or by the release of metabolic products and are thus available for fresh consumption, although this turnover cannot be assessed quantitatively. Lateral and vertical water exchanges will also influence such measurements and these certainly cannot be allowed for, so that any measure of the primary productivity based on this method will only be valid in certain special conditions.

Two experimental methods have, however, led to results which can be regarded as more reliable, although even here there are still difficulties and reservations.

The amount of assimilation can be expressed by the amount of oxygen given off in unit time; if the oxygen consumed in respiration is measured at the same time, it is possible to calculate the net production during a 24-hour period. Investigations of this kind were started by

Gaarder & Gran (1927); water samples from various depths with their content of phytoplankton were placed in transparent and dark bottles and then lowered to the depths from which they had been taken. Determination of the oxygen content (by the Winkler titration method) before and after the bottles were sunk gives the amount of assimilation (transparent bottles) and of respiration (dark bottles): the respiration is determined by comparing the original oxygen content of the dark bottles with the final values recorded, while the difference between transparent and dark bottles gives the amount of photosynthesis. Riley (1939) modified this method by setting up samples taken from the surface on deck and thus obtained a significantly greater number of values during a voyage.

The results obtained by this method have been repeatedly criticized, particularly by Steemann Nielsen (1957), who has pointed out that the oxygen consumption in the dark bottles depends not only on the respiration of the autotrophic phytoplankton but also on the heterotrophs present, and that in the dark bottles the inhibiting action of assimilating algae on bacteria would be suppressed, whereas this would not happen in the transparent bottles. Hence the values obtained would give an inaccurate picture of phytoplankton production.

Steemann Nielsen himself developed a new method (1952) which he used successfully during the Galathea Expedition (Steemann Nielsen & Aabye Jensen 1957). It is based on the following principle: a definite amount of carbon 14 in the form of $NaH^{14}CO_3$ is added to the sample (100–500 cc), the productivity of which is to be measured. The total amount of CO_2 present in the water then contains a known proportion of $^{14}CO_2$. On the assumption that $^{14}CO_2$ is assimilated photosynthetically by the planktonic algae at the same rate as $^{12}CO_2$, then by determining the content of ^{14}C in the plankton after the experiment one can also determine the total amount of carbon assimilated. As ^{14}C has a half-life of more than 5,000 years, samples with a definite content of $^{14}CO_2$ can be kept in stock (details in Steemann Nielsen 1958).

For the determination of the production beneath 1 m^2 of surface, Steeman Nielsen proceeded in the following way. Water samples were taken from various depths of the euphotic layer, treated with ^{14}C, submerged to their orginal depth in sealed bottles and left in situ from sunrise to noon or from noon to sunset; the phytoplankton from the samples was then filtered, treated with concentrated hydrochloric acid to remove carbonate, dried and the content of organically combined ^{14}C determined by a Geiger counter. To avoid excessive expenditure of time, Steemann Nielsen used tanks standing on deck in which the water samples were subjected to light of known intensities (cf. Riley 1939 using transparent and dark bottles for the determination of oxygen). This method has been modified several times.

In general the use of ^{14}C involves several prerequisites, if the experiments are to provide an absolute measure of the productivity. The author of the method makes the following points: '1, No $^{14}CO_2$ must be incorporated in organic compounds except through photosynthesis; 2, the rate of assimilation of $^{14}CO_2$ must be the same as that of $^{12}CO_2$; 3, no $^{14}CO_2$ must be lost by the respiration which takes place simultaneously with photosynthesis; and 4, no organic matter must be lost by excretion' (1958, p. 43). The correction of $+ 10\%$ which is said to be necessary only applies when the respiration rate amounts to 10% of the photosynthesis; with increasing rates of respiration this correction factor will be larger and the method thus becomes less accurate. Strickland emphasizes the point that the interpretation of the values measured in ^{14}C experiments is not so simple as the method itself.

A large number of investigations have been carried out using this ^{14}C method, which has been modified in various ways (cf. among others the Plankton Symposium 1957). In addition, attempts have been made to find correlations between the standing stock and the productivity. In this context, Paasche (1960) showed that there was a closer correlation between the surface area of the phytoplankton (calculated from counts and measurements) and the production capacity than there was between cell volume or cell size and productivity; a correlation factor of 0·74 does not give sufficient accuracy to allow a calculation of productivity from the data measured.

Primary production by photosynthesis is only the first stage in the total production process taking place in the pelagial. It leads to reproduction of the algae by division, which may be so intensive as to lead to plankton blooms, that is, to the mass multiplication of phytoplankton organisms which will then cover the water over large areas. This reproduction exceeds both the natural mortality rate of the phytoplankton organisms and also their consumption by herbivorous animals. The combination of these factors has been expressed in the following formula:

$$\frac{dP}{dt} = P\,(R - M - G)$$

in which t represents time, P is an expression of the amount of phytoplankton (number, weight or volume of the algae or weight of carbon), R the current rate of division, M the current rate of mortality and G the rate of consumption by herfivores.

This, however, does not cover all aspects of the production process. The accumulation of organic matter is not restricted to the formation of simple carbon compounds by assimilation, but also involves the proteins, in particular, the synthesis of which requires nutrients containing nitrogen, phosphorus and sulphur. Thus the processes following on the primary production will be dependent upon the available mineral nutrients; among these phosphorus is of such great importance that it may act as a limiting factor. Even when conditions are very suitable for photosynthesis, production may be restricted by lack of nutrients. This has been confirmed several times in all parts of the ocean, ever since Brandt first drew attention to it. Looked at from the viewpoint of time and space it is true to say that when production has gone on for any length of time without renewal of the water by exchange, the water becomes aged and its productive capacity decreases. The so-called land-mass effect is also of interest, for it has been shown that production is much greater in the vicinity of land masses than it is in areas distant from the coast. This is primarily due to the turbulence above the Continental Shelf, which moves nutrients from near the bottom into the photosynthetically active layers. In oceanic regions great productivity is observed where water rises along divergence lines from moderate depths and contributes its nutrients to the productive layers. In medium and high latitudes the seasonal breakdown of stratification leads to the regeneration of the nutrient content and provides the basis for plankton outbursts.

In areas where there is no permanent or temporary regeneration of the mineral nutrients from neighbouring water masses, as for example in stratified offshore tropical waters, a balanced production may develop, with a more or less stable relationship between phytoplankton and herbivorous zooplankton. Cushing (1959) has drawn an interesting parellel between the scattering layers and the benthos: the animals present in the scattering layers rise at night into the upper layers and bring nutrients, in the form of their excreta, back into circulation; this can be regarded as a kind of turbulence effect caused by animal activity.

In the vicinity of land masses there is also the likelihood that organic matter may be washed down from the land or out of marine coastal vegetation.

The practical use of the mathematical formula mentioned above is further restricted by the varying effects of fluctuating temperatures on assimilation, cell division and animal activity. Local temperature differences may be short- or long-term so that it is difficult to compare areas that are separated from each other. Such comparisons are also made more difficult by the fact that different species adapt to changes at different rates, so that the 'history' of a population will affect its measurable potential, but is not itself measurable. The same applies to light, a factor which is so obviously important in primary production. This has been investigated by Steemann Nielsen & Hansen (1959) and others.

In calculating the primary production beneath 1 m² of surface area it is not only the extent of the euphotic layers and the total amount of its phytoplankton that are of consequence, for the vertical distribution of the latter is also very important. The distribution is more uniform in unstratified than in stratified water, because in the latter enrichment takes place at the discontinuity layer. The light conditions may be more favourable or less favourable for photosynthesis according to the depth of the discontinuity layer, so that noticeable fluctuations in the production per square metre of surface may be due to the stratification, the other conditions being the same. Sorokin (1960) has therefore, emphasized that stratification and the depth of the thermocline should be determined in all investigations on primary productivity.

Steele (1954) has drawn attention to a further difficulty. He has emphasized that the methods used to measure productivity take no account of the particular species of animals and plants involved. The development of the subject has tended away from the usual biological units and uses instead such abstractions as 'carbon content'. The aim of such investigations is to simplify and generalize so that quantitative comparisons can be made between different areas and at different periods. But little consideration is given to the organic complexity of the processes being investigated and to the interspecific dynamic relationships and this greatly reduces the general validity of the measured values and the extent to which they can be compared. The wealth of available results, however, only give a relative picture of productivity in the sea, and it seems doubtful whether a mathematical treatment of the material (e.g. Cushing 1958) really makes the position any clearer. It is only under certain limited conditions that a mathematical model will approximate to the truth.

According to the results of Steemann Nielsen & Aabye Jensen (1957), Sorokin (1960) and others, the highest rate of primary production lies not immediately beneath the surface, but generally at depths of 20–30 m; in the upper layers photosynthesis is inhibited by excessive illumination. Since the red light has already been absorbed in the uppermost few metres this means that there is a relatively low utilization of the sun's energy. A reduction in illumination causes the maximum utilization to move upwards and thus reduces the size of the actively assimilating zone. Stratification also has a modifying effect: if it causes the phytoplankton to be held in the upper layers the light may have an inhibiting effect, but this will depend, of course, on the specific composition of the phytoplankton. Steemann Nielsen (1955) has given a generalized curve for the vertical relationship between light intensity and gross production (Fig. 162).

From the regional viewpoint Steeman Nielsen & Aabye Jensen (1957) have distinguished four different classes of oceanic regions according to the amounts of organic production:

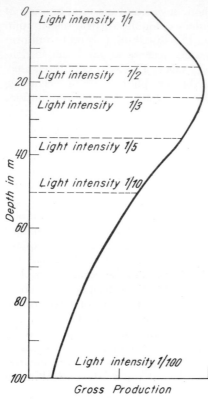

Figure 162. Light intensity and rate of gross production at different depths (after Steemann Nielsen 1957)

1. Those with a very considerable admixture of nutrient-rich water to the photosynthetic zone; the organic production is high and amounts to about 0·5–3·0 g C/m²/day. An example of this class is the southern part of the Benguela Current. The shallow waters of Walvis Bay give a surprisingly high maximum of 3·8 g.

2. Regions with a fairly steady admixture of nutrients, giving production values of 0·2–0·5 g C/m²/day. Areas of this kind were found by the Galathea Expedition in the vicinity of the divergences caused by the equatorial counter-currents in all three oceans.

3. Areas with a small influx of water from medium depths into the euphotic zone, due mainly to turbulence, giving a production of 0·1–0·2 g C/m²/day, as happens for example in most areas of the tropics and subtropics.

4. Areas with typically 'old' surface water without regular or sufficient influx, producing a 'balanced' cycle and a production of only about 0·05 g C/m²/day; the best-known example of this is the Sargasso Sea.

There has been no lack of attempts to estimate the amount of production from the data obtained. In calculating the annual production there are several sources of error; certain assumptions have to be made on the average gross production in g $C/m^2/day$ and on the amount of respiration, and there is also the need to multiply by a factor of 350. Thus even the values first given by Steemann Nielsen (1952 and 1954) were calculated differently in 1957; this author gives an average production of 55–70 g $C/m^2/year$ (corresponding to 0·16–0·2 g $C/m^2/day$), a respiration loss of 40% and the total area of sea surface as 361 × 10^6 square kilometres; from these figures he calculates the annual phytoplankton production in the seas of the world as 1·2–1·5 × 10^{10} tons of carbon, but there is considerable regional variation (Fig. 163).

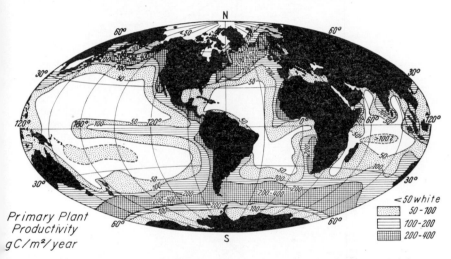

Primary Plant
Productivity
gC/m²/year

<50 white
50–100
100–200
200–400

Figure 163. Distribution of annual production in grams of carbon per m² (after Hela & Laevastu 1961 from Dietrich 1963)

Investigations of the primary production by benthonic vegetation have been less extensive, although the results can be regarded as more reliable owing to the greater accessibility. Apart from the macroflora, the primary producers are the benthonic diatoms, in particular; the heterotrophs also play a part in so far as they feed on dissolved organic matter.

The importance of diatoms can be judged from the dense diatom mats seen in shallow water, although there have so far only been a few estimates of their productivity. Using the ^{14}C method, Grøntved (1960) found the following mean values for the diatom microflora of seven shallow brackish-water areas in Denmark:

in summer (March–October)   861 mg C/m²/day
in winter (November–February)   164 mg C/m²/day

Assuming that in natural environments the potentiality is only half as great as it is under experimental conditions, this gives a mean annual production of 116 g C/m². Simultaneous estimations of the phytoplankton productivity in the same waters yielded significantly lower figures. On the other hand, in a shallow lagoon area in Florida, Pomeroy (1960) found that the production of phytoplankton far exceeded that of the benthonic microflora and of the turtle-grass *Thalassia testudinum* (Fig. 164).

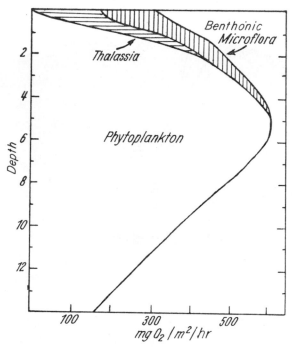

Figure 164. Variation with depth of the relative primary production by the photoplankton, by *Thalassia* and by the benthonic microflora (after Pomeroy 1960)

Grøntved's investigations showed maximum productivity at a depth of 0·5–0·7 m; at the same depth productivity is greater on muddy than on sandy bottoms; it is also dependent upon exposure to wind, and thus on water movements (Fig. 165). From this one must expect great variability in the potential biological productivity of the autotrophic microflora. It is difficult to determine the part played by heterotrophs in the microbenthos and so calculation of the net productivity is even more difficult here than it is in the pelagial.

In 1962 Grøntved published the preliminary results of experiments on the microflora in shallow waters, using the ^{14}C method. His material came from the upper one centimetre of the bottom and the assimilation periods lasted for two hours; he was able to differentiate between a suspended fraction and a sand fraction with adhering diatoms ('psammophytic microvegetation)'.

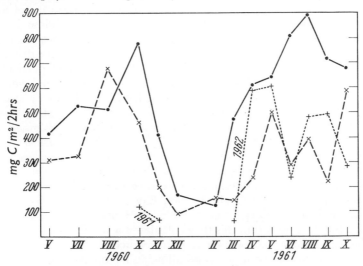

Figure 165. Average total production potential in the upper 1 cm in the Danish Wadden Sea (after Grøntved 1962): *unbroken line* = at 4 sheltered positions, characteristic of the area; *broken line* = at 2 positions sheltered by artificial structures; *dotted line* = at 2 positions continuously covered by water

The summary given in Table 39 shoes that at stations continuously covered by water (A, B), both fractions showed approximately the same productivity, but that at stations exposed from time to time (I–IV, V–VI) the suspended fraction gave much lower values. The productivity of the psammophytic microvegetation at stations I–IV is surprisingly high.

TABLE 39

Average assimilation potential in g $C/m^2/year$ of the microbenthos on Danish tidal flats

(after Grøntved 1962)

Locality	A, B	I–IV	V–VI
Sand fraction	278	833	494
Suspendible fraction	293	59	83

These figures represent the total potential, that is, they include the uptake of ^{14}C during darkness.

In the annual cycle there is a marked winter minimum and a summer maximum. It is remarkable that there is a considerable total potential down to a depth of 3 cm in the bottom. This shows the urgent need for measurements of the illumination present in these depths of sediment, in order to find out whether the potential measured is actually due to the photosynthetic combination of ^{14}C.

Heterotrophic microplants must be regarded as producers of food in so far as they are able to utilize substances such as cellulose, lignin, keratin and chitin, which are inaccessible as food to most animals, and change them into a digestible form; they can also remove dissolved matter from the water and thus increase the amount of particulate organic matter. Such heterotrophs are bacteria, yeasts and various fungi; these have been shown to serve as food for animals by feeding experiments (Zobell & Feltham (1938) on the mussel *Mytilus californianus*, the crab *Emerita analoga*, the gephyreans *Dendrostoma zostericola* and *Urechis caupo*; Kriss & Nowoshiloff (1954) on *Nereis diversicolor*) and also from the analysis of stomach contents. We cannot yet estimate the extent of this potential production, but must agree with Perkins (1958) that it is of considerable importance to the microfauna of the benthos. On a muddy flat in Mission Bay on the Californian coast, Zobell & Feltham (1942) found $2 \cdot 2 \times 10^{12}$ bacteria per cubic foot (0·028 m^3) of mud. From this they calculated a weight of 1.1 g of protoplasm, assuming that each bacterium weighed $0 \cdot 5 \times 10^{-12}$ g and had a water content of about 90% and allowing for the fact that the plate method only accounts for about 10% of the bacteria. This standing crop represents a balance between reproduction and destruction and so assuming 10 divisions per day, the productivity must be 11 g/day/ 0·028 m^3 (approx.). Even allowing for considerable overestimation this is a remarkably high figure, which cannot be overlooked when considering the biology of feeding in the benthal. Using modern methods of continuous culture it should be possible to show experimentally the extent to which bacteria provide food in the benthal.

Micro-organism populations of this kind are also ecologically important in other respects; Zobell & Feltham, for instance, have drawn attention to oxygen consumption, hydrogen sulphide formation and changes of pH and redox potential.

Rees (1940) has given a much simplified diagram of production on a mud flat which shows the qualitative interrelationships but gives no quantitative information (Fig. 166).

There are relatively few data on the actual productivity of the macroflora, whereas the biomass has been measured on several occasions (e.g. McFarland 1959). As a general guide Westlake (1960) gives a

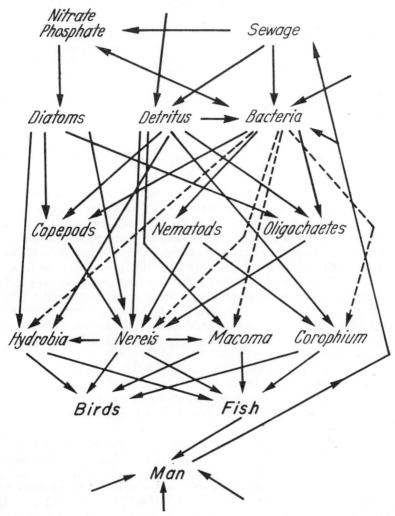

Figure 166. Simplified food cycle for a mud-flat in the eulittoral (after Rees 1940)

figure of 10–20 tons/hectare for the annual production by marine macroflora. Some examples will be given, but these are not strictly comparable because the units of measurement used were very varied and re-calculation will have introduced further inaccuracies.

Blinks (1955) reported the following figures for littoral algae on the coast of California.

TABLE 40

Species	Dry weight g/m² standing crop	Dry weight production in g/m²/day
Ulva	70	3–7·2
Porphyra	300	11–21
Gigartina	750	54
Iridophycus	760	19
Egregia	800	25
Alaria	1,800	14
Fucus	2,130	19–42
Pelvetia	4,240	35
Laminaria (Thalli)	4,400	66

Working on the productivity of the macrovegetation in shallow bays on the coast of Denmark, Grøntved (1958) found that *Zostera marina*, *Ruppia maritima* and *Chara* species produced much less than the other phanerogams and the algae. Grøntved calculated the production potential as the difference between the dry weight/m² at the time of the lowest and highest growth rate of the vegetation. He found that the mean biomass in Dybso Fjord was 379 g/m² and at Naeraa Strand 241 g/m², but that the production potential was lower in the first locality (184 g/m²) than in the second (201 g/m²). The actual potential must be still higher because no account could be taken of assimilates released or of the pieces of dead plant lost during the period between the measurements.

Petersen (1913) estimated the annual production of *Zostera marina* by doubling the weight present in late summer. For the years up to the beginning of this century he arrived at values of 1920, 1120 and 544 g dry matter/m² depending upon the nature of the different localities (according to Grøntved 1958).

In a widely distributed plant community of *Caulerpa* and *Cymodocea* at depths of about 4–6 m in the Mediterranean, Gessner & Hammer (1960) found a production potential in *Caulerpa* of 3·2 and 5·4 g C/m²/day, depending on different amounts of light. In mixed populations of these plants the potential was 10·4 and 18·7 g C/m²/day, so *Cymodocea* evidently has a significantly higher assimilation potential than *Caulerpa*.

Determinations of the standing stock of the macroflora, which have been done several times, do not provide any measure of the productivity which certainly varies considerably, depending on the locality, the latitude and the specific composition of the vegetation. If, however, one takes a value of only 2 g C/m²/day this is some ten times higher than the basic figure used by Steemann Nielsen in his calculation of the total potential of the phytoplankton per square metre of sea surface. If one further assumes that the autotrophic flora occupies a band 10 m wide, one obtains a potential of 7 t C/km coast/year, and taking the total length of coast as 369,000 km (excluding Antarctica, according to the Geographisches Taschenbuch 1953), the total production will be about $2 \cdot 5 \times 10^6$ t C/year. Compared with the equally speculative value given by Steemann Nielsen this is 40% less. This form of vegetation does not, therefore, provide any supply of primary food in oceanic regions, that is, at great distances from the coasts and there are also extensive areas of coast which lack macrovegetation.

The role of the neritic macroflora as a source of food is different from that of the phytoplankton. Whereas a large part of the phytoplankton is grazed while alive by herbivorous animals, the macrophagous herbivores are relatively few in number: among these are the sirenians which, following the extermination of Steller's sea cow, are now restricted to warm-water areas, and the marine iguanas (*Amblyrhynchus*) of the Galapagos Islands. Because of their restricted distribution both these forms are of only local importance in the utilization of macrovegetation production. Of greater importance, particularly in warm seas, are the numerous and widely distributed species of fish belonging to the families Mugilidae, Scaridae and Siganidae which feed, at least in part, on encrusting algae or flat algal mats on rocky coasts, biting or gnawing off pieces. On the rocky coast of Heron Island in the Capricorn Group, Stephenson & Searles (1960) observed that the algal vegetation is controlled by Siganidae and Scaridae in particular. Experimental exclusion of these fishes resulted in the growth form of *Calothrix* changing from flat films or mats to taller mats; this was followed by the settlement of blue-green algae and other green algae and the old mats of *Calothrix* lifted off and were destroyed. In other words, a succession took place which did not occur when the fish were present. It would be very valuable to have similar experimental investigations made in several other places.

Various molluscs (*Chiton, Lacuna, Helcion, Aplysia, Trochus,* etc.) have been observed grazing on algal mats growing on firm substrates and some echinoids also do this. Even nereid polychaetes, which were originally regarded only as predators on account of their jaws, feed on algal fragments or pieces of eel-grass leaves, as do isopods of the genus *Idothea* (Remane 1960). According to Stephenson & Searles a large specimen of the gastropod *Acanthozostera* requires the *Calothrix* production from an area $\frac{1}{3}$ m², which it will keep under control (see p. 230 et seq.).

On the other hand, the grazing of large algae such as *Fucus, Laminaria, Macrocystis* etc. has never been observed to any significant extent. Evidently the production of this kind of neritic vegetation enters the general food cycle in other ways: (*a*) by the release of assimilates into the water, which are then available to the heterotrophic microflora and perhaps also to some protozoans, (b) by contributing to the formation of detritus, dead pieces of the plants being rubbed off into the water where they are utilized by filterers and sediment feeders. Detritus is also of indirect importance because it provides a substrate for bacteria, yeasts and fungi.

In areas near the coast the detritus content is considerable and is usually several times greater than the amount of living plankton present (phytoplankton, zooplankton and heterotrophs). The following figures are based on analyses made by Krey (1952a, 1953):

TABLE 41

	Living plankton	Detritus
Western Baltic Sea	184	814
Central area of English coastal waters in the south-western North Sea	383	4567
Central area of English Channel water of Atlantic origin	255·6	324

These figures must, however, be regarded as subject to considerable fluctuation, so it is not possible to use them in estimating the relative importance of the detritus component and of the phytoplankton for the subsequent links in the food chain.

Coral reefs appear to be of great interest from the viewpoint of biological productivity; the method of investigation so far used, involving the determination of the oxygen content of the water, is very dubious because it does not take sufficient account of the importance of turbulence in the exchange of oxygen between air and water. Some authors are of the opinion that the primary production would itself be great enough for the maintenance of the whole reef community, that a reef is more or less self-supporting (Fig. 167). However, the plant populations on a reef are very scattered and as far as I know there have been no quantitative investigations on the assimilating microflora. Gordon & Kelly (1962) came to the conclusion that one coral reef examined by them was in no way self-supporting, but they suggested that there were possibly basic differences between reefs near to large islands or the mainland and those that are oceanic.

Odum & Odum (1955) analysed the primary producers on a coral reef at Eniwetok Atoll and found zooxanthellae and filamentous algae in coral polyps and other animals (e.g. lamellibranchs), encrusting and bushy algae particularly on the dead parts of the reef, small filamentous algae in and on broken-off coral fragments, and a sometimes dense mat of filamentous algae, particularly inside the coral skeletons. For each of their sample areas of 43·56 m² they calculated that the mean bry biomass of producers was 703 g, but the productivity itself

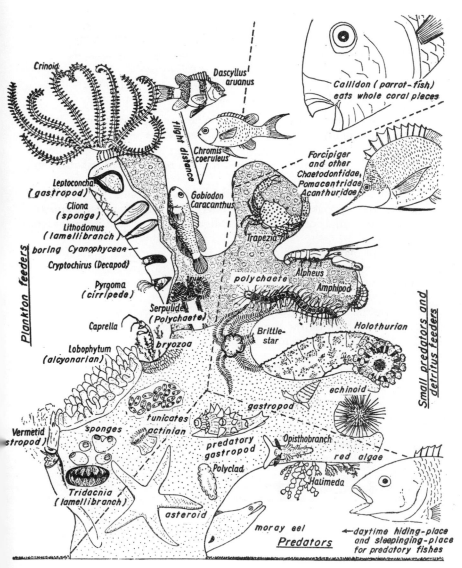

Figure 167. Sketch to show the most important members of the macrofauna living in and on live and dead corals in the Maldives (after Gerlach 1960)

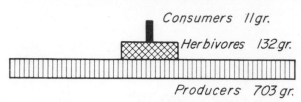

Figure 168. Average biomass (dry weight from a quadrate of 20×20 feet = 43·45 m²) on a coral reef in Eniwetok Atoll (after Odum & Odum 1955)

could not be estimated (Fig. 168). Whether the productivity is sufficient to maintain a biomass of 132 g of herbivores and 11 g of carnivores
cannot yet be assessed. Hiatt & Strasburg (1960) have produced a
diagram showing the food relationships on a reef in the Marshall Islands
(Fig. 169).

It should be mentioned that Siepmann & Höhnk (1962) isolated
yeasts from sponges and holothurians. This might suggest a function
analogous to that of the symbiotic algae but there is no firm evidence
for this.

The primary production of food for the animals is predominantly
in the form of microscopic pelagic algae and detritus and it follows
that the first direct users of this are the filterers and sediment feeders.
The relation between the amount of available food and the growth
and reproduction of the consumers is however in no way linear
(Fig. 170), because the growth of the animals is a process which depends upon several factors. Food is only one of these; another impor-

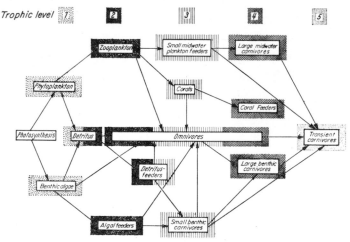

Figure 169. The food web on a coral reef in the Marshall Islands, with special
reference to the fishes (after Hiatt & Strasburg 1960)

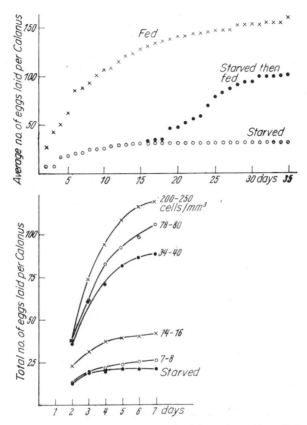

Figure 170. Dependence of egg production in *Calanus finmarchicus* on the available food; (a) comparison of starved and fed animals, (b) effect of feeding with *Chlamydomonas* cultures of different concentrations

tant factor is temperature and there are others which have not so far been fully investigated.

From an analysis of numerous length measurements of calanoid copepods, Deevey (1960 b) found correlations between body length on the one hand and temperature and phytoplankton on the other (Fig. 171). Within their total area of distribution the individual species showed differences both in the order of size of their length variations and in their dependence on the two different factors. These regional differences are determined by the mean annual amplitude of the temperature range, by the number of maxima per year of the phytoplankton and by its seasonal distribution. The mean body length of a given

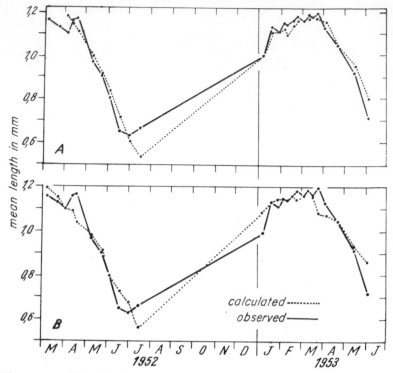

Figure 171. The calculated and observed mean cephalothorax lengths of female *Temora longicornis* in Long Island Sound: (A) calculated lengths based on mean temperature and chlorophyll content in the month previous to sampling; (B) calculated lengths based on the temperature and chlorophyll on the day of sampling (after Deevey 1960b)

population was generally correlated with conditions in the month previous to sampling.

The correlations with temperature are of particular interest and importance because they evidently do not represent a simple variable. According to various evidence the temperatures during the period in which the eggs are laid or the young hatch more or less determine the growth and rate of development of the young.

Taking the lengths of these animals as an indication of body weight, it follows from what has been said that, if hatching and early development takes place when temperatures are relatively high and the phytoplankton is simultaneously poor, then this growing generation will subsequently have lower weights, even if conditions have in the meantime become more favourable. Since reproductive cycles and development can also be correlated with other factors, e.g. light, it follows that the process by which the harbivores are produced is a complex one and not dependent upon the phytoplankton alone. We still lack detailed information on this, although a knowledge of the process is indispensable for any assessment of food chains. From what has been said it follows that investigations on phytoplankton production that are widely separated in space and time do not by themselves tell us very much about the production of the consumers.

Investigations on biological productivity in the pelagial, as in the benthal, are mainly concerned with the problem of the flow of matter, which starts with assimilation and branches out into the numerous links in the food chain of consumers. For the consumers, however, the decisive factor is not so much the weight of food taken in as its content of potential energy. Consequently, the true aim of productivity biology should be the determination of energy flow; this, however, presents great difficulties, not only in the methods employed but also in more fundamental aspects, because measurements of, say, the energy flow alone will not help us to estimate the nutritional importance of substances such as vitamins. Davis (1963) has given definitions of possible terms that can be used in discussing energy flow and productivity.

2. Vertical migrations

Many observations on pelagic animals have shown that their association with a given horizon is only relative. The same species may at one time be restricted to the surface layers, at another it may be more or less uniformly distributed throughout a large water column, or it may be found exclusively in deeper horizons; this may happen at one and the same station and not only in widely separated regions. In freshwater organisms it was recognized as early as the beginning of the 19th century that these changes followed a certain pattern: in bright daylight cladocerans stay in deep water, but in cloudy weather and at night they are found near the surface. There are numerous observations from the marine pelagial which show that vertical movements take place according to a daily and also a seasonal rhythm. These phenomena may be considered from four closely connected points of view:

(1) Their course and occurrence in the most diverse species and groups.
(2) An analysis of the factors causing them.
(3) Their possible significance in the biology of the species.
(4) Their significance in biological processes in general.

(a) *Diurnal vertical migrations*

In any discussion of vertical migration it is the diurnal movements which must occupy the greatest space. These are by far the most wide-spread, can be studied in all parts of the sea as well as in fresh waters and are very suitable for analysis on account of their short periods. The comprehensive works of Rose (1926), Russell (1927) and Cushing (1951) are therefore concerned almost entirely with diurnal vertical movements. It therefore seems unnecessary to give details of the numerous observations which have shown the existence of this phenomenon and I will merely refer to these exhaustive summaries. Franz (1910) questioned the reality of vertical migration; he suggested that during the day the animals see the plankton net and can take avoiding action, so that the different amounts of plankton taken by day or night would only be apparent. Needless to say this view is now known to be completely erroneous.

There are various possible methods of determining a diurnal rhythm in the vertical distribution of single species or of the total plankton. One can either fish a given horizon at regular intervals throughout the 24 hours, using a horizontal net, and determine the quantitative changes numerically, or one can fish different horizons at different times of the day, using vertical closing nets and then compare the numerical results. Both these methods can be used separately or combined. Clarke (1933 b) sampled five horizons simultaneously, using nets set at intervals one above the other and in this way obtained good comparative data on the simultaneous occurrence of species at different levels. Dietz (1948) reported observations in the Pacific and Atlantic Oceans in which echograph traces taken at several stations showed a reflecting layer which shifts up and down in accordance with a daily rhythm.

The general conclusion can be expressed quite briefly: many pelagic animals live by day at greater depths than during the night, that is, they carry out an upward and a downward movement during the 24 hours (Fig. 172). If one takes the daytime depth as the norm, it can be shown that under the same weather conditions in the same place this will be different for different species, and that the same species may remain at different horizons from day to day depending upon the weather conditions. Furthermore, the amount of the vertical movement varies from species to species and within a single species from day to day, from one season to another, and also according to the stage of development and the sex. As a general rule, it has only been possible hitherto to state that the horizon occupied during the daytime will be deeper in bright sunny weather than when the sky is cloudy or overcast, and that juvenile stages generally live closer to the surface than adults. In my view it would be valuable to extend research on this

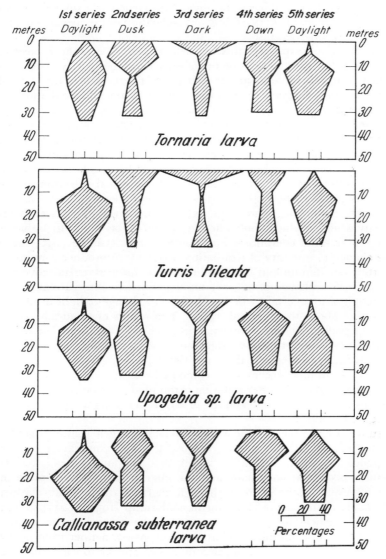

Figure 172. Vertical distribution of zooplankton in relation to time of day on 15–16 July 1924 in the area off Plymouth (after Russell 1925)

problem by carrying out systematic investigations using echo-sounders combined with the different net methods, particularly since the echo-sounder is incapable of providing any indication of the specific identity of the reflecting organisms.

As might be expected, this problem has also been the subject of experimental investigation; plenty of suitable material is available and much of it can be kept in the aquarium. Furthermore, the diurnal rhythm—the change from bright or weak light to complete darkness—lends itself to physiological investigation. Of course, as Russell (1927) pointed out, we must realize that we still have no data on the normal light intensity at the depths in which the animals spend the day. In addition, there are very few experimental investigations which take account of the effect of qualitative changes in light, due to absorption by the water (see p. 51 et seq.), on the physiological behaviour of the animals. And finally, we know almost nothing about the wave-lengths in composite or monochromatic light or about the differences in intensity which produce photokinetic or other effects under the same or varying external conditions. For *Daphnia magna*, Clarke (1933) gave the change in intensity of irradiation necessary to produce a response as 16% per minute and found that *Metridia lucens* starts to move upwards with a change of 4% per minute. Further observations would be required to ascertain whether the rate of change of light in the sea ever rises above this threshold value. The results of Friedrich (1931) on copepods provide no information since they involved horizontal movements. Unfortunately, therefore, the numerous experimental investigations carried out from the time of Loeb (1894) up to Rose (1926) and in more recent years can only be used with strong reservations as a basis for the interpretation of this phenomenon. These investigations do not in fact touch on the problems of comparative physiology that are involved.

In addition to the physiology we also lack any general investigation on the nature of light as an ecological factor. Temperature can be excluded as an effective factor in this field, because water is an extremely poor heat conductor and the migrations may take place in homothermal water or pass through thermal discontinuity layers. Although, in the case of Hydromedusae, Kramp (1959, p. 241) stated: 'Diurnal vertical movements frequently take place but are usually barred by a discontinuity which is not surpassed, neither from below nor from above.' Vertical migrations would appear to show that the organisms concerned are forced to remain within a zone of illumination that is appropriate to them. The pressure, like the amount of light, also changes with depth, but except in fishes we know of no relevant sense organs in marine animals, and only in a few instances have reactions to pressure been found (see p. 78), so that this factor must surely be excluded from any general explanation. Harris (1953) pointed out that the light intensity changes logarithmically with increasing depth (see p. 52 et seq.) and the reaction of the sense organs is also often proportional to the logarithm of the light intensity. As a result the physiological reaction of the animals to light implies a linear relationship to the depth. Light is therefore to be regarded as the decisive factor. In accordance with the hypothesis of Russell, and also of Cushing among others, and following Michael (1911), it can now be said that the different species of pelagic animals require a zone of optimal illumination, which they seek during the day. As the sun goes down the intensity of light in the populated layers is reduced and the optimal zone rises. The animals follow this zone towards the surface and then when the sun rises again and the intensity increases they move down again into deeper water. As far as I know this hypothesis was first put forward by Rose.

At first there were some objections to this interpretation owing to certain apparently contradictory facts. It was observed that the upward movement frequently starts during the afternoon, that in polar regions the zooplankton shows the same rhythm even in summer when it is light the whole time, and that the downward movement frequently starts around midnight, that is, at the time of greatest darkness. Russell (1927) showed convincingly that the first two points did in fact support the hypothesis: as the sun goes down the zone of optimal illumination is already moving upwards during the afternoon, and in polar seas the sun is so low during the night that only a small amount of the light striking the water actually penetrates the surface. Mohn (1905, cited in Russell 1927, p. 256) gives the sum of sun's radiation during June at 83° 6 N as 5·12 J/m²/min at noon and as only 2·47J/m²/min at midnight, so that there are quite large fluctuations in intensity during a 24-hour period. However, the matter is apparently more complicated in the polar regions. Bogorov (1946) has emphasized that the diurnal vertical migrations undergo marked seasonal change, and that there are considerable differences between the Barents Sea and the White Sea. In the Barents Sea during summer the vertical distribution remains almost unchanged, whereas in the White Sea migrations have been observed. There seems, however, to be no information on the vertical distribution during the long period of darkness in the winter months.

Moore (1950) made particularly clear the dependence of the vertical distribution on a given light intensity, because he found the position of the deep scattering layer was correlated with the iso-illumination curves (Fig. 173).

Cushing has shown very elegantly that the downward movements, which often start at about midnight, do in fact support the hypothesis.

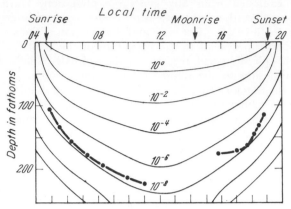

Figure 173. Diurnal movement of the scattering layer, when its depth is dependent solely upon the light (after Moore 1950)

From numerous observations it is known that many pelagic animals in the sea and fresh waters swim faster in light than in complete darkness, that is, they show similar photokinetic reactions to those shown by many terrestrial animals. Consequently, during the time of greatest darkness the animals move slower and therefore gradually sink because of their overweight. At dawn the zone of optimal illumination will lie above the animals and they will swim towards it, so in the morning hours there must be a second ascent, which is then followed, as the light increases, by an active descent associated with the daytime conditions of illumination. In several investigations an upward movement has actually been observed at dawn; these observations were made at widely separated times and places and so would support Cushing's interpretation. Investigations carried out at precise times and depths would presumably confirm this particular aspect of the diurnal ascents and descents. According to Clarke, *Metridia lucens* migrates further upwards during the time of greatest darkness so that here other factors may be involved. In this investigation (l.c., p. 426, Fig. 5) there were, however, no observations between 2126 and 0337 hours, which is the critical period.

Ussing (1938) has postulated that the seasonal vertical migrations of *Calanus finmarchicus* and of other forms involve photokinetic and phototrophic reactions.

In summary, therefore, we can say that many pelagic animals carry out daily vertical migrations, involving an ascent towards the surface during the evening and night and a descent into deeper layers in the morning and during the day. In a number of cases that have been investigated in detail this rhythm shows the following components:

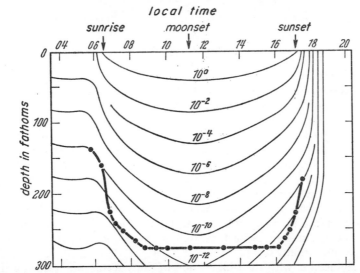

Figure 174. Diurnal movement of the scattering layer, when its rise and fall are dependent upon the light but its depth is dependent upon a temperature layer (after Moore 1950)

ascent with decreasing light during the afternoon and evening, sinking during the period of greatest darkness, a small ascent again at dawn, and a descent to the daytime level during the morning hours. The significant factor in these movements is evidently the light, it being assumed that the animals prefer an optimal zone of illumination with a specific light intensity. Further investigations are needed to show the extent to which these movements may be modified and how far the idea of a zone of optimal illumination can be confirmed.

According to Moore (1950) the depth taken up during the day appears in some cases to be determined by the temperature. He found a deep scattering layer which first sank as the sun rose, but then remained throughout the day at a certain depth in spite of increasing illumination, and then ascended again in the evening. Evidently temperature can be the only factor responsible for the straight line of this deep daytime level. According to Owre (1960) it is probable 'that the level of the upper part of the spread of chaetognaths is controlled predominantly by light and that temperature predominantly controls the lower levels' (l.c., p. 312).

These points are concerned solely with the observational side of the problem. There is an extensive literature on the physiological problems, on how light affects the organisms, but it would not be possible to give a detailed summary of the many different interpretations. Loeb (1904)

M—M.B.

considered that the vertical movement was governed by a simple positive and fluctuating phototropism. He later came to the conclusion that it was also positive and negative geotropism, due to changed light intensity, which determined the direction of movement. Esterly (1911) considered that it was unlikely that the light itself was responsible for the direction, but that the primary cause was the effect of light on the geotropism of the animals.

Ostwald (1902) thought that the nocturnal ascent was due to increasing density of the water at night due to cooling, but overlooked the fact that the daily fluctuations in water temperature are very small (see p. 63). According to this interpretation the organisms were to be regarded simply as physical objects with a given specific gravity. But one of the essential features of living organisms is their ability to regulate, which to varying degrees makes them independent of changes in the physical and chemical conditions. This point should under no circumstances be forgotten when considering the causes of vertical migration.

In connection with the observation of diurnal migration during the summer months in polar regions, and of other phenomena showing a diurnal rhythm it has been suggested that vertical migrations depend upon an autonomous physiological rhythm (see below). Esterly (1917b) reported that under experimental conditions the copepods *Acartia tonsa* and *A. clausi* even show a rhythmic change in vertical distribution when kept in complete darkness (see Table 42).

TABLE 42

Percentage distribution of the copepods *Acastia tonsa* and *A. clausi* in a column of water in darkness to show the effect of time of day.

(after Esterly 1917b)

(The intermediate stages given by Esterly have not been included; so that the sum of the percentages is not 100)

Time	A. tonsa		A. clausi	
	above	below	above	below
08.00–12.00	0	92	31	55
12.01–15.59	13	81	20	66
16.00–17.59	37	62	16	70
18.00–20.00	62	22	45	37

Moreover, in *A. tonsa* this behaviour is different at temperatures above and below 16°C. We do not, of course, know how far the behaviour of these animals in a glass tube 50 cm tall and 3 cm in diameter is comparable with that in open water.

A few years ago Hardy (1953) reported the results of some particularly interesting investigations. He used vertical glass cylinders one metre long which were divided in the middle by a door. This door could be opened by a falling weight and then closed again. The experimental animals, mostly *Calanus*, were placed into the upper or lower chamber of the cylinder under very low illumination and the cylinder was darkened so that the animals would not be affected by the daylight. The cylinder was then submerged to a given depth. After removal of the darkening the door inside the cylinder was opened by the falling weight and after about one hour closed again. The two chambers were then emptied separately and the number of animals in each compartment was counted. From the percentage distribution Hardy obtained data on the upward and downward movement at different depths and at different times of the day. In experiments at depths of 1, 5, 10, 20, 30 and 40 m using *Calanus* caught at even greater depths he observed that 'at each successive depth the members of the population segregated into those which moved upwards and those which moved downwards. The proportions of those moving upwards or downwards, while showing considerable variation between results on different days, on an average increase or decreased respectively in relation to depth. . . .' The idea that *Calanus* has a 'sense of depth' could not however be confirmed by laboratory experiments; Knight-Jones & Quasim (1955) found no reaction to pressure changes in *Calanus*. On the other hand, the importance of light was shown conclusively by an experiment in which the glass cylinder was darkened above and illuminated from below by a mirror: this resulted in a mirror-image distribution of the animals. In my opinion, however, there is no need for the postulation of a 'sense of depth'. The animals had indeed been caught at a greater depth and were therefore adapted to a low light intensity, and they must certainly have more or less retained this adapted condition, especially since the effect of direct daylight was excluded during their manipulation. In their new environment a new adaptation must therefore have taken place. It is not surprising that the animals then moved, some upwards, some downwards, and this may be compared with the reaction of *Daphnia* to an increase or decrease in lateral illumination; some of the animals react positively, others negatively. So far the only possible interpretation is that the animals are in different physiological conditions and therefore react differently. It is true, of course, that the expression 'different physiological condition' cannot be defined quantitatively.

These investigations have in fact only confirmed the importance of light in vertical migration. Nevertheless, one cannot exclude the possibility that other factors may also be involved. One could, for example, test whether the activities of the phytoplankton might play a part as a releasing factor, whether by assimilation during the day or by reproduction which mostly takes place at night. From investigations on *Paramecium* and pluteus larvae (Fox 1925 and others) it is known that fluorescent substances strongly affect the reactions to light and in certain circumstances light may even become a lethal factor. One could test whether the migrations could be determined by feeding, in that the animals ascending into the surface layers at night may find the phytoplankton in a different physiological condition than during the day. If this were true the light would again be of decisive, although indirect, importance. And finally, it is worth noting that, e.g. in *Calanus finmarchicus*, the physiological behaviour of succeeding generations differs, so that it is not possible to produce a clear analysis from a population consisting of the members of two generations.

So far, therefore, it has not been possible to put forward a generally valid theory to explain these diurnal migrations. The position is particularly obscure in view of the fact that such migrations have also been found in animals living at great depths. Thus, Mackintosh (1937) reported that in Antarctic waters, the largest catches of the copepod *Pleuromamma robusta* were at 100–50 m during the day, but at 750–500 m at night. Using a large closing net, Welsh *et al.* (1937) made horizontal hauls at depths of 400 m and 800 m in the Sargasso Sea and found that the greatest abundance of the acanthephyrids *Systellaspis debilis* and *Acanthephyra purpurea*, of sergestids, copepods, *Sagitta hexaptera*, fishes, etc., was at 400 m during the night but at 800 m by day. According to Waterman *et al.* (1939) in the western North Atlantic the decapods *Gennadas elegans* and *Thysanopoda acutifrons* also migrate into great depths, approximately to the region between 100 and 1200 metres. Dietz (1948) has found that the deep reflecting layer, presumably caused by organisms, sometimes lies at 800 metres. Welsh *et al.* (1937) have shown that at this depth light is the only factor which shows fluctuations between day and night. Since the downward movements have already started before the increase in light intensity, these authors considered the idea of physiological rhythms. Welsh (1935) had already shown in four different meropelagic crustaceans (Hippolytidae, Palaemonidae, Penaeidae) that rhythmical displacement of eye pigment takes place even under constant light conditions; there were also differences in behaviour between the different eye pigments in the species concerned. Diurnal rhythms have been confirmed in various marine organisms, particularly in fishes, and Harris (1963) has shown activity rhythms in *Daphnia magna* and *Calanus helgolandicus*. In both species there is an increase in activity during darkness. If an increase in activity takes the form of negative geotropism it follows that an upward movement must take place during darkness.

Here conditions are presumably similar to what happens in many plants and animals, and even man, which have a 'physiological clock' (cf. Bünning 1958). Investigations on other planktonic organisms will certainly provide more detailed information on the physiological basis of vertical migration.

There are few exact data on the distances covered during diurnal migrations. Esterly (1911) found the largest accumulations of *Calanus finmarchicus* in June–July at 360 m during the day and near the surface at night. These animals, therefore, had to travel a distance of several hundred metres in the course of 24 hours, an astonishing performance for such a small organism. Waterman *et al.* (l.c.) (1939) considered that *Gennadas elegans* covered about 40 m in one direction, while *Thysanopoda acutifrons* travelled 600 m in the 24 hours.

The speed at which these animals move is evidently quite considerable. Welsh (1933) reported a speed of 82m/hr in the laboratory for the copepod *Centropages*; in species of *Euphausia* Hardy & Gunther recorded 100–200 m/hr, and Moore (1949a) found speeds of more than 60 m/hr in various other animals.

Hardy & Bainbridge (according to Hardy 1953) placed various planktonic animals in a circular tube and turned this against the direction in which the animals were swimming so that the distances swum could be measured. By means of special opening and closing doors the water was made to move at the same speed as the tube. These authors found that *Calanus* and the nauplii of *Balanus* covered about 16 m/hr, *Centropages* about 30 m/hr and the euphausian *Meganyctiphanes norvegica* about 100 m/hr. From echo-sounder traces, Dietz estimated that the scattering layer descended during the morning at a speed of about 250 m/hr, and that this was exceeded by the ascent for which he recorded a speed of 300 m/hr.

The extent of the daily migrations will be apparent from these speeds. Whatever the factors are that stimulate and control these movements we must regard them as forming an important part of the overall behaviour of these animals. The herbivorous forms, the consumers of phytoplankton, are taken by the upward movements into the euphotic zone with its rich food supply, and then as the light increases they move downwards into depths in which they are less visible to predators that hunt by sight. Carnivorous animals prey on the herbivores and it is conceivable that this rhythm occurs from the surface layers down to the dark deep layers. The feeding of the herbivores can, therefore, be regarded as the initial process and the subsequent participation of the carnivores is an expression of the biological links in the food chain. The ascent of the animals from the deeper layers into the euphotic productive zone also means that mineral nutrients will be carried up again into the productive layers. Gardiner (1937) confirmed that the P_2O content in a closed volume of water rose during the course of a 3-hour experiment when living animals were present. So it is not only hydrographic conditions which are responsible for the routine restoration of the inorganic plant nutrients. This phenomenon may be of considerable importance, particularly from the quantitative point of view, to the biological productivity in areas poor in nutrients.

If we can regard the search for horizons rich in food and a 'need for protection' as the original starting points for these diurnal vertical migrations, we must remember that the extent of the upward movement to reach the zones with food must be restricted by the swimming ability of the organisms concerned. This means that there must be a limit to the depth to be reached and in many cases this is evidently effected by light-sensitive organs and is associated with an optimal light requirement that is presumably innate; in some cases the limit may be set by the animals reaching a given temperature.

Vertical migrations have various side-effects which may in some circumstances increase their biological importance. Hardy & Gunther (1935) have given an example of this, which agrees with Mackintosh's explanation of seasonal vertical migrations (see p. 350). This example is based on the fact that planktonic organisms are only capable of small, localized movements and cannot therefore undertake large-scale migrations, and furthermore the direction and strength of the currents varies at different depths. In Fig. 36 the fine continuous arrows represent the surface currents, the fine broken arrows the currents in deeper water. The heavy lines A–A1, B–B1, C–C1 show the distances which the organisms will be transported if they live by day (heavy continuous lines) in the upper layers, by night (heavy broken lines) in deeper water. Fish (1936 b, p. 197, footnote) has written: 'It is probable that the transfer of adults from one area to another is relatively much slower than that of larvae because, by descending during the daylight hours, they pass beyond the influence of the surface into a slower dominant drift.' It is apparent from such a diagram that the animals may be transported into completely different water masses and that vertical migration gives them a greater independence of the environment.

Many animals appear to have a close association with certain water masses (see p. 356 et seq.) and have therefore been used as indicator species for determining the origin of these water masses and their movements. In some circumstances, however, vertical migration may transport the animals into different water masses and this seems to limit the value of such animals as indicator species. This problem requires further elucidation because it casts certain doubts on the view of Russell (1939, p. 171) that 'these organisms form ideal drift bottles'.

The whole problem of diurnal vertical migration therefore offers a wide field for research, involving direct observation, the recognition of the exogenous and endogenous factors which influence the rhythm and extent of the migrations and the assessment of the general biological significance of these movements. As Hardy has so aptly said, these problems pose 'planktonic puzzle number 1'.

In many ways the analysis of vertical migration in freshwater planktonic organisms is further advanced. Thus, Siebeck (1960) has published a detailed investigation on the planktonic crustaceans of the Lunzer Lake, in which, amongst other things, he comes to the conclusion that the depth of the plankton maximum is determined not so much by an absolute optimum as by an ability to adapt and by the 'previous history of the light conditions' (l.c., p. 451). A more detailed treatment of the whole subject than is possible here would enable a better comparison to be made between freshwater and marine plankton.

(b) *Seasonal vertical migrations*

On the basis of the older observations of Schmidtlein and Lo Bianco, Chun (1888) was the first to point out clearly that in the Mediterranean many forms are absent from the surface layers during the summer months, but occur there abundantly during the winter. By using closing nets he showed that these species were present at greater depths during the summer.

Since then this type of seasonal shift in vertical distribution has been found in several different organisms in many different parts of the sea. Here, however, we can only consider a few of the numerous observations that have been made, particularly as some of these were chance observations rather than systematic investigations. Brandt, Lohmann (1909a) and particularly Rose (1925) made a number of observations in the Mediterranean Sea; Nikitine (1929) investigated the position in the Black Sea as regards *Pleurobrachia, Sagitta,* copepods and cladocerans; Russell (1926 et seq.) worked in the sea area off Plymouth; Sømme (1934) made observations in the Lofoten Islands, and Ussing (1938) has produced data for East Greenland waters which have been extended by Digby (1954). Owre showed seasonal fluctuations in chaetognaths.

Particularly detailed investigations have been made by Mackintosh (1937) on the macroplankton in the South Atlantic in the area of the 80th meridian West, between 55 and 68°S. In December 1933 and in March, September, October and November 1934 he carried out hydrographic and plankton profiles, taking closing-net samples in depths of 50 to 0 m, 100 to 50 m, 250 to 100 m, 500 to 250 m, 750 to 500 m, 1,000 to 750 m and also at three stations in depths of 1,500 to 1,000 m. From the 284 samples taken, some 66,234 animals were counted, some being identified down to the species, others only to the genus or group. The forms present in this wealth of material could be divided according to their habits into:

1. Those found in summer in the surface waters but in winter in much deeper layers (the main species present were *Rhincalanus gigas* (26%), *Eukrohnia hamata* (over 18%) and *Calanus acutus* (over 13%)); this has also been confirmed from other observations in antarctic waters;

2. The species *Sagitta gazellae, Calanus propinquus, Parathemisto gaudichaudi* were found constantly in the surface layers and no diurnal migrations could be established;

3. The species *Sagitta maxima, S. planktonis* and others evidently live only in the deeper layers and they too show no diurnal rhythm in their distributions;

4. Marked diurnal changes in level without any recognizable seasonal rhythm were shown by *Pleuromamma robusta, Euphausia truncata, E. vallentini, E. frigida.*

The investigations of Ussing (1938) in East Greenland waters have confirmed that the various species living within a given area may behave in different ways. During the summer months *Calanus finmarchicus, C. hyperboraeus, Pseudocalanus minutus* and *Oncaea borealis* showed a maximum at depths of 0 to 500 m, whereas for *Microcalanus pygmaeus* and *Metridia longa* this zone showed a summer minimum and a late autumn maximum. Samples taken at depths of 100 to 200 m showed the reverse, so a true seasonal migration takes place (cf. also the findings of Digby 1954).

The work of Nikitine has shown a similar situation in the Black Sea: during the cold season *Calanus finmarchicus, Pseudocalanus elongatus, Oithona similis* and *Sagitta maxima* are present at all depths but during the summer they disappear almost completely from the upper 15 to 25 metres. Conversely, *Centropages kroyeri, Evadne nordmanni* and *E. spinifera* occur near the surface during the warm season, but disappear as the water cools and cannot be found there during the coldest period of the year. Presumably in the three last-named species this is a case of seasonal occurrence, rather than vertical migration.

Any explanation of the factors causing seasonal vertical movements is complicated by the fact that the same species can evidently behave differently in different areas. Thus, Russell (1926) observed that off Plymouth, *Calanus finmarchicus* is at its deepest horizon in June and starts to ascend in July–August. Since the average illumination has already started to decrease in July–August, whereas the water temperature does not reach its maximum until August–September, Russell regarded light as the decisive factor. This interpretation is strengthened by the fact, emphasized particularly by Clarke (1938), that there are very great fluctuations in light conditions during the course of the year, even at a depth of 15 m (see p. 51). Conversely, however, the observations of Nikitine (1929) and of Esterly (1911, 1912) on *Calanus finmarchicus* suggest that these animals move downwards as the water warms up, and both these authors concluded that this species had its upper limit of temperature at about 15°C. In my opinion there is no reason to doubt the accuracy of these conflicting results, and it seems to me that one must draw the following conclusion: if the animals are living in a temperature range that is suitable for them, then light will act as the primary factor controlling seasonal vertical distribution, but if there are fluctuations of temperature which extend beyond the limits of tolerance then temperature will become the deciding factor. This view is supported by the observations of Damas (1905), Sømme (1934) and Ussing (1938) who found that, in the Lofoten area and in East Greenland, *Calanus finmarchicus* and some other species remain in the upper layers during the summer but move down into deeper water in the winter. According to Digby (1954), *Metridia*, which spends the summer below 50 metres, ascends to the surface layers in winter.

Chun considered that in Mediterranean forms the descent into deeper water during summer was due to the increase in temperature, whereas Lo Bianco held that the absence of ascending currents in summer kept the animals in deep water more or less mechanically. Since, however, the appendicularian species behave differently, although they are subject to the same mechanical action of currents, Lohmann (1909a) rejected this interpretation and considered that the migrations were due to the fact that the horizons rich in food varied in position according to the season. This is undoubtedly an important point, but one cannot enlarge upon it because the observational material is still insufficient.

Finally, there is one further point which must be mentioned. During the course of the year many planktonic organisms produce several generations, so in many cases the summer and winter animals belong to different generations. In 1953 Gauld & Raymont measured the respiration of the copepod *Centropages hamatus* and were able to distinguish two groups and to relate the differences between them to the different habits of two generations. Here it can be assumed that the general physiology of the summer and winter animals differs and this may be correlated with the seasonal vertical migrations. According to Harris (1963), in *Calanus helgolandicus* off Plymouth, the endogenous diurnal rhythm of activity is absent during the winter, and this will therefore affect the seasonal behaviour and the vertical distribution. We do not, however, know the factor which causes the biological clock to be altered in rate or switched off.

The behaviour of a species in its own environment may be interpreted not only from its autecology but also from its synecology and this suggests two further possibilities for future investigation:

1. the animals could be influenced by the course of the phytoplankton succession (see p. 397), so that they either escaped from it by downward migration or followed it upwards;

2. it is conceivable that of two or more species with contrasting behaviour only one carries out migrations caused by physical or chemical stimuli, whereas the other escapes from the competition for space and food by migrating in the opposite direction; in other words, one species reacts to abiotic, the other to biotic factors.

In any case it is clear that the seasonal migrations of some species are established facts, that these show very great variations and that the investigation of the factors regulating them, which are possibly of a regional nature, does not present insuperable difficulties. We are also not clear whether and to what extent animals living at even greater depths do not show similar movements. In view of the observations made on diurnal migration I believe that this is quite possible.

Mackintosh (l.c.) has written in detail on the possible effect of these seasonal changes in depth on the distribution of organisms, particularly of *Rhincalanus gigas*, *Calanus acutus* and *Eukrohnia hamata*. In the area of his investigations, the cold, relatively low-salinity surface layer (about 100 to 289 metres thick), moves northwards towards the antarctic convergence. Here it sinks below the subatlantic deep water which is moving in a southerly direction. If then the animals penetrate into this subatlantic deep water by ascending from below they will be again carried southwards towards the antarctic convergence, that is, into the area from which they originally came, although somewhat farther to the east. Boden & Kampa (1953), working in Bermuda, have investigated the possible existence of an endemic plankton population in the region of oceanic islands and have found, among other things, that the winter circulation is a cyclonic, convergent movement about a sinking centre which acts antagonistically to the vertical migrations, so that the zooplankton is again carried into the area of the currents which are converging on the island group.

Sømme (l.c.), like Mackintosh, believes that the downward movement of *Calanus finmarchicus* during winter in the Lofoten area serves to maintain the population. The speed of the currents is lower in the deeper water and this reduces the risk of the relatively small winter population being carried away. As a result, a reasonable population can develop again in the spring in the waters closer to the surface. In my opinion, however, this view is still inconclusive and should only be regarded as a working hypothesis. When all is said and done, diurnal vertical migrations could also have the same effect, provided that they were sufficiently extensive for the two opposing currents to have the same transporting power. On the other hand, it should be remembered that there are also species living in the area in question, for which we know of no migrations of this kind. Evidently therefore there are other factors which must be involved, in order to maintain a population in one area in spite of the transporting action of currents.

3. Fluctuations and lunar periodicity

In certain limited areas, fluctuations in the population-size of individual species have frequently been observed, but these observations have been restricted almost exclusively to regions in the vicinity of the coast because quantitative investigations over a period of years are necessary. The mere record of a fluctuation is not in itself sufficient; one must find out the original causes, and consider the reasons for the increase of the population above its average and its subsequent reduction. Hitherto it has only been possible to carry out a complete investigation in relatively few cases. (In this context see the remarks on 'red tides' p. 399 et seq.)

Segerstråle (1960) has described fluctuations in the Gulf of Finland in the populations of the lamellibranch *Macoma baltica*, of the amphipods *Corophium volutator* and *Pontoporeia affinis* and of chironomid larvae. The original causes are still not completely understood, but some contributory factors can be suggested. The chironomid larvae showed a maximum in 1928; the summer of 1927 was particularly warm and this may have favoured copulation and egg-laying. In *Corophium* there were indications that the reduction in the number of individuals from a maximum in 1928 to a minimum in 1930 was due to infection by a pathogenic yeast (*Cryptococcus gammari*). In the case of *Macoma baltica*, Segerstråle considered that *Pontoporeia* may have been involved, because the maxima and minima of the two species alternate with each other, the crustacean being capable of decimating the young brood of the lamellibranch, particularly during the period of transition from pelagic to benthonic life. He was unable to suggest any reasons for the fluctuations in the population of *Pontoporeia*.

Hagmeier (1930) ascribed a fluctuation in the population of *Mactra subtruncata* off the East Frisian coast to 'a mass influx of free-swimming larvae ready to settle', which found suitable living quarters in the area. An increase in the numbers of predators (asteroids, ophiuroids, gastropods) led to a quick decimation. The entry of large numbers of free-swimming larvae ripe for settlement could have taken place in several different ways. The larvae from several spawning areas could have been carried into the area by different currents and thus become mixed together; they could have been brought in from a

single breeding area by a more or less constant current at the time
of spawning and larval development; epidemic spawning could have
taken place in one spawning area and the water mass containing the
larvae could have reached the area of observation just at the time when
the larvae were ready to metamorphose. In any case this observed
fluctuation was probably the result of several factors.

In the years 1926, 1930, 1933 and 1936 Stephen (1938b) observed
large spatfalls of the lamellibranchs *Tellina tenuis* and *T. fabula* a
Kames Bay near Millport. These occurred in the years in which,
during the period from April to September, the algebraic sum of the

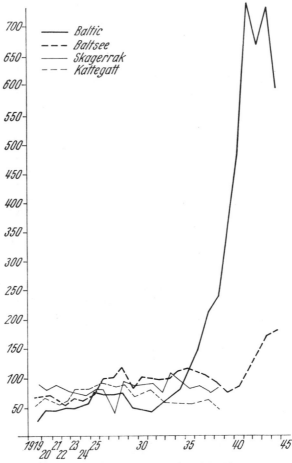

Figure 175.   Cod fishery catches in the Baltic Sea, Belt, Skagerrak and Kattegat from
1919–1944 (after Meyer & Kalle 1950)

deviations of the monthly temperature means from a standard set of means showed sharp rises. Evidently there was, therefore, a correlation with temperature.

According to Coe (1953) there have been seven mass spatfalls of the lamellibranch *Donax gouldi* on the Californian coast near La Jolla during a period of 58 years; when these occurred the populations were 1,000 to 10,000 times as large as the normal. In one case during the year following such a spatfall there were 20,000 animals per square metre in a band 2–5 metres wide and 5 miles long; within 3 years this immense population has been reduced by disease to a small remnant. Coe ascribed the occurrence of this mass spatfall to a combination of the following factors: 1, a large spawning population not too far away; 2, favourable conditions during the larval period; 3, favourable currents at exactly the time the larvae were ready to metamorphose; 4, favourable conditions for growth.

The fishery yields for the Baltic Sea, particularly of the Gotland Deep, provide an example that has been analysed in great detail. Here, during a period of thirty years, there was a marked biological change which was strikingly reflected in the curve for the yield of the cod fishery (Fig. 175). According to Meyer & Kalle (1950), the reasons for this must be sought in hydrographic changes. Data accumulated over a period of several years showed that there was a water mass in the Gotland Deep which had a low content of oxygen but a high value for phosphate (Fig. 176). The change in biological conditions coincided with changes in the chemistry of this deep water; these involved an increase in oxygen content and a decrease in phosphate (Fig. 176), and at the same time there were changes in temperature and salinity. These hydrographic changes were caused by the inflow of high-salinity Atlantic water poor in nutrients, which reached the

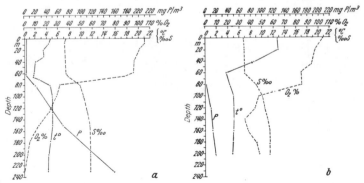

Figure 176. Distribution of phosphorus, oxygen, temperature and salinity in the Gotland Deep: (a) 25.5.1931, (b) 18.10.1934 (after Meyer & Kalle 1950)

Baltic through the Skagerrak and extended along the bottom down into the Gotland Deep, where it pushed the old deep water with its high nutrient content up into the productive surface layers. Thus the hydrographic changes resulted in the enrichment of the surface waters, leading to greater biological productivity, which was reflected in the yield of the cod fishery.

Population fluctuations have had profound and sometimes disastrous effects on commercial fisheries. Only in a few cases is it possible to find out the causes, because the hydrographic and general marine biological background is insufficiently known. Mention will only be made of the displacement of the spawning and fishing areas of the cod,

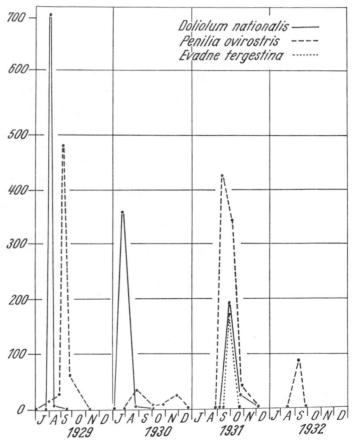

Figure 177. Mean total numbers per 10 minute tow (in thousands) of *Doliolum nationalis*, *Penilia ovirostris*, and *Evadne tergestina*, over the course of four years (after Deevey 1960b)

which is correlated with the increase in the temperature of arctic waters which has taken place during the last decades.

It is to be expected that fluctuations will generally be more extensive and more frequent in the marginal parts of the species' range. Thus, Radovich (1961) found that off the Californian coast, there was a temperature rise in 1957 after nine years of sub-average water temperatures. In 1958, 23 fish species and in 1959 26 species were observed to the north of their previously known range of distribution. A similar situation was found in the invertebrates, in other words, there was a fluctuation in the faunal composition of the area which can be traced back with a fair degree of accuracy to temperature fluctuations. This will naturally cause significant changes in the general biocoenotic structure of the region (Fig. 177). Southward (1963) has drawn attention to this problem in the area of the western English Channel where considerable changes have taken place in both the plankton and the benthos. These have been correlated with the temperature rise in the northern Atlantic already mentioned and with increased water movements associated with it. The most striking change in this area is the disappearance of the herring and the appearance of large numbers of pilchards; the larger populations of pelagic fish lead to a greater consumption of plankton and thus increased decimation of the pelagic larvae of benthonic animals, and so the whole ecology of the area undergoes change.

Several factors or combinations of factors can be suggested for the occurrence of non-periodic population fluctuations of this kind:

1. Changes in nutrient content or temperature produce conditions which, by changing the productive potentiality, bring about increases or decreases in the populations, or allow the immigration or emigration of certain species, thus affecting the biocoenotic structure.

2. Certain current conditions may bring about an aggregation of pelagic developmental stages, which may produce increased populations of certain species from time to time.

3. The developmental stages with the lowest ecological potentiality may encounter particularly favourable conditions during the sensitive phase, e.g. in the transition to independent feeding, so that the survival rate is above average and this will lead to correspondingly large year groups.

4. Epidemic diseases, parasitic infections or predatory enemies can seriously affect a population, and will generally quite quickly reduce one that has built up.

In the marine field the general problems of population dynamics have hitherto been studied almost exclusively by fisheries biologists, whose work must necessarily involve economic considerations as well

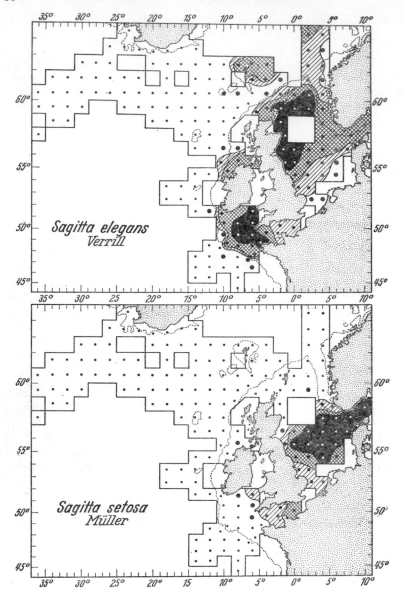

Figure 178. Distribution of four species of Chaetognatha in the North Atlantic and North Sea during 1958–1960. The dots represent the centre positions of the sampled rectangles; *small dots* indicate absence of the species, *large dots* indicate its presence:

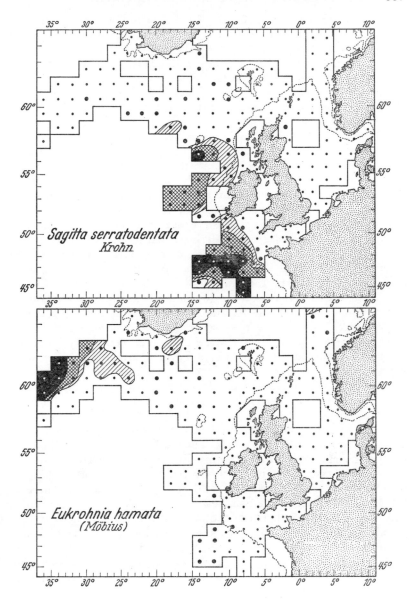

oblique lines, cross-hatching and *black areas* show increased abundance. These species serve as indicators of the water masses in which they live (after Bainbridge 1963)

as the estimation of natural factors. In comparison with work on land, marine research is relatively much more difficult, on account of the greater inaccessibility and the more indirect nature of the observations. At the present time we have almost no knowledge of fluctuations in oceanic areas far from the coasts.

In some parts of the pelagial marked fluctuations take place, which are due to the inflow of water masses which are not normal in the area. By this means, planktonic forms characteristic of these water masses are transported into areas where they are not endemic, giving rise to considerable population fluctuations. This has been shown to occur in the English Channel, the North Sea and in other areas where long-term observations have been made. The species associated with a given water mass indicate that an abnormal inflow of water has taken place and the route it has taken; they have therefore been termed indicator species (Figs. 177, 178). From their occurrence one can frequently draw conclusions on the increased inflow of oceanic water into neritic areas; in the North Sea, for example, they indicate the entry of North Atlantic water. From the fluctuations in the occurrence of pelagic tunicates in the waters off south-east Australia and in the Tasman Sea, Thompson (1948) inferred an increased inflow of warm water masses. These fluctuations are usually not restricted to single species, but are shown by several at the same time.

The biological phenomenon known as lunar periodicity has been known for a considerable time. Certain events, usually connected with reproduction, take place at the time of certain phases of the moon. The classical example is the palolo worm of the South Seas, which appears with such regularity at the last quarter of the moon in October and November (with an interval of 12 or 13 months), that in earlier times the natives of the islands used at these periods to go out and catch these edible animals off the coasts. The palolo is a part of the body of the polychaete *Palola viridis* (family Eunicidae) which contains the gonads. The front part of the body remains in its normal habitat while the remainder is released and swims to the surface where it breaks up and releases the gametes.

A time relationship between the onset of reproductive activity and certain phases of the moon has also been found in several other organisms, both plants and animals. Caspers (1951) has given a comprehensive review of this phenomenon and an analysis of the behaviour of the halobiont chironomid *Clunio marinus* which has also been investigated by Neumann (1963).

The larvae of this chironomid live in the eulittoral in algal cushions on rocky substrates. The imagines copulate and lay eggs immediately after emergence; they do not feed and die after a few hours. This short time span for a very important part of the life cycle presupposes a correspondingly close ecological adaptation to the local conditions. Investigations have revealed a double rhythm of emergence: maxima occur in the 24-hour period, that is, the insects emerge at certain times of the day, in Heligoland in the late afternoon, but in the Atlantic and Baltic populations not until the start of darkness when the periods of light and dark are both 12 hours. There is evidently a genetic difference between these two populations in respect of the relation of emergence time to the diurnal light/darkness change.

During the course of a month the 24-hour maxima occur at intervals of a fortnight, around the days immediately following the new and full moon (Fig. 179 A and B). In this way, the rhythm of emergence of

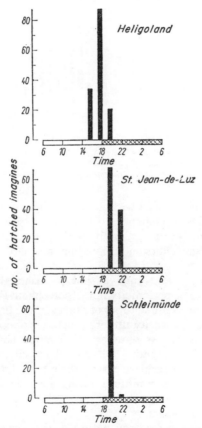

Figure 179 *Clunio marinus:* (A) diurnal distribution of hatching of the imagines in three different populations with artificial light change of 12 hrs light (0600–1800 hrs), 12 hrs darkness (1800–0600 hrs).

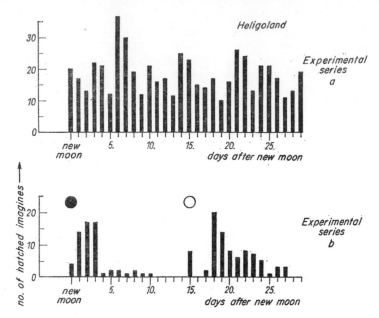

Figure 179 (B). *Clunio marinus:* animals from Heligoland, graphs of the daily hatching numbers in relation to the synodic month; series a = reared in artificial light change (12: 12 hrs, 20°C, 554 imagines in 3 broods in 4 months), series b = reared under natural light (147 imagines, 2 broods in 2 months) (after Neumann 1963).

a population is synchronized with a certain time of the day and phase of the moon which corresponds to low water springs. According to the observations of Neumann, both rhythms are initiated by the influence of light: the endogenous guiding mechanisms have not yet been investigated.

Hauenschild (1955 and 1956 a) has published the results of his investigations on lunar periodicity in the polychaete *Platynereis dumerilii*. These worms live in web-like tubes of mucus in the littoral and during the period of sexual maturity undergo metamorphosis to a free-swimming *heternonereis* form. As soon as this change has taken place (it takes about a week), swarming follows obligatorily, the worms rising to the surface at night and releasing their sexual products into the water. Swarming is at its minimum during the time around full moon, and this is retained even if the worms have been kept in natural light in the laboratory for years.

On account of the obligatory nature of the swarming after metamorphosis is completed, the conditions prevailing at the time of swarming may not be decisive, but the start of metamorphosis must

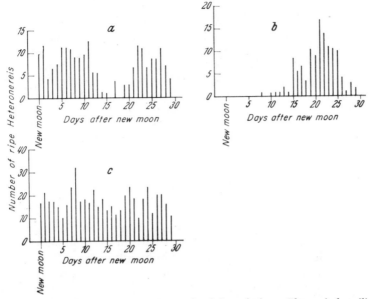

Figure 180. Swarming of the heteronereis of the polychaete *Platynereis dumerili*: (a) natural light, minimum around full moon; (b) continuous artificial light from 3rd day before until 2nd day after new moon, then 12 hrs daily, minimum at the time of strongest illumination; displaced about 14 days relative to (a); (c) continuous artificial illumination: no marked minimum (after Hauenschild 1956a)

determined periodically. In his laboratory cultures Hauenschild found (Fig. 180):

(1) with free access of light: marked periodicity with a minimum spawning at the time of full moon;

(2) with prolonged, uniform artificial illumination: no clear minimum + uniform distribution of swarming dates;

(3) in artificial light with periodic long and short day periods independent of the moon's phases and sometimes the opposite of these: a periodicity with minima at the time of the artificial full moon.

Thus the rhythm of a culture could be altered by changing the time interval between the short-day periods. It was concluded that the shortening of the daily period of illumination, such as occurs when the moon is waning, must be regarded as the factor stimulating metamorphosis and indirectly, therefore, the swarming. A swarming maximum will, however, be stimulated less by the shortening of the daily period of illumination immediately prior to it, but much more by an endogenous rhythm which is induced by the exogenous periodicity which is present throughout the whole period of development.

The physiological basis should presumably be sought in a hormonal system, which according to Durchon (1952, 1953, 1956) and Hauenschild (1956 b) acts as an inhibitor during the development of sexual maturity. Only experiment could decide whether in other cases of lunar periodicity it is possible to prove photoperiodicity. It should be noted that in many cases there is a triple temporal correlation, dependent upon the season, the phase of the moon and the time of day. We are still a long way from fully understanding the interaction of the endogenous processes.

From the ecological viewpoint a rhythm of this kind is of great importance because, by ensuring that whole populations breed within a short period of time (epidemic spawning), it considerably increases the probability that the sexual cells released into the water will be fertilized.

A similar phenomenon occurs in the California grunion (*Leuresthes tenuis*) (see Korringa 1957). These fish spawn in the sand above mean high-water level; the sexually mature individuals assemble along the coast 3–4 nights after full and new moon, move up the shore to high water over a period of 1–3 hours and spawn immediately, so that they are carried back again into the water by one of the next waves. Each individual spawns only once during each spring tide so that there is a fortnightly spawning rhythm. Development is tied to the same rhythm, because the young hatch after 14 days and are washed down by the next spring tide; if development is retarded, then hatching will also be delayed for 14 days.

There have not yet been any experimental investigations to determine whether the rhythm is in some way controlled by hormones, analogous to what has been found in polychaetes.

4. The vertical gradient

Three important abiotic factors show marked vertical gradients: temperature (see p. 59), light (see p. 50) and pressure (see p. 78); to a lesser, sometimes localized, extent oxygen content and salinity also show gradients. The increase in pressure and the decrease of light with increasing depth are continuous gradients, but the other gradients often show considerable discontinuities. Correspondingly, there are marked biological gradients, which have already been mentioned in the preceding chapters; these points will be summarized here and to some extent supplemented, particularly in the case of the pelagial.

In the epipelagial, with its phytoplankton and the animals which

TABLE 43

Vertical distribution of zooplankton biomass in the Kurile-Kamtschatka Trench in May-June 1953

(dry weight excluding coelenterates)

(after Vinogradov 1954)

Depth in m.	mg/m³ mean	Ratio Minimum : Maximum	No. of samples
0– 50	497·6	1 : 10·2	10
50– 100	320·3	1 : 3·2	10
100– 200	264·6	1 : 1·9	10
200– 500	228·0	1 : 2·1	6
500–1,000	59·3	1 : 2·1	6
1,000–2,000	21·8	1 : 1·7	3
2,000–4,000	9·3	1 : 1·6	4
4,000–6,000	2·64	—	1
4,000–8,000	1·73	1 : 4·6	3
6,000–8,500	0·48	—	1

consume it, the vertical differentiation is determined by the gradient in light, the productivity decreasing with depth; this is followed by depths which are characterized almost exclusively by the presence of consumers. This is reflected in the analysis of the vertical distribution of zooplankton biomass made by Vinogradov (1954) (Table 43). if the biomass of the total plankton were taken into consideration the gradient would be even more marked. Taken individually, however, the components which determine the gradients in biomass are in no way uniform. The phytoplankton attains its maximum development at a depth of about 10 metres, and not immediately below the surface.

TABLE 44

Number of copepods caught at different depths by the John Murray Expedition

(after Sewell 1948, p. 238)

Depth in m.	No. of individuals
0– 100	>6,503
100– 200	601
200– 300	220
300– 400	728
400– 500	1,423
500– 600	5,643
850–1,000	3,103
1,500	341
2,000	24
2,800	2

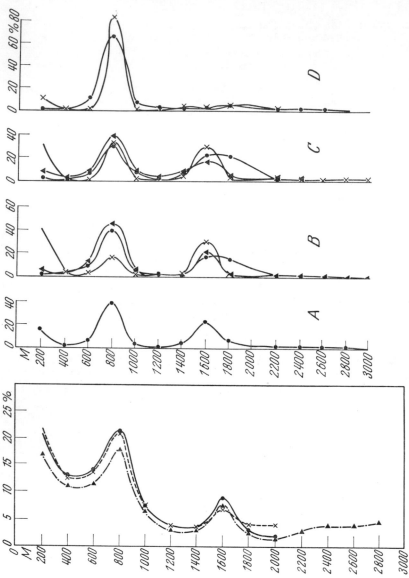

Figure 181. (a) Percentages of total catch at different depths; (b) Percentages of catch at different levels at 38° 16′N 69° 05′W

A: percentage of the total catch; B: ●fishes, — copepods, × euphausiaceans; C: ●decapods, —chaetognaths, × salps;

From hauls taken at different depths, Leavitt (1938) has shown discontinuities in the vertical distribution of the numbers of zooplanktonic organisms (Fig. 181 a and b; cf. also Table 44). Records of scattering layers show that this phenomenon is widely distributed, and there are also important local differences. Discontinuity layers and the boundaries between water masses lying one above the other represent discontinuities in the abiotic factors which are reflected more or less markedly in the biological gradients.

By direct observation from diving spheres and bathyscaphes it has often been possible to confirm the presence of maxima of this type. Thus, off the coast of Portugal, Pérès, Piccard & Ruivo (1957) found that the plankton density decreased from the surface to a minimum at about 300–400 metres; this was followed by an increase down to 800 metres and then by a further minimum between 1,100 and 1,400 metres. Between 500 and 800 metres, Piccard & Dietz (1957) passed through zones showing a considerable density of plankton, made up mainly of salps. Since these animals are filterers feeding on small particles it can be assumed that large amounts of detritus and nannoplankton are present at these depths.

Working in the same area as Leavitt, Clarke & Hubbard (1959)

TABLE 45

Depth distribution of light flashes at a station 150 miles S.E. of New York
(after Clarke & Hubbard 1959)

by day		at night	
Depth in m	No. of flashes	Depth in m	No. of flashes
		50	23
		100	**166**
		170	**135**
		330	36
		400	33
490	7	500	41
550	**27**		
600	63·5*	600	48
700	96	700	60
800	**104**	800	86
900	**118**	900	93
1,000	94	1,000	92
1,100	96	1,100	70
1,200	80	1,200	34
1,300	85	1,300	51
1,400	66	1,400	63
1,500	63	1,500	48
1,600	54	1,600	48
		1,700	60

* The mean of 2 readings of 67 and 60.

recorded the vertical distribution of bioluminescence and were often able to observe a significant maximum of light flashes at depths between about 800 and 1,000 metres (Table 45) which pointed to an accumulation of planktonic organisms. In the Mediterranean, on the other hand, it appears that maxima of this kind have not been observed (Clarke & Breslau 1959). The average duration and intensity of the flashes give some indication of the different species present at different depths and so this appears to provide an interesting supplementary method.

As the amount of phytoplankton and of its sinking dead bodies decreases with depth there must be a reduction in the numbers of phytoplankton consumers and thus a change in the relative proportion of the different types of feeding. The filterers feeding on nannoplankton occur principally in the upper horizons and only constitute a small percentage of the plankton in the deeper parts of the bathypelagial; on the other hand, the proportion of macrophagous predators in the population will increase. In view of this it is understandable that certain groups occur almost exclusively in the epipelagial. Thus, according to Lohmann (1920), the appendicularians are in general restricted to the upper 200 metres, and only a few species can be regarded as true deep-water forms (e.g. *Bathochordaeus chuni*). Among the salps collected by the Meteor Expedition, Krüger found only three deep-water forms (*Doliopsis meteori, Doliolina intermedium, D. resistibile*) which were caught particularly frequently in depths of 400 to 800 metres at 35°S in the Atlantic. Pteropods have also been found mainly in the epipelagial.

Predatory forms, such as heteropods and chaetognaths, are also sometimes widely distributed in the epipelagial: among these Kuhl (1938) regarded only *Sagitta macrocephala, S. decipiens* and *Heterokrohnia mirabilis* as true deep-water forms, which were only found in the lowermost epipelagial and in the bathypelagial, whereas all the others were either stenobathic epipelagial or more or less eurybathic. A similar position has been found in the polychaetes; at Woods Hole, Leavitt (1935) found 15 euphausian species in the upper 400 metres, but only 5 species below 800 metres.

The gradient in impoverishment of species and types of life-form is not however continuous in all groups. In his extensive analyses of the pelagic copepods collected by the John Murray Expedition, Seymour Sewell (1948) gave a number of examples to show that in many areas there was no continuous reduction in the number of species of copepod with increasing depth. As shown in Tables 46 to 50 (and several more examples could be cited), there is in many cases an intermediate species maximum, but the depth at which this occurs varies according to the area.

There are similar findings in the case of the pelagic gammarids. Of the 10 species collected by the Albatross, Schellenberg (1929) names only one (*Synopia ultramarina*) which is stenobathic epipelagial, whereas all the others are bathypelagic or at least prefer the deeper,

TABLE 46

Depth in m	No. of copepod species caught	Area	Author
0	21	Gulf of Guinea	Scott 1894
55	44		
110	15		
293	19		
476	26		
658	47		
841	28		

TABLE 47

Depth in m	No. of copepod species	Area	Author
0	5	North Atlantic	Wolfenden 1904
183	9	50–61° N	
366	10		
549	16		
732	12		
914	12		
1,097	8		
1,280	8		
1,463	8		
1,829	7		
2,195	9		

TABLE 48

Depth in m	No. of copepod species	Area	Author
91– 274	20	Biscay	Farran 1926
183– 274	17		
183– 366	25		
274– 457	30		
366– 549	23		
549– 732	31		
732– 914	32		
914–1,372	39		
1,372–1,820	36		
1,829–2,286	18		
2,286–2,743	14		
2,743–3,658	11		

TABLE 49

Depth	No. of copepod species first taken at this depth	Area	Author
0	109	Indian Ocean	Sewell 1948
100	9		
200	9		
300	7		
400	11		
500	14		
600– 650	16		
850–1,000	24		
1,500	24		
2,000	2		
2,500	1		
2,926	1		

TABLE 50

Depth	No. of calanid species per m^3	Area	Author
0– 25	7	Kurile Trench	Brodsky 1952
25– 50	7		(after Moore 1958)
50– 100	9		
100– 200	10		
200– 500	28		
500–1,000	30		
1,000–4,000	87		

cooler water. The same appears from the work of Birstein & Vino-gradov (1952) in the area of the Kurile Trench (Table 51). An analysis of the data hidden in the literature would doubtless provide further examples.

The long-tailed swimming crustaceans also appear to have a species maximum in greater depths.

The material available on this subject only allows provisional conclusions. It shows that in the pelagic copepods a relatively large num-

TABLE 51

Horizon	No. of pelagic gammarid species	Area	Author
Surface	1	Kurile Trench	Birstein &
Transition zone	1		Vinogradov 1955
Upper Abyssal	13		(after Moore 1958)
Lower Abyssal	12		
Ultra Abyssal	6		

ber is evidently cold or cool stenothermal and therefore prefers greater depths. This conclusion is confirmed by the distribution of copepods in the waters around Iceland, as found by Jespersen (1940) (Table 52). The total number confirms the general picture; in detail, however,

TABLE 52

Regional and vertical distribution of copepod species in the waters around Iceland (after Jaspersen 1940)

Sector	S	W	NW	N	E	Total
No. of surface species	21	16	17	17	16	23
No. of deep-water species	76	22	2	7	6	79

there are certain differences, because the sectors S and W, which are influenced by Atlantic water, have few surface species and many deep-water forms, whereas the sectors NW, N and O which are more under the influence of arctic water show the reverse tendency. There is still, however, no confirmation of this from the results of antarctic investigations. It is also conceivable that deep-water species are tied to special types of water, characterized not only by low and relatively constant temperature, but also by a combination of several different factors. It is also possible that the small amount of light, or its total absence, might be the decisive factor.

From the presence of a relatively large number of species in medium depths it follows that there is a greater variety of ecological niches in this part of the oceans than was at first suspected. Since the abiotic factors are relatively constant, one must look for diversity in the biological factors in these regions; these could possibly be related to the type, amount and distribution of the available food.

The evolutionary problems involved are again of considerable interest. Are most of the species autochthonous, that is, have they evolved at these depths, or are they allochthonous, having come in from centres of evolution in shallower water? There is evidence in favour of both possibilities.

Pelagic nemertines have been found exclusively in medium and great depths (Coe, 1945) and they show very marked speciation, which has called for the establishment of several higher systematic groups. Here there is no evidence of an invasion from higher horizons, caused by species pressure, because the whole group is completely lacking there. It is also very unlikely that the nemertines ever occurred by preference in the epipelagial, and then settled secondarily in the bathypelagial, because the nemertines in general, particularly the errant Polystilifera which are most closely related to the pelagic forms, mostly belong to the infauna. The transition to a pelagic way of life would therefore have taken place at depths with little or no light, so that we must assume that the great variety of species has been evolved autochthonously.

A few examples of evolutionary interest can be cited from the pelagic fishes of medium depths. In the Atlantic, the genus *Chauliodus* is represented by the species *C. danae* and *C. sloani*, the latter being subdivided into two subspecies. The populations of *C. sloani sloani* are separated from each other by areas occupied by *C. danae;* it appears that *C. sloani* has been invaded by *C. danae*. Marshall (1957) regards *C. sloani sloani* as an adaptable form which can avoid the competitive pressure of other species of the genus. There appears to be a similar situation in *Stomias boa boa*, whose area of distribution is divided into two by *S. affinis* and *S. colubrinus*.

Both examples show considerable analogy with the case of competitive invasion in *Gammarus duebeni* discussed on p. 109. We cannot, however, fully understand the picture until the special biological relationships of the competing species have been worked out. The same applies in the case of the genera *Vinciguerria* and *Cyclothone*, the species of which show a marked vertical distribution which cannot be correlated with any definite types of water.

The examples cited on page 366 of groups with mainly epipelagial species and only a few representatives in the bathypelagial suggest a more marked process of evolution in the epipelagial. The few deep-water forms can be regarded as species that have arrived as invaders to avoid the pressure of competition, and have been able to settle in unoccupied niches beneath the epipelagial.

The frequent repetition of this process could lead to an increase in the variety of species in deeper water. This presupposes that the invading species possess considerable plasticity, which allows them to adapt to the differing conditions in their new habitat. The great plasticity of the copepods has already been discussed in Chapter IV, 3, page 149. If then, plasticity in ecology and physiology is correlated with genetic instability one could get autochthonous speciation in a group of species that was originally allochthonous. Presumably the fact that in some groups few or no representatives have evolved in deeper water is the result of genetic stability and reduced plasticity.

One can also think of invasion taking place in other circumstances. If climatic changes occur in areas with relatively constant conditions, which lead to sharp seasonal fluctuations, the greater part of the organisms living in this area must either escape or perish. Thus, forms for which the constantly changing conditions are more fatal than a rather constant reduction of light and temperature could escape into deep water. Repeated long-term climatic changes, such as occurred for example in the Pleistocene, could then also lead to an enrichment in species and give rise to a diversity of species that was not autochthonous.

In general, these ideas must be regarded as hypothetical because the available evidence is still very limited. In medium and great depths the abiotic environmental conditions are relatively very constant and so in any future investigations special attention should be paid to the dynamic relationships between the organisms.

5. Diversity in relation to latitude

Biological differentiation in relation to latitude is a phenomenon well known in the sphere of terrestrial biogeography and it is also found in the sea. In the warmer parts of the oceans there is a significantly greater variety of forms than in cool temperate and cold seas. This is very striking in groups that are largely restricted to areas of warm water, such as the reef-building corals, which have a particularly large variety of forms in warm seas having a surface isotherm of 21°C or above during winter. But this type of gradient is also shown in groups whose total area of distribution includes both higher and lower latitudes; this is shown by the examples given in Table 53, and further analysis of data in the literature would certainly yield other instances.

If one considers this gradient in relation to biological habits it appears that it is the animals of the pelagial and the epifauna that show

TABLE 53

Distribution of plants and animals between warm- and cold-water areas of the oceans in the surface layers (down to 200 m), calculated as a percentage of the figures for warm water

	Warm water	Cold water	Author
Algae	100	6·4	Zenkevitch
oceanic Hydromedusae	100	14·4	Kramp 1959
pelagic polychaete genera	100	20	Friedrich
Calanids (pelagic copepods)	100	15·6	Brodsky 1959
Brachyura Decapoda	100	3·1	Thorson 1952, 1957
Pteropoda	100	14·3	Tesch 1946, 1948, 1950
Prosobranchs	100	22·2	Thorson 1952, 1957
Nudibranchs	100	6·3	Thorson 1952, 1957
Gastropods excluding nudibranchs, pelagic species and Pyramidellidae	100	26	Fischer 1960
Lamellibranchs	100	32·2	Fischer 1960
Appendicularians	100	9	Lohmann 1914a
Doliolum	100	8·5	Neumann 1935
Tunicates	100	6·6	Hartmeyer

the greatest diversity in warm seas; this phenomenon is less conspicuous or even absent in the infauna. Thus, for example, the lamellibranchs on the north American coast, which include several infaunal
elements, show the highest percentage of cold-water species in Table
53. Thorson (1952 and 1957) stated that the ophiuroids, holothurians,
cumaceans and among the prosobranchs the Naticidae, which live
in soft bottoms, are groups which show no gradient in the number
of species (Fig. 182). The findings of Riedl (1956 and 1960) appear to

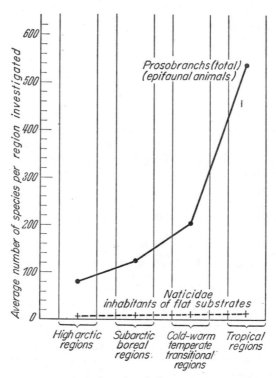

Figure 182. Average number of species of all prosobranchs from large areas of the
same size in different latitudes (———): below; the corresponding curve for the Naticidae (×—×—×) (after Thorson 1952)

point in the same direction; in pure muddy bottoms in the Adriatic
he found a number of turbellarians, a nemertine (*Callinera burgeri*) and
an enteropneust (*Saccoglossus mereschkowskii*) which were previously
only known from northern areas. In this context one should also mention that a number of species living interstitially in sand have a wide
latitudinal distribution.

There are only a few groups which show a greater diversity in higher latitudes. According to Fischer (1960), the brown and red algae appear to be best developed in cold temperate latitudes; according to a quotient produced by Feldmann, viz., $\dfrac{\text{Rhodophyceae}}{\text{Phaeophyceae}}$ the brown algae account for a higher percentage in the cold seas: the quotient is 1·5 in arctic seas, 2 in the North Sea, 3 in the Mediterranean and 4·6 on the coasts of tropical American (cf. Pérès 1961). Whales, seals and sea-birds also occur in greater numbers in colder seas than in warm-water areas. Further examples of this could probably be found.

It is, however, a general rule that there is a greater diversity or organisms in warm seas than in regions with cold water. At the moment this only applies to the upper water layers (about 200 metres).

The presence of a larger number of species in warm waters than in cold waters presents problems of evolutionary biology and also of ecology which may be discussed here.

From the viewpoint of ecology one could ask whether the number of ecological niches is greater in lower than in higher latitudes. From the purely physical and chemical side this question must be answered in the negative; on the contrary it could be argued that the seasonal changes in several of the abiotic factors might offer a greater diversity in ecological niches dependent upon physical and chemical factors. If, on the other hand, one takes the organisms into account the question must be answered in the positive. For example, in warm-water areas the precipitation of calcium is considerably facilitated by the temperature conditions (see pp. 95, 239); this results in much biogenic calcium precipitation, in the form of various kinds of reef formation, thus providing members of several animal classes with niches which are lacking in temperate and cold latitudes.

There are also several algae, in particular green and red algae, which contain precipitated calcium, and this is most marked in warm waters. Indeed the calcium-precipitating green algae are entirely restricted to the tropics and subtropics, and among the red algae there are only a few temperate and cold-water genera that do this (*Lithothamnium, Lithophyllum, Melobesia, Corallina, Iania*), but these too have their main centre of distribution in warm-water regions. These two groups of algae behave somewhat differently ecologically, because in contrast to the red algae the green algae tolerate stagnant water; usually however representatives of both groups are present alongside each other. In his review of the littoral calcareous algae, Lemoine (1940) points to their role as builders of sediment; the ooliths and pisoliths occurring in fossil marine sediments have been ascribed to the activity of Cyanophyceae in warm seas.

One very striking phenomenon is the formation in the Mediter-

N—M.B.

ranean of what Quatrefages described as 'trottoirs'; according to Deboutteville & Bougis (1951) the thalli of calcium-precipitating algae, particularly of *Tenaria tenuosa*, are collected by wave action into a bank, which projects at the water surface as a shelf-like formation (Fig. 122).

The various growth forms of calcareous algae provide diverse substrates for a rich macro- and microfauna, and at the same time they also supply food (see pp. 259 et seq.).

Magnesium carbonate is deposited in addition to calcium carbonate. In oceanic water the ration of Ca: Mg is 0·32:1, but in algal depositions the calcium is about 8–40:1. This proportion varies in the individual species according to the time of year; comparative regional investigations would therefore be interesting. Gessner (1959) gave the following figures (based on the data of Prat & Hamackova 1946):

	Ca : Mg
Corallina officinalis	8·8 : 1
Corallina rubens	8·3 : 1
Vidalia volubilis	15 : 1
Lithophyllum cristatum	11·6 : 1
Peyssonelia rubra	7·8 : 1
Halimeda tuna	26·5 : 1
Dasycladus clavaeformis	25 : 1

A higher mean temperature may therefore be reflected in an increased number of ecological niches for the epifauna of the warm-water littoral and thus provide opportunities for a greater number of species. In this example there is no need to postulate a direct relation between the animals concerned and the temperature.

This diversity of species in equatorial water is not, however, so evident in the case of pelagic and especially of holopelagic organisms. Richter (1961) expresses this in the following words: 'The heteropods as true warm-water planktonic organisms live in an environment which is marked by the diversity of its fauna, and which is rich in ecological niches of various kinds; it is therefore scarcely likely that selection is very severe. Finally, the "population density" of heteropods is comparatively low in all oceans, as is characteristic of highly specialized predators. So, since there is a plentiful and varied supply of prey one can exclude food competition as a selection factor, especially as the heteropods, in spite of very similar habits, show great differences in size between the species and so take prey of varying sizes and thus also occupy different ecological niches. All this must at least have contributed to maintain the diversity of the evolutionary forms in this group of gastropods' (l.c. pp. 207–209).

Fischer (1940) has discussed diversity in relation to latitude from the viewpoint of evolutionary biology. He pointed out that one must take into consideration the temporal and climatic changes in the areas concerned as well as the direction and speed of organic evolution. A warm-water girdle, whose position may have changed, is therefore a relatively constant characteristic of the seas of the world, whereas the higher latitudes are subject to greater changes associated with transgressions, regression and polar displacements. The climatic conditions of the polar and adjoining temperate regions will vary according to whether the poles are sited in areas of land or of sea and according to the relative size of the water or land masses surrounding the poles. Since polar displacements have been demonstrated it is certain that the cool- and cold-water regions of today have had a changeable history, which has possibly prevented the evolution of greater biological diversity.

By assessing the evolutionary factors in the warm and colder regions, Fischer came to the conclusion that genetic variability need not necessarily be affected by these differences. He also estimated that the rate of generation succession is small, since in many tropical forms this is no greater than in their relatives living in higher latitudes. There are, however, forms in which significant differences exist in the succession of the generations, so that a generalization in the sense of Fischer does not appear permissible. Maturity is frequently reached more rapidly, thus giving a larger number of generations with a smaller average body size (see pp. 248 et seq.). Thus for the pelagic copepod *Temora longicornis* five generations have been recorded in the English Channel (Digby 1950), two in Loch Strive (Marshall 1949) and one main breeding season in Oslo Fjord (Wiborg 1940). Bogorov (1959) gave the following varying weights and lengths for female *Calanus finmarchicus* from different parts of the Barents Sea:

	Temperature in °C	Wet weight in mg	Length in mm
Southern region (69°N–75°N)	4·7	0·79	3·22
Central region (75°N–77°N)	1·08	1·44	4·78
Northern region (77°N–80°N)	−0·5	2·46	4·57

In agreement with Fischer and other authors (e.g. Schmalhausen 1949, Dobzhansky 1950), one must assume that there are fundamental differences in the selection factors. In higher latitudes the abiotic factors will have a greater decimating action, whereas in warmer areas biological competition will be more effective; but the abiotic factors have a more random action and are therefore less selective than the biotic (see pp. 150, 337). Also in higher latitudes there is stronger competition between members of the same species, whereas in lower latitudes there is greater interspecific competition. This difference is partly due to the differences in the diversity of the forms, but it must also be important in the evolutionary process. Finally, it is probable that the uniform high mean temperature of tropical seas is more advantageous to a wider range of mutation within a species than the temperatures of temperate and polar regions which fluctuate markedly between the upper and lower limits of tolerance.

The points discussed here provide material for assessing the significance of latitude in relation to the diversity of forms. It is quite possible that further valid evolutionary points of view could be made so that it is necessary to test the variour theoretical views on a large number of objects.

There are other instances of a gradient related to latitude. In the prosobranchs, Thorson (1950) showed that proceeding from low to high latitudes there is a percentage decrease in the occurrence of pelagic developmental states and a corresponding increase in direct development (Fig. 127, p. 249). Crinoids show viviparity particularly among the species living in the antarctic (11 out of 19 species), whereas out of more than 600 unstalked crinoids from other parts of the sea only five are known to be viviparous (John 1938). The percentage of antarctic ophiuroids showing viviparity is similarly high (Mortensen 1933).

This phenomenon can be regarded as an adaptation to different ecological conditions. Pelagic larvae are dependent for food upon the presence of nannoplankton. The latter shows few or no seasonal fluctuations in the warm-water areas but has marked maxima and minima in the temperate and cold regions, which may vary in time. For polar animals with pelagic larvae it is an important prerequisite for the conservation of the population that breeding should coincide with the annual cycle of the nannoplankton, because the pelagic stages represent a sensitive phase in development. In some cases, there evidently is such correlation. Usually, however, the sensitive phase of development is reduced by shortening or completely suppressing the pelagic stages; the organism thus becomes adapted to fluctuating conditions. 'In high arctic seas pelagic development is suppressed in up to 95 per cent of all the species of marine invertebrates' (Thorson 1946, p. 434).

Brood protection also removes the developmental stages from the action of the less selective abiotic factors. Thorson (1946) distinguishes between different types of pelagic larvae in marine invertebrates:

A. (a) Planktotrophic larvae with a long pelagic life; feeding on plankton, producing large numbers of eggs and a rapid succession of generations (one year cycle). Capable of being widely distributed.

    (b) Planktotrophic larvae with a short pelagic life; not so

common, the pelagic stage evidently only serves to find a suitable substrate for metamorphosis.

B.   Lecithotrophic pelagic larvae; independent of an external food supply, good chances of being distributed, but only a few, large eggs produced.

At first glance the third type appears to be particularly suited to high arctic regions, because it favours the distribution of the species and is also independent of food supply. Thorson regards this type as a compromise and continues: 'The absence of lecithotrophic larvae (in high arctic regions) may probably be due to the fact that the competition in high arctic areas is so keen that no compromise methods, but only an "either–or" will suffice if a species is to survive. A species living here must either reproduce in a totally non-pelagic way—"the safety first principle"—i.e. a slow method with relatively few young per mother animal, with small chances for spreading, but with good chances for recruiting the stock in a stable way, or it must use the planktotrophic pelagic method with larvae which are able to make the most of the phytoplankton produced during the short summer time, to develop quickly at very low temperatures, and in this very hazardous way to take their chance for a good spatfall' (Thorson, 1946, pp. 437–438).

Knudsen (1950) working on the prosobranchs of the area between Sierra Leone and the Niger delta found that the ratio between pelagic and non-pelagic development was 31:69%, which appears to be incompatible with the figures given by Thorson. The contradiction is only apparent, because Knudsen points out that in the narrow region of Shelf there is an active exchange between coastal and oceanic water, by means of which pelagic larvae would easily be transported away from the areas suitable for their settlement. There may, therefore, be a few exceptions to the general phenomenon recognized by Thorson, due to special local conditions.

It is evident that here too there is a close relationship between ecological problems concerned with survival and certain evolutionary problems. At the same time it shows the general importance of the ecological work on the larvae of benthonic animals carried out in recent decades, particularly by Thorson.

Gradients due to latitude may also be expected in physiological characters, whether in comparing related species within a differing north–south distribution, or populations of a single species from different latitudes. Bullock (1955) has given a comprehensive review of the vast literature on this subject which has accumulated since about 1936. The problem may be stated as follows: in poikilothermal animals the rate of many living processes (e.g. movement, development, respiration) is dependent upon the temperature; many poikilotherms live under differing temperature conditions, particularly if they have a wide distribution; in general, however, individuals of the same species from areas differing in temperature do not show the expected differences in the rate of functioning; that is, they are adapted to the special conditions of the area in which they live.

Some of the investigations carried out in this field may be mentioned: Takatsuki (1928), Spärck (1936), Thorson (1936), Fox (1936, 1938, 1939), Wingfield (1939), Peiss & Field (1950), Dehnel (1955), Tashian (1956), Démeusy (1957), Roberts (1957) and Vernberg (1959). The animals investigated included molluscs, crustaceans, polychaetes and fishes, and observations were made, for example, on their growth, heart beat, respiration frequency, oxygen consumption. It is not possible to discuss the individual works, but the most important results may be mentioned:

1. In general, the rate of functioning of animals from cold regions is similar to that of warm-water animals, measured at the temperature of their normal surroundings.

2. At lowered temperatures the rate of functioning of cold-water animals is reduced less than that of the animals from warm waters; cold-adapted animals have lower temperature coefficients than related warm-adapted species.

3. From 1. and 2. it follows that at a given temperature the rate of functioning of northern animals is, in general, higher than that of the southern animals.

4. In northern animals there is a greater resistance to cold, in southern species a greater resistance to heat (e.g. Vernberg 1959 on *Uca pugnax*).

5. Animals become less sensitive to temperature fluctuations from south to north (e.g. Tashian 1956 on *Uca pugnax*, Demeusy on *Uca pugilator*).

6. Species from temperate latitudes show a considerable ability to adapt to lower temperatures; in general, tropical species show less ability to adapt.

7. In species and in populations of widely distributed species from temperate regions there is often a seasonal change in the basic metabolism, so that animals from the summer and winter months, measured at the same temperature, show different values (e.g. Marshall, Nicholls & Orr 1935 in *Calanus finmarchicus*, Edwards & Irving 1943a in *Emerita talpoida* and the same authors 1943b in *Talorchestia megalophthalma*).

Thus, it appears that the metabolism of the animals is correlated with the ecological conditions under which they live, and that corresponding to the ecological gradients in temperature there is also a latitudinal gradient in the metabolic processes.

When examined in detail, certain exceptions, and even apparent contradictions, have been observed, e.g. by Fox (1938, 1939), Fox & Wingfield (1937), Wingfield (1939), Tashian (1956) and Roberts (1957). Bullock (1955 p. 323), however, considered it possible that if size and season are taken into account, most of these cases show some ability to compensate.

There are several related problems which should be briefly mentioned. The mechanisms of adaptation and regulation are almost unknown; presumably they are partly on the level of tissue and cell physiology (e.g. according to Peiss & Field (1950), Roberts (1957), Masacchia & Clark (1957)) and include enzymatic processes in particular (Precht et al. 1955). There may be various separable processes within a level which, according to Bullock (l.c. p. 332), may account for the existence of adaptations: 'We may propose that it is the factor of balanced alteration which is the crucial one limiting the extent of acclimation in a species and its frequency among species.'

There are also evolutionary aspects, which Tashian (1956), for example, emphasized in his investigations on the fiddler-crab *Uca pugnax*. This species is represented by the subspecies *Uca p. pugnax* from Cape Cod to central Florida, and by *U. p. rapax* from south Florida to Brazil: 'During its evolution and distribution from tropical regions, *Uca pugnax* succeeded in migrating to the temperate zone without being accompanied by any great morphological changes. The fact that it was successful physiologically is more important, from an evolutionary viewpoint, than any structural features that may have been related to these adaptive processes' (l.c. pp. 44-45)

From the evolutionary viewpoint it is important to know whether and to what extent the different physiolcgical characteristics are genetically fixed or have an exclusively adaptive character, that is, are not passed on from one generation to the next. In some cases there appear to be populations of individual species with genetically fixed differences, but in others there are exclusively adaptive characters, because the characteristics change with the time of year. An adaptive condition may be reached very early in the life of an individual, if a 'conditioning' to the special conditions of the environment takes place during the sensitive phase which is often very short. In such a case even long-term (several months) pre-treatment before the experiment would cause no change in the response, and yet there is no genetic fixation. Individual cases of this kind are known (Démeusy (1957) on *Uca pugilator*, Kinne (1962) on *Cyprinodon macularius*), but the available material is still far from being sufficient for the formulation of any general ecological rule. The question whether there is genetic fixation or not, can only be settled on the basis of hydridization experiments.

There are various related ecological aspects: the specific abilities to adapt explain the substitution of related species in parallel animal communities (see p. 282); the greater productivity in high latitudes is understandable when we know that the low mean temperature does not actually slow down living processes (see p. 378); a reduced ability to adapt to changed temperature may be compensated by modification of behaviour in relation to activity and choice of habitat (Roberts (1959) on *Pachygrapsus*) (Fig. 183).

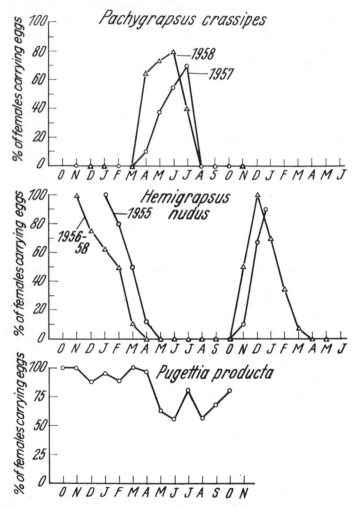

Figure 183. Annual reproductive cycle of three species of crabs on the west coast of N. America, as measured by the percentage of females carrying eggs in monthly samples (after Boolootian *et al.* 1959)

There is a further ecological problem. Experimental investigations on gradients related to latitude are almost exclusively concerned with temperature as an isolated factor. But the light also changes according to season and latitude (see p. 51). Since different living processes are simultaneously dependent upon several external factors we must in many cases investigate the combined effect of temperature and light. In discussing the reproduction of the echinoid *Strongylocentrotus purpuratus*, Boolootian & Giese (1959) put forward the following hypothesis: 'Temperature affects the duration of reproductive activity and of spawning. Light has approximately the same periodicity throughout the range of the purple sea urchin. Hence, the identical peaks of spawning by minimal day length. Food determines the stimulation and the production of gametes and thereby controls the amplitude and duration of gonadal activity.' In this case food supply is also involved as one of several factors involved. More attention should be paid to this aspect in future investigations.

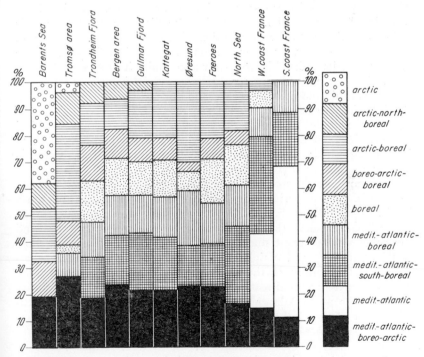

Figure 184. Composition of the echinoderm fauna of European seas according to zoogeographical region (after Brattström 1941)

6. Marine biogeography

Problems involving the geographical distribution of marine organisms have been discussed in the last chapter and also in the preceding section. In spite of this a special section is now devoted to 'marine biogeography' for the following reason: classical biogeography seeks to provide an outline of the spatial distribution of systematic units, whether species or higher categories. The material for this is derived primarily from the investigation by specialists of collections from restricted areas; this yields a varied mosaic of areas some of which can be well differentiated while there are large areas that have been little examined; this depends on the density of the hauls taken and on the specialists involved. The distribution of a number of species can be plotted from these data and this provides a basis on which we can recognize geographical regions, provinces and so on. Statistical analyses of this kind raise problems regarding the origins of such units/ (Fig. 184); these will primarily involve geological events, the actual distribution of ecological factors and the ecology of the organisms. Biogeography thus becomes a constituent part of ecology and changes from its classical position of studying the distribution of systematic units to dealing with the distribution of ecological units and systems. Ecology is, however, dependent upon the exact identification of the systematic units and so biogeography also retains its importance in the classical sense; naturally its most reliable results have been obtained from the neritic and littoral regions, and these will be the ones to which the following discussion mainly refers.

A few examples may now be given of proposed biogeographical systems; Zenkevitch (1949) put forward a generalized scheme in which he distinguished: arctic—northern temperate—tropical—southern temperate—antarctic (Fig. 185).

Bogdanov (1963) proposed a similar classification but with more zones, viz. an equatorial zone was followed to the north and south by the tropical zones characterized by the trade winds, the subtropics, the temperate zones, the subpolar (subarctic and subantarctic) and the polar regions. These marine zones are largely parallel with those of the continents and mainly parallel with the lines of latitude, although as they approach the continents the boundary lines are more or less deflected in the direction of the poles. The greatest irregularity occurs in the north Atlantic where, because of the south-west–north-east course of the Gulf Stream, the boundary between the temperate and subpolar zones ends in a curve in the area of the White Sea.

On the basis of the occurrence of the Heterosomata (flatfishes), a group containing almost 600 species, Chabanaud (1949) divided the neritic region into the following five zones proceeding from north to

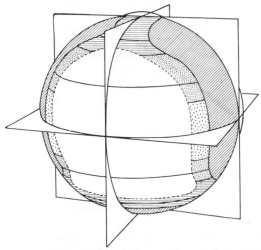

Figure 185. Geographical divisions of the littoral fauna in an ideal ocean (after Zenkevitch 1948): *oblique lines* = continents; *coarse stippling* = tropical; *fine stippling* = temperate; *horizontal lines* = arctic and antarctic

south: arctocryozone, arctopsychrozone, thermozone, notopsychrozone, notcryozone. These zones extend latitudinally and occur in three different super-regions: America, Afro-Asiatic and Australo-New Zealand. The American super-region is divided into two sectors (Pacific and Atlantic), the Afro-Asiatic into three (Atlantic, Indian, Pacific). The super-regions are further divided into a total of 29 regions, which are themselves subdivided into sub-regions.

As is usual in biogeography this classification is based on positive and negative characters: the positive being based on the limited occurrence of systematic groups, such as an area showing characteristic endemism, the negative being based on the absence of groups that are otherwise widely distributed. There is still no generally recognized standard for evaluating such characters, whether endemism or the absence of certain groups, although several authors have attempted this (cf. Hedgpeth, 1957, pp. 362–366). There is still no terminology for the units of the classification.

In his biogeographical classification of the neritic region, Ekman (1935, 1953) distinguished the following main divisions:

Arctic
Boreal   European Atlantic, Baltic Sea
                North American Atlantic
                Temperate Pacific
Warm waters   Indo-West-Pacific

                          Atlanto-East-Pacific   tropical and subtropical
                                                      America
                                                  tropical and subtropical West
                                                      Africa
                      Mediterraneo-Atlantic Sarmatic
                      South coasts of Africa
                      South-west Africa
                      South Australia and Tasmania
                      New Zealand
                      Peru and North Chile
    Antiboreal   South America
                      Oceanic islands
                      Kerguelen
    Antarctic

For comparison we may give the classification of the temperate
and cold areas of the southern hemisphere drawn up by Knox (1960):
this corresponds with Ekman's divisions from the south coasts of
Africa southwards, but in the reverse order:

Antarctic Province	(within the antarctic convergence)
South Georgia Province	(within the antarctic convergence, similarities with the Falkland Islands and the antarctic province)
Kerguelen Province	(in the west wind drift, close to or within the antarctic convergence, strong subantarctic element)
South America with	
Magellanic Province	(56°–42°S, cold temperate)
Central Chilean Province	(42°–c.30°S, mixed water)
Peruvian Province	(c. 30°–2°S, warm temperate)
uncertain: Tristan Province	(Tristan da Cunha and Gough Islands)
South Africa	
West coast Province	(from the Cape to c. 18°S)
Cape Province	(warm temperate)
South Australia	(with several provinces not yet properly defined)
New Zealand	(with several provinces: warm water in the north, cold water centre in the south, transition zones in between particularly on the east coasts)

These few examples of large-scale classifications show a certain
conformity, but also some differences; these are due partly to the
different values given to the characteristics of the geographical units,

partly to the heterogeneity of the material under consideration, and partly to the fact that the classifications are based not only on the actual distribution of the organisms but also on the distribution of the water masses. It would be better in the first place to establish only the distribution of the organisms, but on as broad a basis as possible. One might subsequently seek correlations between the exogenous factors and the limits of distribution of the organisms.

Temperature must be regarded as the principal external factor responsible for large-scale distribution. This can be explained by a few examples.

In the arctic and antarctic, the supralittoral and eulittoral zones are very poorly developed or may be completely absent. Here the formation of ice prevents the establishment of a flora and fauna; in the arctic, extensive dilution with fresh water from melted ice may also be a further exogenous factor. There are, moreover, considerable differences in faunal composition between the arctic and antarctic (cf. Uschakov 1963).

Supralittoral animals of temperate latitudes, such as the talitrid amphipods, are probably limited towards the poles by the severe shortening of the active phases of the life cycle, and towards the tropics by the hazards of desiccation and strong light. Amphipods in the supralittoral are presumably favoured by brood protection, which renders the sensitive juvenile stages independent of fluctuating food supplies and protects them against injurious external factors. In tropi-

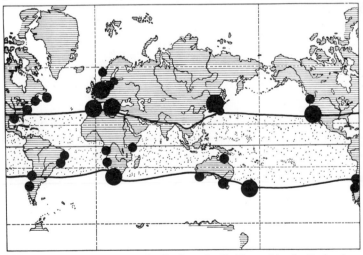

Figure 186. Map showing the distribution of talitrid amphipods digging in sandy beaches (black circles indicating number of species 1–4) and the distribution of the crab genus *Ocypoda* (dotted areas between black lines) (after Dahl 1952–53)

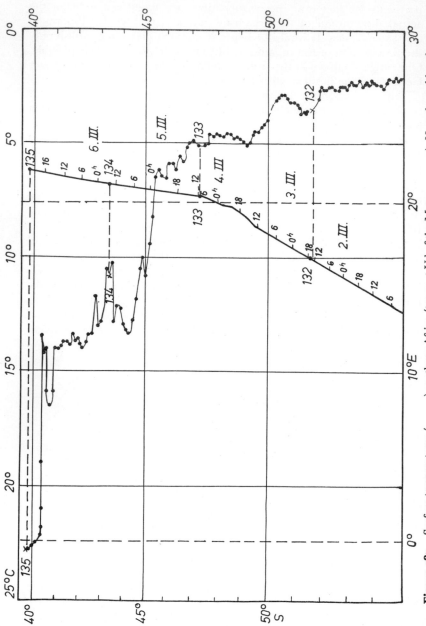

Figure 187. Surface temperatures (———) southern Africa (transect Vd of the Meteor voyage). Note the sudden change in temperature from 6° to 13°C in 45°S and from 14° to 23°C at 40°S

cal regions, on the other hand, brachyuran decapods have been able to occupy this zone, because their pelagic larvae are less exposed to such unfavourable factors (Fig. 186).

The concentration of reef-building corals with their accompanying fauna is a characteristic of the warm-water girdle; other littoral animals have a similar distribution, and the green alga genus *Halimeda* (with 21 species) is circumtropical (Hillis 1959). For these organisms the safe limits are the isotherms around 24°C. The occurrence of the coral general *Lophohelia* and *Amphihelia* outside the warm-water region should be noted.

The extent to which temperature limits may arise in the oceanic pelagial is shown in Fig. 187, which depicts two marked boundary zones encountered during the *Meteor* voyage south of Africa. The euphausiaceans, for example, show distinct boundaries which are comparable to this (Fig. 188). Einarsson (1945) has shown the same for the north Atlantic species of this group (Fig. 189).

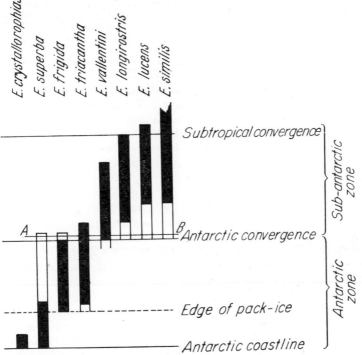

Figure 188. Distribution of species of *Euphausia* in the surface waters of the antarctic and sub-antarctic zones; *black* = normal range, *white* = occasional occurrence (after John 1936)

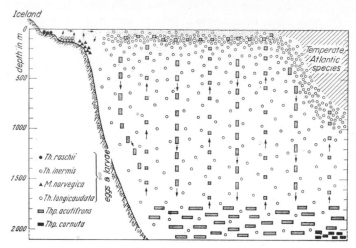

Figure 189. Schematic illustration of the vertical and horizontal distribution of the north Atlantic euphausiids in a section going southwards from Iceland. For *Thysanopoda acutifrons* the different sizes of the rectangles denote respectively: eggs and larvae, adolescents and adults. All stages of *Thysanopoda cornuta* are confined to deep water

In many cases the restricting role of temperature on the distribution of organisms is due not so much to its effect on the adults as to the greater sensitivity of the developmental stages. As a rule the sensitivity of the life stages is: adults < germ-cell maturity < spawning and development. The survival of adult animals in a sterile area of distribution is, therefore, understandable. At the same time, however, adaptations may occur, because in the echinoid *Paracentrotus lividus* at Naples, Hörstadius (1925) found normal and abnormal cleavage distributed in the following way:

	Summer	Winter
13°	abnormal	normal
26°	normal	abnormal

Runnström (1936) found similar conditions in several ascidians and in the echinoid *Arbacia*; observations of the same kind have been made in other places, so the dominant role of temperature in the control of geographical distribution is certain.

The fact that reproduction and development are more dependent on higher temperatures than the vegetative functions means that in many species the populations will breed the whole year round in areas with uniformly favourable temperatures, but only seasonally in unfavourable regions. In the cooler parts of their range, warmth-loving species will breed during the summer months, but species from cool regions living in the warmer marginal areas of their range will breed in spring and

TABLE 54

Geographical relationships of the Mediterranean fauna

(after Pérès & Picard 1958)

Species from the Mediterranean	Hydroids (192 species)	Crustaceans (Decapoda reptantia, 129 species)
and:		
North Atlantic (in the wide sense)	41·6%	56·6%
Senegal	0%	17·9%
Subtropical Atlantic coasts	3·6%	2·3%
Circumtropical	10%	2·3%
Cosmopolitan	17·2%	4·6%
Indo-Pacific		
(immigrants through the Suez Canal)	0%	3%
Endemic	27·1%	13·2%

	Echinoderms (107 species)	Ascidans (132 species)
North Atlantic (in the wide sense)	50%	31·8%
Senegal	14%	2·2%
Subtropical Atlantic coasts	4·6%	1·7%
Circumtropical	0·9%	4·5%
Cosmopolitan	2·8%	5·3%
Indo-Pacific		
(immigrants through the Suez Canal)	0·9%	3·2%
Endemic	26·1%	50·4%

autumn, or in the winter. Statistical confirmation of the occurrence or absence of a species in an area will provide a good method of assessing whether it belongs to the fauna of a given region (Figs. 190 a and b).

Table 54 provides an example of the biogeographical analysis of a single region; it was drawn up by Pérès & Picard (1958) to show the geographical provenance of some of the animal groups that occur in the Mediterranean. This sea, together with the adjoining Black Sea, has often been discussed from the viewpoint of classical biogeography (e.g. Ekman 1935, 1953; Kosswig 1950, 1955; Slastenenko 1959). Of the many and varied problems connected with it, some mention may be made of recent changes caused by the immigration of fishes from the area of the Red Sea and Indian Ocean through the Suez Canal. As Kosswig has emphasized, this invasion of tropical forms has taken place against the gradients in salinity and temperature and is probably possible because the immigrating forms are fairly euryoecous. Presumably, however, invasions of warmth-loving forms from the Red Sea area took place before the opening of the Canal, after the Tertiary Tethys fauna had largely died out or been displaced when the climate deteriorated at the end of the Tertiary and in the Pleistocene.

Over the course of several decades numerous observations have been made in the northern margins of the Atlantic (West Greenland, Iceland, Spitsbergen, Bear Island, White Sea), and the recorded

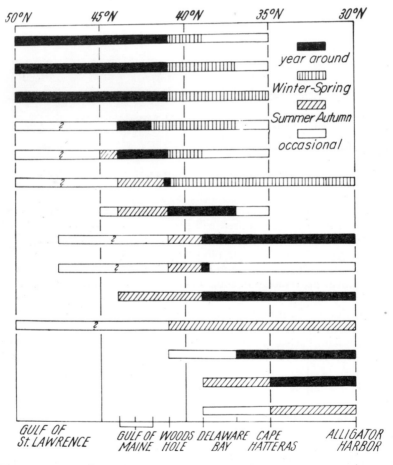

Figure 190a. Latitudinal and seasonal occurrence in eastern American waters of 14 copepods

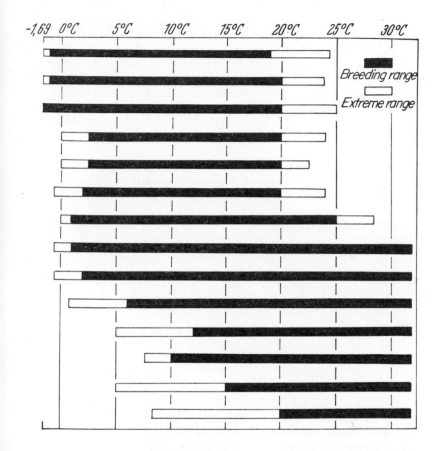

Figure 190b. Known temperature ranges of 14 copepods from eastern North American waters (after Deevey 1960a)

temperature rise has been reflected in the invasion by Atlantic forms; Blacker (1957) has given a summary for the area around Spitsbergen.

A further example of the historical importance of biogeographical relations between different areas is the relationship of the warm-water species on the American Pacific coasts with those of the West Indies region. These two faunal areas are separated from each other by central America, but in almost all classes they show greater similarities to each other than the west coast of America does with the central and west Pacific. For the echinoderms, excluding the crinoids and holothurians, Ekman (1946) has given the following outline (Table 55): The close relationship between the faunas of western America and

TABLE 55

Amphi-American and amphi-Pacific relationships of the warm-water echinoderms of western America

(after Ekman 1946)

Common species		Closely related species		Common genera	
W. America	W. Indies	W. America	W. Indies	W. America	W. Indies
2·3%	2·3%	40·5%	36·2%	50%	63%
W. America	Central & W. Pacific	W. America	Central & W. Pacific	W. America	Central & W. Pacific
3·7%	0·9%	12·2%	3%	21%	11%

the West Indies is understandable on the basis of a connection between the two areas, across the central American sill, lasting into the Tertiary. The significant differences between the east and west Pacific suggest the existence of a barrier to distribution, which has for a long time hindered exchange between these two regions; this barrier is the extensive area of the east Pacific which has no islands.

Geological history and distributional ecology are also involved in the interpretation of disjunction, that is, of cases where one or even two closely related systematic units (species, genera or even families) are found in two spatially separated areas. This type of distribution may occur in relation to longitude or latitude; in the latter case it is known as bipolarity and is found both in benthonic animals (e.g. *Priapulus caudatus—P. tuberculatospinosus*) and in pelagic forms (e.g. the hydromedusa genus *Botrynema*, the forms of the appendicularian *Fritillaria borealis*). In assessing such cases of disjunction it should be remembered that the observations in the intermediate area may have been rather scattered and in the pelagic forms there is the possibility of equatorial submergence. Nevertheless there is scarcely any doubt of the existence of disjunction. From the presence of numerous circumglobar species living alongside forms with a restricted distribution it is evident that the interpretation of geographical distribution requires a consideration of geological aspects as well as of the ecology of the individual species.

In an area such as the Mediterranean which is more or less enclosed, with a relatively well-known fauna, recent changes in the fauna can be recorded quite easily; this, however, is not so simple in the case of more open areas which have only been the subject of sporadic investigations. A number of recent immigrations have taken place in the area of north-west Europe; these, however, are due to introduction rather than random immigration. Some of these may be mentioned here:

The American slipper-limpet *Crepidula* was introduced to the coasts of England with oyster spat; it spread and was introduced to the North Sea coasts of Holland and Germany. Ankel first found this mollusc at Sylt in 1934 (see Werner 1948).

The cirripede *Elminius modestus*, a native of Australian waters, first appeared on the coasts of England after the war (1939–1945) and since then it has increased its area of distribution (see Crisp & Southward 1959, Barnes & Barnes 1960).

The introduction of the mitten crab *Eriocheir sinensis* into central Europe took place considerably earlier; it has caused damage to inland fisheries.

The diatom *Biddulphia sinensis* suddenly appeared in large numbers in the North Sea during 1903; since then it has maintained itself in this area and occurs regularly in plankton blooms. Ostenfeld considered that it has been introduced, but Müller Melchers (1952) thought that its appearance was due to transport by strong currents and repeated sporadic occurrence in marginal areas.

A novel approach to this subject has been made by Szidat (1955, 1961) who investigated the parasite fauna (primarily copepods, nematodes, trematodes, cestodes, acanthocephalans) of gadid fishes and flatfishes from the coasts of Argentina and compared them with the parasites of related fishes from the north Pacific and Atlantic. He came to the conclusion (Fig. 191) that some of the Argentine gadids were not related to the north Atlantic species, but were more closely connected with north Pacific forms, so that distribution must have taken place along the west coast of the American continent and round Cape Horn. Only one of the gadids investigated must have succeeded in crossing the warm-water region in the west Atlantic, because its parasite fauna agreed closely with that of the north Atlantic species.

On the other hand, the flatfishes have spread out north and south from the centrally positioned warm Sea of Tethys of Tertiary times and they show no differences between their parasites.

This method of zoogeographical investigation and interpretation, which goes back to the early years of the 20th century and the ideas of von Ihering, is certainly very promising; it seems to me, however,

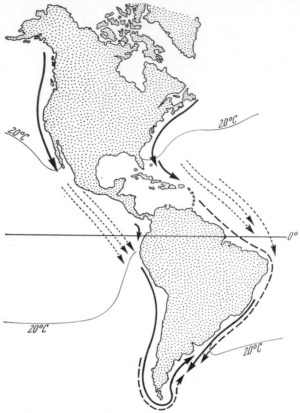

Figure 191. Migration routes of the gadids (——) and flatfishes (— — —) occurring on the coasts of Argentina, based on the relationships of their parasites (after
Szidat 1961, somewhat modified)

that there is still no adequate evidence to allow any firm conclusions,
first, because in the present state of knowledge the problem of the
host specificity of parasites is not sufficiently clear, and secondly,
because in the parasites discussed here, which have alternate hosts,
the specificity of the intermediate hosts has yet to be investigated.
But the answer to these and other questions will depend on whether
the parasites do in fact migrate so exclusively with their hosts. This
seems, therefore, to be an almost unrestricted field for future research.

7. Interspecific relationships of marine organisms

In the sea, as on land and in fresh waters, there are many examples of close or less direct relationships between individuals of different species. These go beyond the associations of animals due to similar environmental requirements and extend from facultative epizoism to highly specialized parasitism and symbiosis. Some examples have long ago found their way into the textbooks; summaries have been given by Caullery (1922), Dales (1957, 1966) and Hopkins (1957, including an extensive bibliography), so that only a few causes need be cited here.

Commensalism, in which one partner derives a definite advantage from the cohabitation, without the other being seriously injured, is widespread in the form of epizoism. Numerous sessile animals settle facultatively or even obligatorily on others, particularly on those with hard shells; thus, ciliates, hydroids, bryozoans, some sedentary polychaetes and cirripedes settle on lamellibranchs, gastropods, crustaceans or on the calcareous tubes of other sessile forms. In many cases the epibiont only settles on the firms substrate provided by the adult forms, but the part of the body on which it settles may also be important, for it may even exploit the water currents produced by the host for its own respiration and feeding.

Polychaetes of the family Polynoidae frequently occur in the tubes of other polychaetes and also in the ambulacral grooves of asteroids; the gastropod shells occupied by hermit-crabs contain the polychaetes *Nereis fucata* and *Harmothoe coeliaca*. A lamellibranch with a sucker-like foot was found by Ohshima (1930) in the burrow of a holothurian. Members of the fish genus *Fierasfer* hide in the cloacal cavity of holothurians or in the mantle cavity of lamellibranchs; similarly the small crab *Pinnotheres* is found in lamellibranchs, and related forms occur in the hind-gut of echinoids and holothurians and in the pharynx of tunicates. Prawns live as commensals in sponges, lamellibranchs, tunicates and crinoids, and an alpheid has been observed, usually in pairs, between the spines of an echninoid. It is probable that, with the exception of *Fierasfer*, these commensal forms obtain food as well as shelter from their hosts, by picking up excess or rejected food particles or by exploiting the faeces. Wells & Wells (1961) found bryozoan colonies living hypercommensally on the carapace of a pinnotherid that was living in a holothurian; these authors considered that both the crab and the bryozoan were filtering off particles from the respiratory water current of the holothurian. In some pinnotherids the relationship appears to be very close to parasitism, and the invasion of holothurians by *Fierasfer* causes considerable damage.

The association of *Echeneis remora* with whales, sharks, turtles and large fishes is well known, the remora attaching itself by means of a sucker formed by the modified dorsal fin. *Echeneis* presumably feeds on fragments of the food of its transporting animal, although it has also been found in the branchial cavities of *Xiphias, Istiophorus* and *Mola*.

There are several cases of the association of various animals with cnidarians. The peculiar relationship between the pelagic gastropod *Phyllirhoë* and the medusa *Mnestra* has already been mentioned (see p. 179), and so has the association of the fish *Nomeus gronovii* with *Physalia* (see p. 172). The fish *Amphiprion* finds protection among the extended tentacles of large actinians (see p. 264). Numerous associations exist between crustaceans and actinians, for quite a number of crabs fix actinians on to their carapace; hermit-crabs shift 'their own' actinians across when they move into a new gastropod shell, and the Indian Ocean crab *Lybia* carries actinians around in its pincers, using them as weapons. In the latter case it has not been proved that prey is actually caught by the actinians, but this is conceivable; in any case the food will be seized by the first pair of walking legs, because the pincers are occupied in holding the actinians.

Some mysids (Clarke 1955) and prawns (Limbaugh *et al.* 1961) are also associated with actinians. Here the prawns appear to act as 'cleaners', collecting detritus, fragments of cast epithelium, food particles and parasites from the actinian. Limbaugh *et al.* showed experimentally that an association of this kind also has an effect on a third species: they removed all the snapper prawns (*Alpheus armatus*) from an area, and after a few days the actinians with which they had been associated had all been eaten, presumably by stomatopods. Evidently, therefore, the snapping shrimps keep the stomatopods away from the actinians.

Prawns often act as cleaners of fish, and there are also fish which clean other fish. This highly interesting pattern of behaviour (see Eibl-Eibesfeldt 1955, von Wahlert 1961) is somewhat analogous to that of the well-known oxpeckers on land.

Symbioses between animals and unicellular algae are particularly widespread in tropical seas (summary in Yonge 1957). From experimental evidence, Yonge (1944) concluded that the algae profit by the direct exploitation of the animal's metabolic products and that the animals profit from the removal of metabolic products and in some cases (e.g. foraminiferans, radiolarians, lamellibranchs of the family Tridacnidae) by actually consuming the algae.

Some ciliates which live in the gut of various animals should probably also be regarded as symbionts. Thus, Beers (1963) found that the gut of the echinoid *Strongylocentrotus droebachiensis*, which feeds mainly on *Laminaria*, regularly contains three holotrich species, and occasionally four other species. Here the ciliates eat bacteria, and their division is dependent upon the feeding activity of the sea-urchin and on other still unknown factors. It is possible that it is these endocommensals which have enabled the echinoid to make the transition to a vegetarian diet.

On the other hand, the fungi (Eccrinales) found in the gut of arthropods have been regarded as neutral commensals (Lichtwardt 1961).

Groups which are all parasitic or which contain parasitic representatives are also widespread in the sea, e.g.

Sporozoans in nemertines, polychaetes, crustaceans, echinoderms, lamellibranchs, gastropods, cephalopods, fishes;

Trematodes particularly in molluscs and fishes;

Cestodes in crustaceans, molluscs and fishes;

Acanthocephalans in mammals and birds with developmental stages in crustaceans and fishes;

Nematodes, leeches, acarines and among the ceopepods particularly the Caligoidae, Arguloidae, Lernaeopodidae etc, occur in various hosts, including invertebrates.

More or less extensive parasitism has evolved in several classes and orders (synopsis in Hopkins 1957): among the turbellarians there are parasites in all the orders and in the most diverse hosts; parasites occur among the nemertines (*Carcinonemertes*), polychaetes (*Ichthyotomus* and the eunicids *Haematocleptes*, *Labrorostratus*, *Oligognathus* and *Ophiuricola*); the myzostomids are endoparasites and ectoparasites on echinoderms; the prosobranch gastropods contain every transitional stages from suckers of blood and tissue fluids to endoparasites; parasitic forms are found among the tardigrades, pycnogonids, cirripedes and isopods; the parasites known as whale lice are amphipods of the family Cyamidae.

In some cases, parasitism may be of considerable ecological importance, as when the host population is seriously affected, sometimes causing castration or mass mortality. Parasitic castration has been observed, for example, in gastropods parasitized by trematodes, in crabs by cirripedes, in copepods by Elliobiopsidae. A particularly serious case of mass mortality was the extermination of *Zostera marina* by the rhizopod *Labyrinthula macrocystis* in the coastal regions of the North Atlantic. This involved not only the destruction of the *Zostera* plants, but also of the whole fauna endemic in the *Zostera* beds.

The importance of interspecific competition as an ecological factor has already been mentioned (pp. 109, 200), so this needed not be discussed in further detail. It should, however, be emphasized that relatively little is known about this. Competition could be the cause of certain phenomena which give the impression of succession in the occurrence of various species. Several long-term investigations have shown that over more or less extensive areas a single species of a class or order occurs in the largest numbers. This species will show

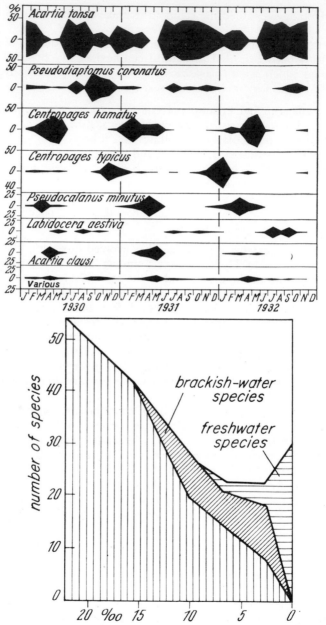

Figure 192.   Relative percentages, by count, of the species of copepods in Delaware Bay from January 1930 to December 1932 (after Deevey 1960 a)

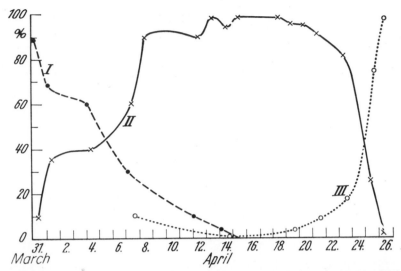

Figure 193. Percentages of *Thalassiosira* (I), of *Chaetoceros* (II) and of peridinians (III) in the total phytoplankton off the coast of Massachusetts, March–April 1938 (after Lillick 1940)

minima and maxima depending upon the time of year; other species only appear at the time when this dominant species is at its minimum. The occurrence of species whose maximum coincides with that of the dominant species, seems to be dependent upon the relative abundance of the main species at a given time (Fig. 192). The extent to which this involves direct interspecific relationships or the different ecological requirements of the species can only be clarified by further investigations. Frequency graphs, as given for example by Störmer (1928) for two copepod species or in Fig. 193 give some idea of this problem.

In the pelagial, there have been numerous observations and experimental investigations during the last thirty years on the various predator-prey relationships between phytoplankton and zooplankton. There is no need to discuss the history of this as it has been dealt with several times by Lucas (e.g. 1938, 1947, 1949, 1955).

One of the most striking phenomena is the mass mortality of fish associated with blooms of certain phytoplanktonic organisms. Under special hydrographical conditions, certain individual planktonic plants may either increase phenomenally in the vicinity of the coast or in oceanic areas. These may impart a striking coloration, usually reddish or brownish, to the water, which has given rise to the terms 'red water' and 'red tide'. Observations on this phenomenon have been summarized by e.g. Brongersma-Sanders (1948) and Hayes & Austin

TABLE 56

Number of individuals of *Gonyaulax monilata* inside (a, b, c) and outside (d, e) an area of coloured water off the coast of Florida

(after Slobodkin 1953)

a		b		c		d		e	
Depth in metres	Millions Indiv./l	Depth in metres	Millions Indiv./l	Depth in metres	Millions Indiv./l	Depth in metres	Millions Indiv./l	Depth in metres	Millions Indiv./l
0·15	3·60	0·21	4·60	0	8·20	0·15	0·80	0·15	0·57
0·45	2·15	0·37	1·00	0·30	5·70	0·60	0·17	0·45	0·46
0·78	0·9	—	—	0·45	1·45	1·53	1·19	0·99	0·07
—	—	—	—	2·60	0·45	2·60	0·94	—	—
—	—	—	—	—	—	3·29	0·73		

(1951). The principal organisms causing these water blooms are: blue-green algae of the genus *Trichodesmium* and a number of dinoflagellates, such as *Prorocentrum micans*, *Gonyaulax polyedra*, *G. polygramma*, *Gymnodinium brevis*, *G. splendens*, *G. mikimotoi*, *Cochlodinium catmatum*, *Glenodinium rubrum* and others. Table 56 gives an idea of the number of individuals producing a 'red tide' off Florida. Mass outbreaks of diatoms and green algae may also occur (cf. e.g. Tsujita 1955).

The fish mortality frequently associated with these outbreaks has been traced to certain toxins produced by the phytoplanktonic organisms. (Investigations are still needed to find out whether these are true exotoxins, that is, substances given off into the water, or endotoxins which are only released when the phytoplankton cells decompose. There have only been a few studies on the nature of these poisonous substances and their action (Abbott & Ballantine 1957, Starr 1958).)

Gunter *et al.* (1947) reported an extreme case in which the waters off Florida contained enormous numbers of a species of *Gymnodinium* which caused serious injury to the mucous epithelium of several coastal animals.

Damage to animals from mass outbreaks of phytoplankton has also been observed in freshwater organisms (e.g. Prescott 1948, Lefèvre & Nisbet 1948, Ryther 1954); benthonic organisms may be damaged by outbreaks of certain planktonic plants (e.g. Nightingale 1937, Loosanoff & Engle 1947). This is evidently, therefore, a widespread phenomenon. There have been some particularly extreme cases in which mass outbreaks have resulted in the accumulation of enormous amounts of metabolic products causing damage to animals. But even lesser concentrations of such phytoplanktonic organisms may produce substances, the effect of which is less obvious owing to the lower concentration: these would require more detailed investigation.

Hardy (1935) developed the theory of animal exclusion, according to which water masses gradually acquire from their inhabitants certain chemical constituents which prevent further colonization by some animals. Lucas developed the more comprehensive theory of biological conditioning, which has since been confirmed by some experimental evidence. Thus, Wilson (1951), for example, reared parallel cultures of the eggs and larvae of the polychaetes *Ophelia bicornis* and *Sabellaria alveolata* in water from the Irish Sea and from the English Channel near the Eddystone. Development was normal in the Irish Sea water, but the larvae reared in the English Channel water were abnormal and died prematurely (Fig. 194). The biological effect of the two types of water was, therefore, very different, in spite of the fact that the conditions were otherwise the same (salinity, temperature, feeding); this is doubtless a case of biological conditioning.

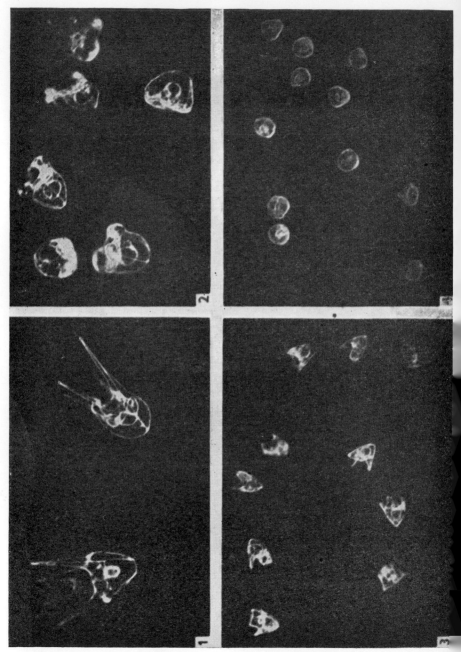

Figure 194. Plutei of *Echinus esculentus*: 1 and 3, normal development in water from the Celtic Sea; 2 and 4, abnormal development in water from off Plymouth (after Wilson 1952)

From a series of experimental investigations, Bainbridge (1953) found that plankton animals might be inhibited or stimulated by phytoplanktonic organisms, or their behaviour might be neutral; in several experiments effects were also obtained from the culture liquid after the algal cells had been filtered off. Allen (1956) showed that filtrates of *Chlamydomonas* cultures encouraged the growth of *Chlorella*, even under conditions were this organism did not otherwise thrive.

Sea water has long been considered to have a bactericidal action (e.g. Cassedebat 1894, cited in Benecke 1933, Vaccaro *et al.* 1950, Burkholder 1963), which has been ascribed to the presence of marine plants. Devèze (1955) demonstrated an inverse relationship between the development of the biomass in the pelagial and that of the bacterial cells, the latter showing a minimum at the time of maximal phytoplankton production.

These ideas on the occurrence of inhibiting and stimulating substances have, in the meantime, received confirmation from culture experiments and from investigations in the sea itself. Attention has been paid particularly to vitamins of the B_{12} complex and to carotenoids (see pp. 100, 238; Provasoli & Pintner 1953, Droop 1954, 1959, Lewin 1954).

In view of these physiological relationships it is understandable that neighbouring water masses, in spite of tiny differences in their physical and chemical characteristics, may show considerable differences in their populations, which are often characterized by special species. Several authors have reported the occurrence of this kind of indicator species, belonging to various animal groups, e.g. Russell (1935), Fraser (1952).

Thorson (1946) showed that in northern seas many benthonic animals spawn soon after the outburst of phytoplankton before the temperature starts to rise. Similar observations were made by Miyazaki (1938) on oysters on the coast of Japan. This suggests that the animals react to specific plant substances.

Benthonic animals are evidently involved in the phenomenon of biological conditioning in another way. Many species inhabit quite specific substrates; at the time of metamorphosis and the transition to benthonic life their pelagic larvae must be carried purely at random to this substrate, unless a 'guiding mechanism' could be demonstrated. Now it has been shown, for example, by Wilson (1948, 1953, 1954) in the polychaete *Ophelia bicornis* and by Jägersten (1940) in the archiannelid *Protodrilus rubropharyngeus* that specific substances reach the water from the substrate and that these stimulate metamorphosis; thus the larvae carried by currents into the vicinity of such a substrate will mostly find a place to settle. The nature of these substances and the mechanism of their action are not yet known; according to Jägersten (l.c.) the larvae become photonegative in the presence of the relevant substance and are thus led to the bottom.

Bourdillon (1954) confirmed that the metamorphosis of the larvae of *Alcyonium* was speeded up by a water-soluble substance from the skeleton of horny corals, which also had an attracting effect.

The work of Valera & Kylin (1942, 1943) and of Levring (1945) has shown that littoral algae also have an effect, that is sometimes inhibiting, sometimes stimulating. Further investigations on antibiotic substances in marine algae have been carried out by Pratt *et al.* (1951), Vacca & Walsh 1954, Chesters & Scott (1956), Ross (1957), Nigrelli (1958), Burkholder (1958) and Burkholder *et al.* (1960). According to these studies this phenomenon is very widespread in the sea and is of general validity, as has been emphasized particularly by the work of Burkholder (1958) on gorgonarians. The whole subject is still, however, by no means clear; we still do not know the extent to which substances with antibiotic action pass into solution and condition the medium, how long they remain active and which organisms react to them.

There is also the possibility that apparently simple cases may be more complex than they seem to be. Thus, Bein (1954), for example, investigated a sea water in which there was a mass outbreak of *Gymnodunium* accompanied by fish mortality; from this water he was able to isolate *Flavobacterium piscicida*. Twenty-four-hour cultures of this bacterium in a 0·1% peptone solution, when added to an aquarium containing various fishes, caused their death. The specific action of the bacterial culture was tested by appropriate control experiments. It is, therefore, an open question whether the observed fish mortality was caused directly by *Gymnodinium* or by the bacterium accompanying it. This may also happen in other cases.

Stohler (1959) has drawn attention to another possibility. A red tide in the Baja California (Mexico), associated with mass mortality of fishes, molluscs, *Sipunculus*, etc., was caused by *Goniaulax polyedra*, which was present at the rate of 15·4 million individuals per litre. Owing to the red pigment of the flagellates, infra-red rays were absorbed more strongly than normal, so that the water temperature increased to 29·4°C. The heat together with the oxygen consumption of the *Gonyaulax* population led to a reduction of the normal oxygen content and the sensitive species died. The subsequent decomposition of the dead organisms caused further oxygen deficiency, which together with the toxicity increased the mortality; there was therefore a kind of chain reaction which requires further study.

All the cases mentioned so far have been concerned with interspecific reactions. From some investigations, e.g. Pratt (1943), on the unicellular alga *Chlorella*, we learn of self-inhibition, when the population density has reached a certain value. Possibly the simultaneous outbreak of plankton blooms is also reducible to this kind of phenomenon, although it is not always safe to apply the results of laboratory investigations to conditions in the wild. In the same way the so-called allelokatalytic affect, in which the members of a species stimulate themselves mutually, is still not clear; it could play a role in the increased settlement of pelagic larvae at certain places or in simultaneous mass spawning.

In general, it should be emphasized that interspecific effects caused by chemical agencies are absolutely selective and in no way affect all the sympatric species. Inhibition or extermination of populations of certain species in a restricted area probably creates new conditions for the species not affected, perhaps providing them with improved conditions. On the other hand, one would scarcely expect any long-term after-effects, following, for instance, a red tide, because the original conditions will soon be re-established by the free exchange of water. This raises the question of the direction, speed and method of recolonization.

Chapter VII

LIFE ON THE MARGINS OF THE SEA

The preceding chapters have been concerned principally with the primary inhabitants of the sea, that is, with the organisms which have either existed in the sea from the beginning or have lived there so long that it is now scarcely possible to recognize where they originally came from.

However, there have been many exchanges of organisms between sea and land as well as between fresh and sea water; there has in fact been emigration from the sea to the other two major habitats and immigration from them into the sea. The forms which have taken part in these moves have to a considerable extent retained biological or ecological relationships with their original habitat and are therefore more or less restricted to the marginal area of the seas; the expression marginal area can be taken literally in the spatial sense and it also denotes that salinity is the most important factor in comparing sea and fresh water or sea and land. The numerous biological and physiological phenomena connected with this exchange cannot be considered in greater detail here, but there is an extensive body of literature on this subject. Here we will mention a few examples in order to illustrate the general process. Remane (1963) has given a detailed account of biological evolution in the sea and fresh water.

1. Emigrants

Various crustaceans have moved from the littoral zone on to the adjacent land. Amphipods of the genera *Talitrus* and *Talorchestia* live in the supralittoral, either in sand or under stones and jetsam. Isopods of the genus *Ligia* are found on hard substrates in the supralittoral, *L. oceanica* being active at night in cool temperate climates while the tropical *L. exotica* runs around in strong sunlight. These two species still have to go down to the water at intervals to wet their gills and so they are tied to its immediate vicinity, but the closely related *Ligidium* no longer lives near to the coast but in marshland. Some of the fiddler-crabs which invade grassy mounds in the shore zone behave in a simi-lar way (Altevogt 1957). Hermit-crabs of the genera *Coenobita* and *Birgus* penetrate farther inland and are even more independent of the water. *Coenobita* is still tied to its protective 'house', but the robber crab (*Birgus latro*) lives without such protection; the hind part of its body is strongly chitinized and has become symmetrical. The degree of adaptation to terrestrial life is well shown in the modified structure of the gill cavity which enables these crustaceans to breathe air. On the other hand they are still tied to the sea because the development of the larvae takes place in the water.

The same applies, for example, to the opisthobranch gastropod *Alderia modesta* which in the North Sea and Baltic Sea lives more or less amphibiously on algal growths up into the supralittoral, but its veliger larvae develop in the water.

Among the polychaetes, representatives of the family Nereidae have invaded the land; in Indonesia several species of *Lycastis* live in holes in the earth, and in laboratory cultures it has been possible to rear them on a vegetarian diet (Pflugfelder 1933, Harms 1948). *Parergodrilus* can also be included here (Reisinger 1960).

Among the nemertines, too, there are several species which live a terrestrial life completely independent of the sea. There have probably been different emigration routes, independent of each other.

Numerous investigations have been devoted to the invasion of fresh waters by marine animals. In many cases these quite obviously concern relicts, such as the amphipods and polychaetes of Lake Baikal; Seger-stråle (1957) has dealt in detail with the relict amphipods *Pontoporeia affinis* and *Gammaracanthus lacustris*, the isopod *Mesidothea entomon* and the fish *Myoxocephalus* (*Cottus*) *quadricornis* and others in the marginal areas of the eastern Baltic Sea.

In the European area macrofaunal immigrants into fresh waters in-clude, for example, the medusa *Craspedacusta sowerbyi*, the prawn *Atyaephyra desmaresti* and the mitten-crab *Eriocheir sinensis*. In addition, there are several forms which penetrate into brackish waters (see p. 419). Thienemann (1950) mentions a range of forms which have in-vaded the waters of caves and also draws attention to species found in inland subterranean waters. The geographical distribution of these animals remained a mystery until the more recent investigations on the interstitial microfauna in coastal areas and rivers demonstrated their origin from marine forms (summaries, etc., in Delamare Deboutteville 1960, and Ax 1956).

Ax found identical species in marine and freshwater interstitial sys-tems and this suggests that immigration is not complete but is still continuing in temperate latitudes. This points to astonishing physiological capabilities, in so far as the species concerned have been able to overcome the difference in salinity between sea and fresh water. Here the absence of light in subterranean sites in the mesopsammon and in caves evidently plays an important role, but the functional mechanism of this is still not clear (cf. amongst others Friedrich 1961).

Very few members of the Porifera, Cnidaria and Bryozoa have penetrated into fresh waters.

In warm-water areas the passage from the marine to the freshwater régime has become more complete than in temperate and cold areas of the sea, but this will not be discussed in further detail. The dominance of these lines of emigration may be due partly to the higher mean temperature and to the greater constancy of temperature and light as ecological factors, and partly to the smaller changes which have taken place in geological time in warm-water regions; in areas nearer the poles such changes must have been greater owing to the ice ages.

2. Immigrants

Remane (1950) has given the first general account of the freshwater immigrants into the sea, using data from the area of the North and Baltic Seas. The relatively recent investigation of this problem is due in part to the fact that the material consists mainly of small organisms, knowledge of which has only increased in recent times. The following points are taken from Remane's work.

It is only in a few animal groups that we can recognize a large number of marine species which can be regarded as immigrants from fresh waters; this occurs particularly in the phyllopods, rotifers, oligochaetes (but not lumbriculids) and hirudineans; groups with a few species or even single representatives which extend into brackish water or marginal areas with higher salinities include the gastrotrichs, tardigrades, ostracods and the Hydracaridae among the mites. Freshwater molluscs and copepods must be regarded as hostile to sea water; the nematodes appear to find this migration route difficult or even impossible, although even here there are a few immigrants. In contrast to the Hydracaridae, the Halacaridae are predominantly marine animals which occur in a wealth of species and to some extent also in great numbers of individuals; they, however, have been derived from terrestrial forms and so are not directly comparable with the Hydracaridae.

It is only the phyllopods and rotifers that show immigration into the pelagial (see p. 190); the other forms are tied to the benthal or to the phytal where they occur in shore pools with salinities showing greater fluctuations than other habitats.

The hirudineans are remarkable in so far as the marine representatives all belong to the family Piscicolidae, and these are classified in a number of specifically marine genera; no genus in this family contains species from both sea and fresh water. This suggests a considerable amount of speciation and structural change. In contrast to this the other groups of immigrants into the sea have produced few new species, while there are scarcely any examples of the development of new forms of organization warranting generic rank. The peculiarity of the Piscicolidae in this respect is probably correlated with their parasitic habits; it is possible that a comparable analysis of the trematodes and cestodes which also occur in large numbers in the sea might also establish certain general principles, which are known from the land and fresh waters.

There are two groups of freshwater insects which are noteworthy as immigrants; Corixidae are often found in large numbers in shore pools and have also occasionally been collected in oceanic waters far from the coasts (*Trichocorixa verticalis* according to Gunter 1959), although in this case one cannot exclude the possibility that they have been carried there. The hemipterans of the genus *Halobates* (see p. 190) can, however, be regarded as permanent inhabitants of the open ocean. Among the Diptera there are some flies which extend far into the sea, where their behaviour shows rhythms dependent upon the tides (see pp. 356 et seq.).

Marine vertebrates are generally regarded as immigrants from fresh waters, even if their freshwater origin can in some cases be traced back to the Palaeozoic. During the course of geological history there have probably been sevral migrations from fresh waters, and the Recent fish fauna also shows numerous examples of return migration (Romer 1955, Remane 1963).

3. Terrestrial forms in the marine region

(a) *Insects, arachnids, myriapods*

Within these groups there are only a few truly marine animals and only one small genus which has invaded the open sea, namely *Halobates*, the hemipteran genus mentioned above (see p. 172). From the viewpoint of tolerance to salinity the insects occurring on the coast and in the littoral zone have been divided into those tied to salt—the halobionts—and those that are salt-loving—the halophils; these terms, however, include forms which only live at inland salt-water sites, and so here I would prefer the alternative use of the terms thalassobionts (those tied to the sea) and thalassophils.

Thalassophilic forms are those which prefer the marine coastal habitat but which are not restricted to it and whose preference for this environment may vary according to the locality. Several different factors may be involved in this preference for the coasts: humidity, abundance of food, specialization for particular foods, suitable substrates for burrowing forms and so on. The diversity of these forms, particularly in regard to their biology and behaviour, prohibits a more detailed treatment.

Thalassobiont forms occur in various groups of arthropods, and a few examples should be given. Schuster (1962) has described the ecology and distribution of small littoral arthropods based on collections made in the Mediterranean and Brazil, from which it appears that some genera and even species have a worldwide distribution. The reasons for the association of thalassobiont arthropods with the marine littoral are still not at all clear but they appear to be not entirely dependent upon the salinity of the substrate.

Several genera of lice are parasitic on pinnipedes. They mostly live on the head of the host, but some species also occur on the body and extremities. Their biology and physiology are insufficiently known. They are air-breathers which can accumulate pockets of air between their bristles and scales, thus enabling them to survive for a long time under water. The degree to which the individual species are specialized for living on certain hosts is not yet clear (Freund 1927).

Among the thalassobiont beetles (v. Lengerken 1929) there are a few submerse forms such as the chrysomelid *Haemonia* (*Macroplea*) *mutica*. Both the larvae and imagines of this species live in coastal brackish waters, the imagines on plants such as *Ruppia* and *Zostera*, the larvae in mud. For this submerse way of life the only special adaptations shown are those convected with respiration: in older larvae the hind pair of stigmata is provided with tiny hooks which are used to bore into the air-containing tissues of plants. With the help of their hairy antennae the imagines collect the gas bubbles formed by plant assimilation or they cut the plant tissues and take the gas which emerges when assimilation is low. These beetles will die if the antennae are removed.

Several species belonging to different families live in the tidal zone, where both larvae and beetles are completely covered with sea water at high tide. At this time they remain in excavated tunnels, crevices, small cavities, under stones, etc., where they probably use air held by capillary action for respiration (Alluaud 1926); I do not know of any special adaptations for respiration. At low tide most of them move around in search of prey. An exact analysis of their activity rhythms would be interesting because the periods of light and submergence will fluctuate with the tides.

From the spray zone down to the level of low water springs, there are species of thalassobiont Diptera Nematocera, in which the larvae have very diverse habits. In those living in the lowermost eulittoral an epidemic hatching of the very short-lived imagines has been observed (see pp. 356 et seq.), and this is correlated with the onset of the lowest tides. Diptera Brachycera are found in great numbers on the seashore; the degree to which they are tied to marine biotopes is extraordinarily varied; many of them are closely associated with the masses of seaweed washed up on the beach.

There are only a few species of thalassobiont Lepidoptera. The phytophagous larvae of some species feed on halophytes and are therefore more or less tied to coastal areas, but most of them appear to occur also at inland sites with salt. Submerse larvae living on eel-grass usually occur in brackish and fresh waters.

A few species of Diplopoda and Chilopoda occur in the tidal zone on rocky coasts, where they find shelter under stones and in rock crevices.

In the supralittoral and to some extent in the eulittoral, thalassobiont insects, in particular, have an important share in the composition of the community and thus play quite a considerable role in the biology of these zones. Like the arachnids and myriapods they are of very little importance in the open sea.

Spiders of the genus *Desis*, which has about twelve species, are regular inhabitants of the eulittoral in coastal coral-reefs; according to Berland (1940), however, they have not reached coral islands lying far distant from the continents. Their habits are somewhat comparable to those of the European freshwater spider *Argyroneta aquatica*; they spin chambers in crevices in the coral-reef or among groups of serpulids, in which they store air, even at high tide. At low tide they go hunting and are evidently not selective in their choice of prey.

There are a few species belonging to other spider families in which similar habits have been described.

Among the pseudoscorpions, the species *Obisium littorale* amongst others has invaded the marine region.

(b) *Marine iguanas, chelonians and sea-snakes* (cf. Neill 1958)

Marine iguanas (*Amblyrhynchus cristatus*) feed entirely in the sea. These lizards, which are up to at least a metre in length, live on the Galapagos Islands in the immediate vicinity of the water. During the night they lie hidden in caves and crevices, but by day they feed exclusively on seaweeds which they eat in the sea. They only take the tender tips which they apparently swallow without chewing. They swim by snake-like movements in which the laterally compressed tail plays an important role; the legs are not used during swimming. The dorsal crest which is best developed on the nape may also assist the snake-like movements. As adaptations for diving the nostrils are surrounded by a raised ring of skin and the small ear-drums are protected by thick pads of skin. These reptiles, which were once present in enormous numbers, lay their eggs on the shore (cf. Mertens 1927). When threattened they flee towards the land, not into the water, which is therefore only used for feeding.

The marine turtles can be regarded as inhabitants of the margins of the seas in so far as they lay their eggs in pits dug in the sand on certain islands and coastal stretches; apart from this they spend their lives swimming in the sea and are sometimes encountered far out in the ocean. Their limbs are modified to form paddles in which the fingers and toes are not articulated. Movement takes place principally at the elbow joint, the most important muscles being attached to a ridge on the powerful but very short upper arm bone. These reptiles swim fast and elegantly, but they can also drift motionless and asleep at the surface. The leathery turtle (*Dermochelys coriacea*), the hawk-billed turtle (*Chelonia imbricata*) and the loggerhead turtle (*Caretta caretta*) are carnivores, which feed on medusae, crustaceans, cephalopods, lamellibranchs and fish, while the green turtle (*Chelonia mydas*) feeds largely on seaweeds and eel-grass. As a further adaptation to this long sojourn in the water they have developed a form of buccal respiration, in which the highly vascularized mucous epithelium takes up oxygen from water in the mouth. 'Specimens (leathery turtles) kept on land appear to have trouble in breathing. At any rate they only survive out of water for a few days (not more than five at the most)' (Wagner 1959, p. 72).

During egg-laying, marine turtles secrete great quantities of tears; the suggestion that this is due to pain, as suggested in popular films, is much too anthropomorphic and it is far more likely that it is a method of removing sand grains from the eyes (Wagner 1959). It is also possible that the tear glands have a function similar to that found in birds (cf. Schmidt-Nielsen 1959), but this has not been investigated.

There are several snakes living in the sea, the vast majority of which are endemic in the Indo-Australasian region; only one species, *Pelamis platurus*, also lives on the west coast of central America. These reptiles do not reach the lengths sometimes claimed, although specimens three metres long have been confirmed. There are two subfamilies, the Hydrophiinae and the Platycaudinae, both characterized by the laterally compressed rudder-like tail and the very narrow or completely reduced ventral plates. The latter character suggests that they would be helpless on land. Little is known of their habits. Most of the species are believed to be viviparous, the young probably being born on land. They feed mainly on fish; the statements on their venom are contradictory, and it is said that these these reptiles will only bite when actually hunting for prey. According to recent investigations the mucous epithelium of the buccal cavity is supplied with numerous capillaries which enable these snakes, like the turtles, to take oxygen from the water. This would explain the records of sea-snakes seen resting on the bottom and remaining submerged for several hours.

(c) *Birds of the ocean and coast*

The bird life of the seas and coasts is so numerous and biologically so varied that only a few general remarks can be made here. For further information the reader should consult Alexander (1959), Fisher & Lockley (1954); both works are fully illustrated. In the strict sense, marine birds are those such as penguins and many petrels which spend the greater part of their life in or on the sea, and only go ashore to any extent during the breeding season. There is, however, no sharp line of demarcation between these and the sea-birds which are more closely associated with coastal areas.

Penguins occur entirely in the southern hemisphere, with the exception of one species which has reached the Galapagos Islands. They live chiefly in open water and show a number of important adaptive characters: the wings form flippers which can beat at the rate of 120–200 strokes per minute, giving the bird a speed of up to ten metres per second. The feathers are short with flattened shafts and reduced vanes, so that they cover the body like a layer of scales. The bones lack air spaces so that the body is relatively heavy and when diving the bird does not have to overcome a great amount of buoyancy. The characteristic erect posture is due to the very posterior position of the legs; these have large webbed feet which are used in steering when the bird is swimming.

Penguins feed exclusively on fishes, pelagic crustaceans (Euphausiacea), cephalopods and other pelagic animals. The upper surface of the tongue and the gums are covered with hard, posteriorly directed horny denticles which help to grip the prey. One of the principal enemies of the penguin is the leopard seal which, on account of its greater speed, can even chase and catch these birds in the water; sharks and large bony fishes will also sometimes attack penguins.

Penguins breed on land in colonies which are usually large; most species build a very simple form of nest, but the Emperor penguin, *Aptenodytes forsteri*, carries the single egg on the webbed feet, enclosed in a fold of skin between the body, tail and feet. Large feeding crèches are formed in the breeding colonies: a number of young, up to 150, congregate in one place and are guarded by 12–20 adults, while the remainder go off in search of food; the young are fed in the colony.

Many birds of the order Procellariiformes spend a large part of their life on the water or sailing in the air above; this group includes the albatrosses and petrels. In contrast to the majority of terrestrial gliding birds, the wings of the petrels are long and narrow; terrestrial gliding birds use vertical air-streams or up-currents, whereas those that glide over the sea mainly make use of horizontal winds, the speed of which may change very rapidly. With narrow wings the angle of attack can be changed more quickly. Other petrels, such as the shearwaters (*Puffinus*) and diving petrels (*Pelecanoides*) spend more time swimming and diving; these have a whirring flight.

Some of the Procellariiformes are predatory and more or less omnivorous and their beak is powerful and hooked; others sieve plankton and have lamellae on the upper mandible and a very extensible floor to the mouth. These plankton feeders, e.g. in the genera *Pachyptila* and *Halobaena*, are small and nocturnal; their nocturnal habits are correlated with the vertical migration of plankton during darkness.

Some species breed in large colonies, e.g. the albatrosses, usually building simple nests of mud and pieces of vegetation; others such as Wilson's petrel (*Oceanites*) and the shearwaters (*Puffinus*) nest in burrows. On account of their short legs some petrels are extraordinarily helpless on land.

The order Pelicaniformes contains some quite distinct biological types, including the frigate birds, tropic birds, gannets and boobies, pelicans and cormorants. The flightless cormorant (*Nannopterum*) lives on some of the islands in the Galapagos group; all the forms mentioned are in general rather helpless on land and many of them are incapable of taking off from the ground. They therefore nest on steep coasts, cliffs, trees, etc., where they also settle to rest. The tropic birds and cormorants, for example, have pigeon-like flight with fast wing-beats, whereas the frigate birds are essentially soarers. There are marked differences in their feeding habits: cormorants swim at the surface from which they dive and follow their prey underwater; this is why they can be trained to catch fish, as was formerly done by the Chinese. The others mostly dive from the air for prey sighted in the water. The development of the bill is correlated with the feeding habits; the cormorants have a hooked tip to the bill, most of the others having a powerful, conical bill which acts rather like an awl when they dive. Frigate birds, on the other hand, take their food in flight from the water surface, although they also chase other birds and force them to jettison their prey.

The auks to a certain extent form a geographical analogue to the penguins of the south. With the exception of the extinct giant auk (*Pinguinus impennis*), they can, however, fly although the wings are short and only permit a kind of whirring flight. The wings are used as paddles in swimming. As in the penguins, the legs are positioned posteriorly so that the body is held erect; in walking and sitting the whole of the leg rests on the substrate and the tail acts as a third support. They feed chiefly on fish, but some species also take crustaceans (auks) and molluscs (puffins). The considerable diversity in feeding habits is reflected in the very varied development of the bill.

The auks live on cliffs in northern seas, where they form large breeding colonies. The clutch usually consists of a single egg, only a few species laying two eggs; the egg is laid either on the naked rock, or on a very simple nest mound or sometimes in a crevice. Unlike the penguins, the auks do not form crèches for rearing the young.

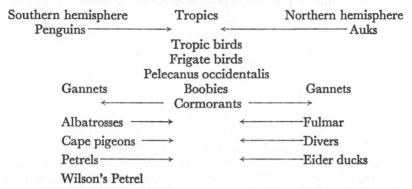

Diagram showing geographical distribution of sea birds

Southern hemisphere	Tropics	Northern hemisphere
Penguins ⟶		⟵ Auks
	Tropic birds	
	Frigate birds	
	Pelecanus occidentalis	
Gannets	Boobies	Gannets
⟵	Cormorants ⟶	
Albatrosses ⟶		⟵ Fulmar
Cape pigeons ⟶		⟵ Divers
Petrels ⟶		⟵ Eider ducks
Wilson's Petrel		

Only brief mention will be made of the coastal birds. Biologically, these are not sharply differentiated from the sea birds. They show various differences in their ability to dive (which is very well developed for example in some ducks), in their ability to swim, fly and soar, in the position of the breeding site relative to the coast, and in the extent to which they obtain food from the sea. Among coastal birds mention may be made of the gulls and terns, sandpipers, plovers, oystercatchers, redshank, avocet, ducks and geese. Biologically these show great differences in the type of food and methods of feeding, in breeding biology and in social and other behaviour. It is not possible, within the scope of the present work to discuss such interesting points in detail. Future investigations should include work on their behaviour, the ability to dive and its physiological and anatomical requirements, the excretion by special salt glands of salt taken up in food or drinking water (Schmidt-Nielsen 1959), sensory mechanisms used in feeding underwater, migrations and so on.

It is difficult to assess the role of oceanic and coastal birds in the total economy of the sea. They doubtless play an important part as scavengers, particularly in coastal regions, and they are also considerable consumers of fish; the extent of this consumption is shown by the large amounts of guano deposited in some places. In view of the enormous size of many colonies and the quantity of food they require one should not underestimate the transport to the land of organic matter built up in the sea, although it should be remembered that these colonies are restricted to relatively few coastal areas so the effects of this movement of matter will be small in relation to the total area. These birds do, however, play a part in the general structure of the communities. In certain localities they may even have an effect on fisheries. In general, the position of sea-birds in the economy of the sea demands a more detailed treatment, based on their distribution and on the records of the numbers of individuals observed in different oceanic regions.

(d) *Coastal mammals*

Among the mammals certain members of the Carnivora and Sirenia have become aquatic and live along the sea coasts. The polar bear and sea otter show relatively few adaptations to aquatic life and are mainly to be regarded as land mammals. On the other hand, the Pinnipedia (regarding these as a suborder of the Carnivora) use coastal areas and the sea-ice as places for resting and for breeding; both morphologically and physiologically they are very well adapted to remain for long periods in the water where they are also able to sleep while floating. Scheffer (1958) has given a comprehensive account of the eared seals (Otariidae), walruses (Odobenidae) and true seals (Phocidae), which are the three families of Pinnipedia.

Pinnipedia occur in all geographical regions but they are particularly well represented in the cold-water areas of both hemispheres, e.g.

Arctic:	walrus (*Odobenus rosmarus*), bladder-nosed seal (*Cystophora cristata*), harp seal (*Pagophilus groenlandicus*), bearded seal (*Erignathus barbatus*), Alaska fur seal (*Callorhinus ursinus*),
Boreal:	common seal (*Phoca vitulina*) ringed seal (*Pusa hispida*), grey seal (*Halichoerus grypus*),
Warm waters:	monk seals (*Monachus monachus, M. tropicalis, M. schauinslandi*), Californian sea-lion (*Zalophus californianus*), northern elephant seal (*Mirounga angustirostris*),
Antiboreal:	Kerguelen fur-seal (*Arctocephalus gazella*),
Antarctic:	Weddell seal (*Leptonychotes weddelli*), crab-eating seal (*Lobodon carcinophagus*), leopard seal (*Hydrurga leptonyx*), southern elephant seal (*Mirounga leonina*).

Pinnipedia show adaptations for moving in a watery medium, for catching food by diving and to salinity changes. Heat is conserved by a layer of blubber (usually several centimetres thick) beneath the skin. The elongated, torpedo-shaped body, the short hairs and the development of the limbs as flippers are all characters adapted for living and moving in the water; the limbs are in many cases unable to support the body, so that on a firm substrate these animals are only capable of a shuffling movement. In connection with the ability to dive a high content of nitrogen and carbon dioxide is tolerated in the blood; the muscles contain a high concentration of myoglobin which probably allows a certain amount of oxygen to be stored. Dives of up to twenty minutes have been observed in several species and, as in whales, these require special respiratory mechanisms and certain structural modifications of the circulatory system. During diving, the rate of heart-beat decreases, the body temperature falls and the peripheral blood vessels become constricted.

Although much salt is taken in with the food the salt content of the blood is scarcely higher than in land mammals. The related problems of salt excretion have not yet been explained.

The seals are mainly fish-eaters; walruses feed chiefly on benthonic animals, particularly molluscs; the crab-eating seal (*Lobodon carcinophagus*) is so named because it feeds mainly on crustaceans; the leopard seal (*Hydrurga leptonyx*) is a predator which hunts other seals and penguins as well as fish. There are considerable differences in the dentition and the shape of the teeth which are correlated with the different feeding habits (Fig. 195).

Most seals live in large groups within which there are family parties

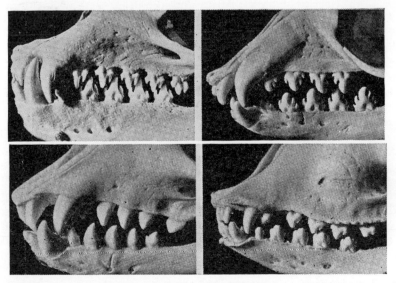

Figure 195. Dentition of seals: (a) grey seal (*Halichoerus grypus*), (b) bearded seal (*Erignathus barbatus*), (c) leopard seal (*Hydrurga leonina*), (d) crab-eating seal (*Lobodon carcinophagus*) (photo by S. Welleershaus, with the assistance of the Institute of Marine Research, Bremerhaven)

consisting of several females and a single bull. The communal life results in special behaviour patterns connected with the formation and maintenance of the harem, defence of the territory, etc.; one example is the inflatable proboscis of the elephant seal, the ethological meaning of which is still not clear.

Herd formation, their difficulty in moving on land and the lack of large enemies in their habitat have led to extensive exploitation of seal populations for the production of pelts. Some seal stocks are already seriously threatened.

In some forms the young are reared in the main living area, but the Alaska fur seal (*Callorhinus ursinus*) undertakes long migrations between its feeding grounds and its breeding area. This species breeds on islands in the Bering Sea, lives during winter between 30°N. and the edge of the ice and has extensive feeding grounds off the coast of California.

The sirenians are vegetarians, which are now restricted to warm-water areas, following the extinction of the small population of Steller's sea-cow (*Rhytina gigas*) on the Commander Islands in the Bering Sea; this took place in the middle of the 18th century shortly after the species had been discovered. The genus *Dugong* Lacépède (=*Halicore* Illiger) has one species in the coastal areas of the Indian Ocean, while *Trichechus* Linne (=*Manatus* Storr) has species on the Atlantic coasts of north America and West Africa and also in the fresh waters of the Amazon and Orinoco basins.

As in the whales, the hind limbs are reduced in the Sirenia; the fore limbs are modified to form flippers which are relatively flexible in their movements. The tail fluke, which is similar to that in the whales, is rounded in *Trichechus* and bilobed in *Dugong* and *Rhytina*. In contrast to the seals, the sirenians do not normally come out on dry land although the flippers are capable of supporting the body to a certain extent.

Sirenians feed on seaweeds and eel-grasses which are torn off and shaken to remove the sand. On the skull the premaxillae are elongated and ventrally tilted to give the impression of a beak. On the underside of the skull and also on the long symphysis joining the two halves of the lower jaw there is a rough horny plate which helps in gripping the food. The upper lips which have mobile bristly pads at the sides, are also used to seize drifting vegetation. The teeth are replaced in the same way as in the elephants; new teeth are formed behind and these move forwards, while the front teeth are resorbed at the roots and eventually fall out.

There is therefore great similarity between sirenians and elephants in the form of the skull and also in the structure and function of the jaws and teeth; as Mohr (1957, p. 14) wrote: 'The sirenian is a relative of the elephant adapted for aquatic life.'

Seals, sirenians and whales all have a similar arrangement of the superficial musculature which enables them to move the hind part of the body somewhat like a snake. In whales and sirenians the tail fluke is horizontal so that the movement is up and down; in the seals, on the other hand, the two hind limbs come together to form a vertical surface which is moved from side to side. In contrast to the condition in whales the neck vertebrae in seals and sirenians are not fused with each other and the skull can still be moved up and down. Seals and whales have a voice whereas sirenians appear to be mute; seals have a dense covering of hair (except for example the walrus) while the whales and sirenians have practically hairless skin.

4. Brackish waters, estuaries and coastal subterranean waters

Many bodies of water of varying size have salinities which are intermediate between those of oceanic water and true fresh water. Waters with intermediate salinities are known as brackish or mixohaline; under this designation Remane (1958) would also include inland waters containing salts, in which the relative ionic composition deviates considerably from the brackish waters of marine origin. On account of this difference in chemistry which produces physiological effects that are often profound, I would remove inland waters from the field under discussion here, even although there are some types there which are close to those in the marine region. One can differentiate between surface brackish water with little or no current (ponds, ditches, inland seas such as the Baltic and Black seas), surface brackish waters with strong currents (in river estuaries) and subterranean brackish waters (coastal subterranean waters).

There has been much discussion on ways of defining brackish waters, as opposed to the sea and fresh waters, and on the classification of brackish waters according to their salinities. To the biologist, a definition based on salinity presents special difficulties, because it gives a classification in which the behaviour of the organisms is related to the salinity. This behaviour is, however, extraordinarily variable in individual cases; many forms do not tolerate any reduction in the salinity of $> 30\%_0$ which is characteristic of oceanic water, others will not tolerate any small increase above the amount present in fresh water; some will tolerate different amounts of dilution or enrichment of the salt content, while others live exclusively in intermediate salinities and are incapable of living either in the oceanic or the freshwater region. Also, because salinity and temperature have a reciprocal effect on the physiology of the organisms and at the same time change the saturation value of oxygen, waters of the same intermediate salinity in different geographical areas with different climates may show

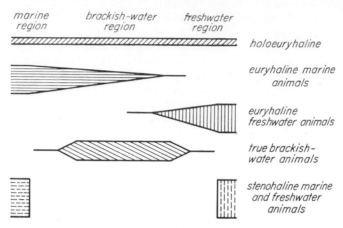

Figure 196.   Diagram to show the relationship to salinity (original)

different biological phenomena. Definitions based only on salinity will therefore scarcely be valid in all parts of the world. On the other hand, some points have been shown to be generally applicable; Remane & Schlieper (1958) have summarized the ecological and physiological problems of brackish waters with special reference to the North and Baltic Seas. In the following brief account of the more general problems use has been made of this summary and also of the results of a symposium on brackish waters held in Venice in 1958.

The distribution of organisms in relation to salinity may be represented somewhat as in Fig. 196, in which the different thicknesses of the bars only give an approximate indication of the relative number of species. On the basis of several investigations on different phyla from separate localities, Remane (1934) has constructed the ideal curve shown in Fig. 197, which has, in general, been confirmed by numerous investigations over a number of years (Fig. 197), but which has in some cases been shown to be not valid owing to exceptions caused by special local conditions. It shows that, in general, brackish water is poor in species when compared with the sea and fresh waters, but that the species minimum lies much closer to fresh water than to the region of oceanic water. This asymmetrical position of the minimum indicates the great sensitivity of freshwater life to salt, which raises problems in so far as fresh water represents an extreme environment on account of its poverty in ions and nevertheless shows much greater differentiation and speciation than brackish water.

Among the special biological problems we may mention: a great decrease in size from higher to lower salinity in a number of mollusc

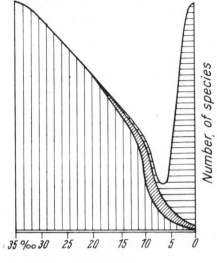

Figure 197a

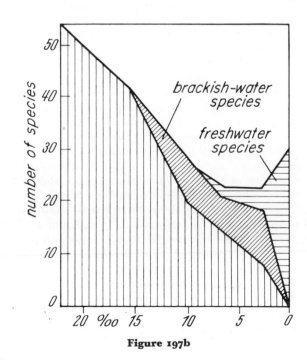

Figure 197b

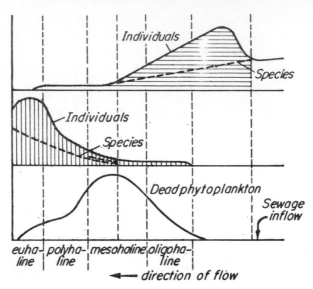

Figure 197c. Number of species in relation to salinity: *horizontal lines* = marine species; *vertical lines* = freshwater species; *oblique lines* = true brackish-water species; *unhatched* = holeuryhaline species
(a) idealized curve (after Remane 1934), (b) dependence of nematodes on salinity (after Gerlach 1954), (c) curves from the lower reaches and mouth of a river (after Kuhl & Mann 1961)

and fish species, which is also seen to a lesser extent among the echinoderms, cnidarians and plants but does not occur in the polychaetes and the microfauna; structural changes have been observed, for example, in the development of calcareous parts of the body, in the growth form of algae and of the hydroid polyp *Cordylophora caspia* and in anomalies in reproduction; attention has already been drawn (p. 284 et seq.) to the phenomenon of brackish water submergence and to changes in the structure of communities. The development of races with different tolerances to salinity has been observed repeatedly, but should be the subject of further investigation (cf. Höhnk 1953, Kinne 1956, 1963).

Genuine brackish-water species as such are not always clearly recognizable, because in geographically separated areas some of them may occur as pure freshwater or pure marine species. According to their habits they may be relatively easily defined as 'limnogenic' or 'thalassogenic' forms; there is a great preponderance of thalassogenic forms in brackish waters. From this it appears that the regulation of the water and salt content in the body fluids must be easier for marine forms than for forms of freshwater origin. The following point may perhaps help us to understand this phenomenon: since thalassogenic species, in so far as invertebrates are concerned, are generally isotonic with their external medium, great demands are made on their regulatory mechanism when living in dilute sea water. They are not, however, forced to a marked reversal of direction such as occurs in the limnogenic species, in which the organs at a certain hypertonicity of the body fluids had originally to contend with water inflow and salt outflow, but at higher external salinities must balance or tolerate water outflow and salt intake. This is doubtless largely a problem in the field of cell physiology (Schlieper in Remane & Schlieper 1958).

From this ability to reverse it is perhaps also understandable 'that freshwater organisms, once they have moved across the critical brackish-water zone, acquire an extreme insensitivity towards high and varying salinities, whereas marine organisms seldom attain such a high degree of independence of the salt concentration' (Remane in Remane & Schlieper, 1958, p. 128).

Several observations confirm that repeated fluctuations in salinity subject the organism to considerable stress, and are more lethal than a permanently low salinity. In tropical coastal waters the marine fauna is often very seriously damaged as a result of heavy rainstorms, but after a few months and a return to the normal salinity this will have been made good by new settlement. This applies particularly to the animal and plant life of the upper littoral (Webb 1958, Goodbody 1961).

Tidal estuaries are brackish-water areas showing regular fluctuations in salinity. The less dense river water lies above the sea water which moves upstream along the bottom, but there is rarely a clearly marked stratification. Usually the river water, heavily laden with particles, mixes with the clearer sea water to produce a more or less extensive brackish-water zone. In some estuaries, particularly at the transition from fresh to brackish water, particulate matter and muddy sediments are stirred up, so that brackish water is often characterized by a particularly high content of substances producing cloudiness. It was originally assumed that the cloudiness was mainly due to the precipitation of colloidal substances in the fresh water when mixed with salt water (cf. Lüneberg 1953, Postmas & Kalle 1955).

Furthermore, the brackish-water zones of estuaries are characterized by considerable instability; the upper and lower limits will be periodically displaced by the normal tides (Fig. 198). Fluctuations in the outflow of rivcr water and in the inflow of sea water due to winds will produce non-periodic changes; and finally such areas will differ from stationary brackish waters owing to the presence of currents which change in direction and speed in relation to the tides.

In recent times the biology of many estuaries has been affected by man-made salination and enrichment; salination may shift the

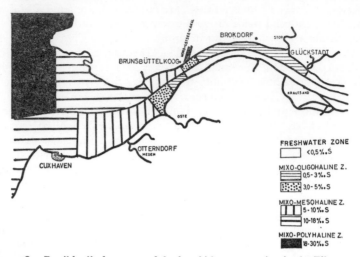

FRESHWATER ZONE
< 0,5 ‰ S

MIXO-OLIGOHALINE Z.
0,5 - 3 ‰ S
3,0 - 5 ‰ S

MIXO-MESOHALINE Z.
5 - 10 ‰ S
10 - 18 ‰ S

MIXO-POLYHALINE Z.
18 - 30 ‰ S

Figure 198. Possible displacement of the brackish water region in the Elbe estuary (after Caspers 1959)

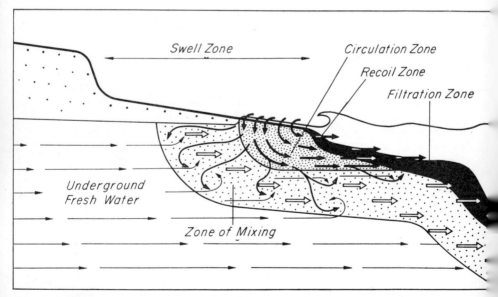

Figure 199. Diagram of the water circulation on a sandy tide-less coast (after Delamare Deboutteville)

natural limits of the freshwater region upstream and the enrichment of nutrients, which usually takes place in the freshwater area, strongly favours the development of phytoplankton.

In addition to the work of Thiemann (1934), investigations on the biology of estuaries has been carried out in various places (e.g. Caspers and Kuhl & Mann in the Elbe, Day et al. in South Africa, Gunter in the Gulf of Mexico, Milne in England, Odum in Florida and so on). The following phenomena are widespread: in the lower reaches of the rivers the diatoms show a pronounced maximum which is followed by a minimal area; this is then replaced seawards by a further maximum, due to incoming euryhaline marine species. The minimal zone is the area where the freshwater phytoplankton dies out (Fig. 197). Marine zooplankton organisms penetrate the brackish-water zone to varying distances, but only rarely extend as far as the true fresh water; among those found at certain seasons are *Noctiluca*, *Sagitta*, Hydromedusae, *Pleurobrachia*, appendicularians, copepods and polychaete larvae. True brackish-water species are only present in the plankton in small numbers, but they may have large numbers of individuals, as for instance *Eurytemora affinis* and species of *Acartia* among the copepods.

Benthonic animals penetrate into the marginal areas (particularly in shallow waters), sometimes reaching the fresh water or its lower limits; animals that do this are the flounder, *Cordylophora caspia*, the barnacle *Balanus improvisus*, *Neomysis vulgaris*, the polychaete *Nereis succinea* and others. With a salinity of 0·08 to 2·7‰ Gunter (1961) gave the ratio of animals of marine origin to those from fresh water as about 2:1 for species and 25:1 for individuals.

The biology of coastal subterranean waters was first studied by Remane & Schultz (1934), and more recently several other investigators have worked on this subject in various areas of the sea. This habitat is characterized by a great number of peculiar and specific forms, whose numbers are continually being increased. As they live in interstitial water they belong exclusively to the microfauna (summary in Delamare Deboutteville 1960) and some have a cosmopolitan distribution, as for example *Protohydra leuckarti*, some nematodes, the tardigrade *Batillipes pennaki* and others. The cosmopolitan distribution of many crustacean genera is also very striking. The diagram (Fig. 199) given by Delamare Deboutteville illustrates the marginal position and the mixed character of this habitat.

Summaries of the history of the classification of brackish waters have been given by Remane (1958) and Segerstråle (1958), amongst others. After agreement at the Venice Symposium in 1958 the following classification was adopted:

Euhaline region $= > \pm 30\%_0$ salinity

Mixohaline region $\begin{cases} \text{mixo-polyhaline zone } = \pm 30 - \pm 18\%_0 \\ \quad \text{salinity} \\ \text{mixo-mesohaline zone } = \pm\ 18 - \pm 5\%_0 \\ \quad \text{salinity} \\ \text{mixo-oligohaline zone } = \pm\ 5 - \pm 0 \cdot 5\%_0 \\ \quad \text{salinity} \end{cases}$

Freshwater region $= < 0 \cdot 5\%_0$ salinity

Gunter (1961) has criticized the wisdom of setting the limit at $0 \cdot 5\%_0$, because a hard fresh water may itself contain $0 \cdot 3 - 0 \cdot 5\%_0$ of salts; he proposed to set this limit where the ratio of the total salinity to the chloride begins to change and becomes greater than $1 \cdot 8$. In view of the numerous references in limnological literature to the importance of calcium and of carbonate in ecological processes (cf. Schlieper in Remane & Schlieper 1958) this proposal in my opinion deserves serious attention, and particularly as a subdivision of the meso- and oligohaline zone is already being considered as a result of experience in the Baltic Sea, which has been studied in such detail.

CHAPTER VIII

ECONOMIC ASPECTS OF MARINE BIOLOGY

As in all fields of human activity the results of research in marine biology can be applied to everyday practice, and a few examples will be
given here. More detailed information is given in F. Pax's handbook,
Meeresprodukte (1962).

Fisheries
Until the second half of the last century, marine fisheries were restricted to areas close to the coast. Apart from the technical equipment
used, they were based on an empirical knowledge of the seasonal occurrence of the species fished, of certain correlations with weather conditions and of fluctuations observed in abundance and breeding periods,
etc. The causes of these phenomena were not understood because there
were no basic facts to work from. These involve a knowledge of the
autecology and ethology of the species exploited by the fisheries, and
their synecology; the latter naturally implies a knowledge of hydrographic conditions (cf. Buckmann 1956, Fleming & Laevastu 1956).
Practical fisheries are interested in large catches of marketable and
valuable fish. Intensive fishing of inshore fishing grounds, and its direct
consequences, has increased the demand. Advances in the technique
of constructing ships and machinery and present-day extension of
fishing limits has forced commercial fishermen to operate in increasingly distant waters. At the same time efforts must be made to control
the catches in order to avoid overfishing. This requires knowledge on
the following points:

(a) The quantity of young fish produced each year.
(b) The proportion of young fish lost before they reach commercial
    size.
(c) The age at which this size is reached, that is, the rate of growth.
(d) The subsequent growth and the fluctuations from year to year
    in the composition of the population.

It is quite obvious that these are very complex problems which have so far been only partially solved. Since most species have pelagic developmental stages the size of the brood can be estimated by taking quantitative plankton hauls and counting the known eggs and larval forms, especially if spawning takes place in restricted areas and is associated with particular water conditions. Considerable caution should, however, be taken in estimating the number of young, even from large numbers of samples, because the distribution of the plankton is frequently very irregular.

After consuming the yolk-sac the developing fish is dependent upon the available food, and at the same time it will be decimated by planktonic predators. It is, therefore, quite essential to have a basic knowledge of the composition and amount of the total plankton in the area where the young fish are developing.

As a result of periodic growth, which is largely dependent upon the season, otoliths, scales and bones develop growth rings (Fig. 200).

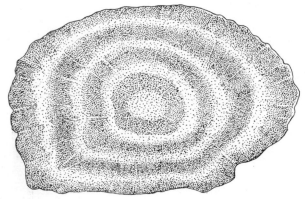

Figure 200. Growth rings (diagrammatic) in the otolith of a four-year-old plaice (based on Wimpenny)

Analysis of these will give the age of the animals and taken together with measurements of weight and length they provide an estimate of the growth rate. Examination of representative samples from actual catches will then enable us to form an idea of how the proportion of the age classes fluctuates from year to year. These fluctuations are due to the growth of the younger year classes to a size at which they can be caught by net, to natural mortality, to the effect of the fishery itself and to immigration and emigration. Comparison of the data provides information on population dynamics, from which one can predict the possible yields in subsequent fishing seasons (on population dynamics see Beverton & Holt 1957).

The natural mortality is due to various factors, mostly of an ecological nature: some of the animals fall prey to larger predators, others may be killed by parasites (see p. 397), and in some areas mass mortality may be caused by outbursts of phytoplankton or by severe climatic changes (see p. 399).

Some of the migration routes of fishes have been worked out from the results of numerous marking experiments and it has been found that in certain cases great distances are covered between the grazing and spawning areas (Fig. 201 and 202); recaptures have shown that several

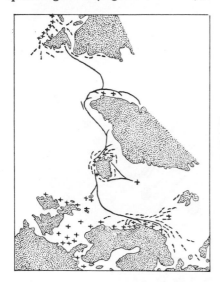

Figure 201. Migration routes of the cod in the North Atlantic; + spawning areas, – – – indicates normal spawning migrations; —— indicates distant migrations (after Lundbeck 1955)

species of tunny carry out transatlantic and transpacific migrations (Mather 1960). A cod marked in the North Sea was recaptured 4½ years later on the Grand Banks off Newfoundland; this migration had presumably taken place by way of the Scotland–Faeroes–Iceland–Greenland ridge (Gulland & Williamson 1962). Transpacific migrations of tunny have been demonstrated on several occasions (Anonymous 1961). There appears therefore to be an exchange between widely separated populations, at least in the case of good swimmers such as tunny, unless one is in fact dealing with a more or less circum-oceanic population. It is even more likely that immigration and emigration take place between neighbouring parts of populations, but we are not yet in a position to assess the extent of this. This constitutes a further uncertain factor in the assessment of population dynamics. Parts of populations may sometimes also form spawning communities, possibly also leading to divergence.

The development of new fishing grounds must be based on sound

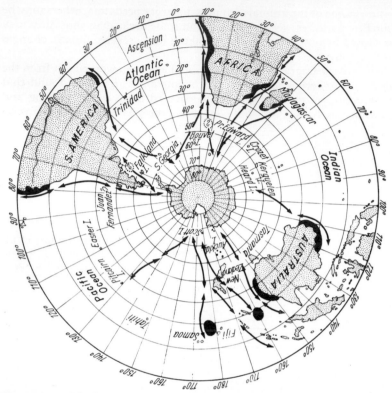

Figure 202. Distribution and migrations of the humpback whale in the southern hemisphere (after Slijper 1961)

economy, in terms of time and money and with some prospect of success, if the appropriate environmental requirements of the species are known to be present.

In practical fisheries it is very important to know the behaviour patterns of the species to be caught, such as the diurnal and seasonal behaviour of shoals (see pp. 335 et seq.), their reaction to obstacles and enemies and to various external stimuli. These are problems which have so far been little investigated and which may require extensive research in aquaria. Bull (1961) drew attention to the investigations of Brawn, according to which *Gadus callarias* shows certain behaviour patterns associated with sound production before and during spawning, so that 'the ocean where the cod shoals are congregated must be full of the sound of cod grunts. This suggests that the use of hydrophones to find cod would be of value, but only seasonally' (p. 250). It has been shown that some fish species (*Blennius pholis, Pleuronectes platessa, Gobius minutus, Gadus virens, Gadus callarias*) perceive the sound waves from echo-sounders and associate them with other stimuli. From this Bull (1961, p. 251) has concluded: 'This means that the suggestion that commercial trawl fishes might, by conditioning, become more wary of capture and, hence, lead to reduced catches cannot wholly be ruled out of consideration.'

Some fisheries have suffered catastrophically as a result of fluctuations in the abundance of the species being caught, as for example the herring fishery off Bohuslän (Sweden) in the 19th century and the Californian sardine fishery; after a peak production of $\frac{1}{2}$ million tons in the middle of the 1940s, the latter suffered a complete collapse (Lundbeck 1958). The displacement of the spawning areas of Icelandic cod to West Greenland had far-reaching consequences for the fishery. It is obvious that events of this kind are not isolated phenomena, but can only be understood in the context of hydrographic and general biological processes; they emphasize once more how necessary it is to base commercial fisheries on the results of general marine research.

Fishes and whales are to a large extent the end links in the food chain, or expressed in other words they stand at the peak of the food pyramid (see pp. 327, 332). The reduction in the amount of living organic matter from link to link in the food chain and the rough estimates made, for instance, of the amounts of plankton in the North Sea have raised the question of harvesting the plankton instead of the fish. So far, no economically feasible method of directly exploiting the plankton has been developed; this is due to the small size of planktonic organisms, their water content, the spatial and temporal irregularities in their occurrence and to fluctuations in the composition of the plankton as a whole (Jackson 1954). To a large extent we also still lack experimental investigations on the nutritional potentialities of different types of plankton. Using plankton containing 75–90% copepods from the Georges Bank region, Clarke & Bishop (1948) only obtained a weight increase in rats if they fed a mixture containing two parts of meal to one part of plankton; if the proportion was reversed there was a drop in weight after seven days. Apart from the nutritional value assessed in calories there is also the problem of toxicity or at least of the extent to which accumulated trace substances can be tolerated. This subject has been discussed in detail by Walford (1958) in his *Living resources of the sea.*

Walford believes that it would be possible to produce large-scale cultures in brackish- or sea-water ponds, in areas with the appropriate conditions. This would provide much valuable information both on the autecology of the individual species and on biological productivity in general.

In addition to fishes, the crustaceans and molluscs also play an important role in world fisheries, but other groups are mostly of local interest only as a source of food. Here, too, the problems of fisheries research are very closely connected with those of general marine biology; crustaceans and molluscs can in many cases be more easily caught, because they are less mobile; these animals are also more amenable to experimental study than the majority of marketable fishes. Thus, it is possible to produce cultures of oysters, mussels and lobsters; the culture of pearls and the commercial rearing of lamellibranchs provide good examples. Only future research and technical advances can show whether the scattering layers with their probable wealth of crustaceans and cephalopods can be regarded as profitable places to fish.

The exploitation of the macro-vegetation along the sea coasts (the green, red and brown seaweeds) has been carried out for some time (Chapman 1949, Walford 1958). In some areas these algae play in a not unimportant role as sources of food, fodder and fertilizer and also as raw materials for the production of iodine, potash, agar-agar, carrageen and alginates; the production of iodine and potash from this source is, however, now much reduced. The establishment of large commercial undertakings using algae as a raw material is impeded by the variations in the chemical compositions of the species, so that common processing offers little prospect of success; further the density of algal vegetation fluctuates, and its content of important chemical substances changes according to the time of the year.

Nevertheless, a more detailed knowledge of, for example, the content in vitamins, amino acids, antibiotics and stored trace elements could lead to a more intensive exploitation of algal vegetation. From this biochemical aspect there are possibilities of utilizing other plants and animals which have so far not been considered, leading eventually perhaps to their production in cultures. This line of approach is suggested, for instance, by the results of Dietrich & Hohnk (1958) who found that the marine fungus *Ceratostomella* produces a very valuable oil with a high content of vitamin E (cf. Burkholder 1963, Volbehr 1963). A series of papers on this subject entitled 'Biochemistry and pharmacology of compounds derived from marine organisms' has been published in the *Annals of the New York Academy of Sciences*, **90**, 615–950.

Land reclamation

In some coastal areas the eroding action of waves, currents and tides is counteracted by the process of sedimentation which results in deposition. This can be promoted by various kinds of construction so that usable land can be reclaimed from shallow-water areas. Operations of this kind have been carried out for several years on the North Friesian coasts and in the Netherlands.

In this work, marine biological problems are often involved in various ways (Wohlenberg 1939, 1954, Muller 1960). The composition of living communities in shallow waters is dependent upon the nature of the substrate and will thus act as an indicator of the sediment being formed and of the future arable soil. The abundant construction of tubes by the infauna (see p. 258) helps to bind the sediment, and allows the entry of water containing oxygen; in this way the layers below the surface become aerated, leading to oxidation of the organic matter present in the bottom. Mass outbursts of diatoms, which frequently cover large areas with a slimy coating, provide some protection against erosion, albeit a weak one, and at the same time they produce organic and (from their siliceous shells) inorganic sediments. Lamellibranchs, such as *Mytilus edulis*, that live in dense populations give some protection against erosion and their filtering activities and production of pseudofaeces will, at least temporarily, bind any finely suspended material. Dissolved substances, particularly calcium, are deposited in large quantities in the shells of molluscs, which often occur in enormous numbers.

Plants living in shallow water, such as dwarf eel-grass (*Zostera nana*), glasswort (*Salicornia herbacea*), Townsend's cordgrass (*Spartina townsendi*) and halophytic plants of the shore zone, such as sea starwort (*Aster tripolium*) and sea club rush (*Scripus maritimus*) help to accelerate sedimentation by reducing the currents and binding the sediments.

The animals and plants of these eulittoral areas show conspicuous zonation which is dependent upon the degree to which they are covered with water and on other factors (see p. 210 et seq.). With progressive sedimentation, which is equivalent to an increased period of desiccation, there is a succession in the settlement which is of importance for the practical process of land reclamation and also involves a number of interesting ecological and ethological phenomena.

For the time being we cannot yet estimate the part played by the micro-organisms living in shallow-water sediments (Höhnk 1952–1956, Siepmann 1959) in processing dead organic matter or in promoting sedimentation. Höhnk found that the roots of *Aster tripolium* were covered with a mat of fungus mycelium, but was unable to decide whether this was a case of true symbiosis (verbal communication).

Economic and scientific aspects are therefore closely bound up with each other in these problems of land reclamation.

Pests

Another economic aspect of marine biology which can be briefly mentioned here concerns pests, that is, the organisms which damage the property of mankind. Ray (1959) has given an extensive summary of this subject in 'Marine boring and fouling organisms'.

The timber-destroying shipworms (lamellibranchs of the family Teredinidae) and boring crustaceans (the amphipod *Chelura terebrans*, the isopods *Limnoria lignorum* and *Sphaeroma terebrans*) have long been known and frequently investigated. In the Teredinidae intracellular digestion of the wood by a cellulase has been established. The role of symbionts in the digestion of the wood is still not apparent. These animals satisfy their nitrogen requirements partly from the wood itself and partly by the supplementary intake of nannoplankton. Their growth is hindered by very cloudy water and also by lowered salinities, so that they are often incapable of surviving in harbour installations in estuaries.

The crustaceans that bore into timber make burrows several centimetres long, close beneath the surface of the wood; in this way they often prepare the substrate for settlement by Teredinidae. The nutrition of these animals is of special interest because timber-destroying fungi are involved. These crustaceans can only use timber as food after it has become infected by fungi; they feed partly on the fungus mycelium itself and partly on the wood exposed to its enzymes (Becker *et al.* 1957, 1958). It is not certain to what extent the relationship between fungus and crustacean can be regarded as mutual symbiosis.

Barnard & Reish (1960) observed the uptake of wood by the copepod *Tisbe gracilis*. No detailed investigation has yet been made on whether these animals take in the wood as food or only accidentally when feeding on the micro-organisms attached to it. So one cannot yet say whether this is actually a pest.

Considerable damage can be done by timber-destroying fungi themselves. Dissolution of the intervening lamellae causes a loosening of the cell structure and thus softens the outer layers of the timber which then become mechanically abraded. This may cause considerable damage to ships and installations, particularly in warm-water areas (Becker & Kohlmeyer 1958).

A knowledge of the physiology of the fungus is essential before effective measures can be taken against this type of damage. Since timber is an allochthonous substrate in the sea the question arises how this particular way of life arose. Barnard & Reish (1960) have pointed out that some species of *Limnoria* bore into the holdfasts of brown algae and a harpacticid eats into algae; substrate borers are common among the isopods, amphipods and harpacticids. Although wood is allochthonous in the sea it must to a certain extent have always been present there and must have played a part ever since woody growth has existed on land and in fresh water.

There are indications that micro-organisms may be involved in the corrosion of building materials (metals and even concrete), but this requires further investigation.

Growths on ships' bottoms, moored buoys, etc., cause destruction of the protective paint and thus assist corrosion, and also reduce the speed of vessels by increasing frictional resistance. These growths show successional stages, which vary according to the physical nature of the surface and exposure to light and currents.

Methods of combating these growths are based partly on changing the structure of the surface coating so as to impede the first settlement and partly on killing the settled organisms with poisonous paints.

The sea as a 'dumping ground'

The flow of domestic and industrial waste effluents into rivers increases the amount of organic and inorganic matter entering the sea. This waste matter is, to a considerable extent, remineralized during its passage down-river to the estuary, even if the effluent was originally cloudy. When this water enters the sea a certain degree of enrichment takes place which is not only noticeable in river estuaries (Kuhl & Mann 1958). This can be traced by the persistence of the river water in the sea (Krey 1956), but it is also noticeable, even in larger areas, e.g. over wide areas of the southern North Sea, by increased turbidity due particularly to increased development of plankton (Kalle 1953 1956).

When cloudy effluent is fed directly into the sea, which happens in many places, it has been observed that *B. coli* and other bacteria die relatively quickly. On the other hand, the breakdown of polluting organic substances appears to take place more slowly in sea water than in fresh water; it has been stated that the breakdown, measured as the biological oxygen demand (BOD) at 20°C requires about twice the time. The number of competent investigations in this field, and thus the chance to make comparisons, is still however very few (bibliography in Friedrich & Luneburg 1962).

One explanation of these phenomena is that the *B. coli* and other organisms causing breakdown in the effluent have been killed by the changed osmotic conditions; so the cause would lie in the salinity of the sea water. On the other hand, Muller (1953) alleges that sterilized or chlorinated sea water is less effective in killing micro-organisms than untreated sea water. On this view it would appear to be not so much the salinity as such but rather the 'biological content' of the sea water which is responsible for these phenomena. More detailed analyses are still lacking, but this theory would need to be tested by an investigation of the part played by marine bacteria, phytoplankton and zooplankton; this would involve a study of the seasonal fluctuations in quantity and species composition of these organisms and also of the relevant hydrographic conditions. In this connection the bacteriostatic and bacteriocidal action of sea water is noteworthy (see p. 403).

In coastal waters the release of polluted water may have direct economic effects on the fisheries; in some coastal areas fish mortality has been correlated with pollution (Kandler 1953); in the U.S.A. in 1949 there was a legal case between the oyster industry and the oil industry which involved a claim of 40 million dollars (Gunter 1959). Consumption of uncooked or insufficiently cooked molluscs from polluted areas is not only unappetizing but may also be dangerous to health. In bathing resorts the fouling of coastal waters may have serious consequences.

We therefore need to know more about the ways in which fouling substances break down in sea water and also to find indicators which point to the relevant coastal areas being influenced. It seems that in the sea, polychaetes of the genus *Capitella* take over the role of the Tubificidae in fresh waters (Reish & Barnard 1960). Investigations of this nature have, however, only just begun and no general conclusions can yet be drawn. The preliminary nature of most of the contributions to the first international conference on waste disposal in the marine environment (Pearson 1960) clearly indicate that this subject is still in the early stages.

In Europe, industrial pollution has so far played a relatively small part but in the U.S.A. this is a subject of far-reaching importance. Legal measures have already been taken against oil pollution which occasionally causes the death of large numbers of sea birds; the sinking of atomic waste has become the subject of international conferences. Acid industrial effluent ought to be less damaging in the sea than in fresh water because, apart from the greater dilution, the buffering action of sea water is considerably greater.

Conditions in the open ocean ought not to be affected by pollution, but in coastal areas increasing attention must be paid to this problem. In view of the relative paucity of our knowledge this is a field where biological research should be undertaken in close co-operation with hydrographic investigations. There is no doubt that this would also throw further light on the breakdown of organic matter in the true oceanic regions.

CHAPTER IX

DRAFT FOR A GENERAL BIOLOGICAL PICTURE OF THE OCEANS

Having discussed several different phenomena and drawn attention to various problems, it now seems desirable to attempt some kind of synthesis of the general biology of the oceans. At the XIII International Congress of Zoology in Paris, Zenkevitch (1949) presented such a synthesis under the title 'La structure biologique de l'ocean'. At that time he went into abstract concepts far beyond anything that had been postulated by, for instance, Chun in 1896. New knowledge and fresh points of view now make it possible to extend this kind of abstract picture and in doing so the first task is to attempt a representation of the static relationships.

The spatial distribution of marine organisms involves two phenomena, namely the distribution of taxonomic diversity and the distribution of biomass; attention must be paid to distribution in relation to latitude, the gradient from coast to open sea, and the vertical distribution from the surface to the bottom of the deep sea (Figs. 203, 204). The continuity of the environment makes such an attempt much easier for the oceans than it would be for terrestrial or freshwater regions.

In the pelagial and benthal there is a very striking latitudinal gradient in the distribution of taxonomy diversity, with a great wealth of species in the warm-water regions which decreases as the latitude increases and reaches a minimum in the areas of sea covered by ice. This minimum is not, however, so low as it is in terrestrial or freshwater regions, because the relative stability of the ecological factors provides a large number of forms with a chance to adapt, so that living populations can exist even under the ice. In addition, currents bring into these areas quantities of allochthonous food so that the animal life is not exclusively dependent upon locally produced food.

Parallel with the changes in taxonomic diversity there is a latitudinal gradient in the differentiation of habitats and living communities. For example, the temporary or permanent formation of ice hinders or

P—M.B.

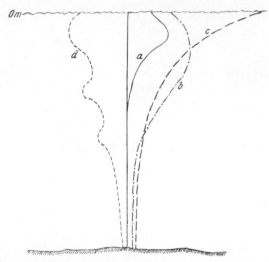

Figure 203. Diagrammatic illustration of the vertical distribution of (a) the phytoplankton, (b) diversity in the zooplankton, (c) diversity in the zoobenthos, (d) the amount of the zooplankton. The curves are not drawn on the same scale) (Original)

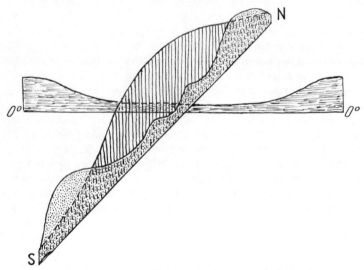

Figure 204. Spatial distribution of biological phenomena in the surface layers of the pelagial; distribution of biological diversity in relation to latitude (vertical lines) and of population density (dots), distribution of population density from Shelf to Shelf along a line of latitude (0°-0°); diagrammatic (Original)

completely suppresses the development of the supralittoral and eulittoral along polar coasts, whereas in warm-water regions special vegetation types (e.g. mangrove) and increased calcium deposition by algae and animals produce very complex habitats with numerous ecological niches.

Certain biological phenomena also show latitudinal gradients (Fig. 205); here we may recall the occurrence of forms with viviparity, lecithotrophic or planktotrophic larvae in certain animal groups, e.g.

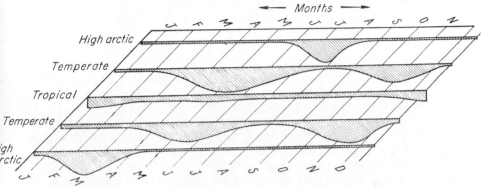

Figure 205. Production cycles in different latitudes (Original)

prosobranch gastropods and echinoderms. The preponderance of planktotrophic larvae in warm-water areas can be correlated with the continuity of light in the lower latitudes and the resulting high production of plant food, whereas viviparity and lecithotrophic larvae can be regarded as adaptations to periodic and non-periodic fluctuations in food supply.

The gradient from coast to open sea also shows changes in distribution: in the neritic region meropelagic forms play a large and sometimes major role; the larvae of benthonic animals in particular, may dominate the plankton. In oceanic regions, on the other hand, holopelagic organisms are in the majority, and they often have a shortened period of development (e.g. in Hydrozoa and Scyphozoa). Large pelagic animals, such as whales, basking and whale sharks, sword fish, etc., are predominantly oceanic.

In both the pelagial and the benthal there is a marked gradient in the distribution of organisms between the surface and the sea bottom. The upper, well-illuminated layers have phototrophic plant life, which decreases rapidly with depth, whereas in deeper water the ecological factors are less differentiated in space and time and the habitat is to a large extent monotonous, although not as much as is sometimes supposed. Water pressure is an ccological factor with a marked vertical gradient; the physiological and ecological importance of pressure in the distribution of organisms is still not completely understood, but it may well be a contributory factor in vertical distribution. In the upper waters of the sea there is great diversity of species and of higher systematic groups, but this decreases with depth. The deep sea is inhabited by stenobathic deep-sea forms, which are specifically adapted to the peculiarities of their environment, and also by eurybathic forms with a greater vertical distribution.

The diversity of life forms decreases with depth, and at the same time there is a change in the relative numbers of the different life forms: in the benthal, for example, the deposit feeders increase in number, whereas in the pelagic a greater proportion of the animals, at least among the fishes, feed on large fragments.

However, the vertical gradients are evidently not a simple linear functions but show stages and discontinuities depending upon a variety of factors and on local conditions. A number of forms which live in deep water in lower latitudes appear in shallow water in higher latitudes, so that dependence upon temperature seems to be an important factor.

The development of large populations is dependent upon various exogenous factors which determine the extent and timing of the primary production of organic matter by photosynthesis. Consequently, the distribution of biomass follows certain general rules, with numerous local variations. In the oceanic pelagial, it can be said that there is, in general, less production and therefore a lower biomass in higher latitudes. In the middle latitudes there are marked seasonal changes and corresponding seasonal fluctuations in productivity, which sometimes lead to the production of very large standing crops; these have a high average biomass. Productivity is lower in regions closer to the equator, although it is higher in the area of the north and south equatorial currents, on account of the larger amounts of nutrient present, than it is in the true equatorial region.

Such wide generalizations cannot be made in the case of the neritic region of the pelagial; here upwelling of deep water and exchange of matter between the bottom and the water cause marked local fluctuations; some areas with upwelling water are amongst the most productive in the sea.

In the littoral benthal the biomass conditions are more difficult to assess and have been little investigated; when one considers the development of coral reefs and mangrove swamps one has the impression that, on average, there is a greater biomass in lower than in higher latitudes, but the difference is not nearly so marked as it is in the case of the taxonomic differentiation.

Comparisons of biomass in neritic and oceanic areas generally show a higher figure for the former; this is because the supply of mineral nutrients is greater and more constant in neritic regions, giving a higher primary production, to which the large plants also contribute, whereas in areas distant from the coasts primary production relies entirely upon the phytoplankton.

For similar reasons there must also be a vertical gradient in biomass: primary production is tied to the illuminated upper layers; the amount of particulate organic matter is reduced as it passes through the successive links in the food chain, because it is oxidized to produce energy. Therefore, as the depth increases there is less available food, so that biomass production is reduced. There are several indications that autochthonous food is available for the benthonic fauna even in deep water, because heterotrophs (bacteria, yeasts, fungi) are widely distributed, sometimes in very large numbers. This is reflected in the relative abundance of deposit-feeders in the abyssal benthal.

The large-scale distribution of marine organisms is to be regarded as closely related to the dynamic physical phenomena of water exchange by currents, upwelling and convection, and these also determine the small-scale biological processes.

The occurrence of organisms in communities or associations is related to the specific environmental requirements of the species. These are extraordinarily diverse, because most species only occur in regionally restricted areas. On the other hand, the same type of sediment at the same depth may support parallel communities, in which closely related species and the same kind of life form predominate, so that from this point of view it is possible to recognize a limited number of associations.

In the littoral region these associations often occupy narrow bands or zones, which are dependent upon exogenous factors, such as the period of submersion, light, exposure to waves, type of substrate and so on. We do not yet have a general, worldwide classification of these zones, although a considerable amount is known about rocky coasts.

Naturally, interspecific relationships exist within the associations, but it is only possible in a few cases to estimate whether these determine any ecological balance, to the extent that the term biocoenoses could be used.

The distribution of taxonomic diversity with its maximum in the tropical-subtropical littoral and the existence of geographical and vertical gradients can be regarded, on the one hand, as statistical distributional phenomena, and, on the other hand, as the result of dynamic biological processes. As an example of the distributional phenomena, we may take the existence and occupation of numerous ecological niches in warm-water regions, the number of which decreases with latitude and depth. Mention has already been made of productivity which is a dynamic biological process. Furthermore, the regional or vertical differences in the diversity of forms may be the result of exchange with neighbouring land or freshwater areas and of evolutionary processes. The exchange of faunal elements appears to take place more easily in regions with high temperatures than elsewhere, but the amount of immigration into the sea is too small for this to be a very important factor.

From the viewpoint of evolution the following point can be made regarding the gradient that exists between the equator and the poles: during recent geological times the warm-water region has not suffered any marked change of climate, so that its fauna and flora have had a relatively long period of time available for undisturbed speciation; in the polar regions, on the other hand, the warmth-loving fauna and flora of the Tertiary were destroyed or pushed back by climatic changes in the Upper Tertiary and Pleistocene, so that in the post-glacial there had to be fresh colonization by organisms suitably adapted for the changed conditions. This time factor must therefore be taken into account when considering that the scale of taxonomic differentiation decreases towards the poles. There are also other evolutionary factors that have a corresponding geographical gradient: the rate of growth and development, and thus the number of generations, is greater in warm waters than in cold; the smaller number of offspring resulting from the development of large lecithotrophic eggs or from viviparity reduces the probability of mutations in higher latitudes as compared with lower latitudes; the greater taxonomic differentiation associated with palaeoclimatic conditions implies a greater number of ecological inches which themselves provide increased selection possibilities for the survival of mutant forms.

Similar points can be made regarding the vertical gradient in differentiation. In the littoral benthal there is great diversity within a limited area, associated with a combination of exogenous ecological factors; these are primarily the fluctuations in currents and water levels. In this vertically arranged environment the formation of zones, each with its suitably adapted organisms, gives a stepwise distribution which is comparable, from the viewpoint of evolution, to the marginal zones of organisms that are distributed horizontally. As the depth increases the sediments and the hydrographic conditions become more widely distributed, and they change less frequently and to a smaller extent, so that they become less effective as selection factors. In view of the completely different feeding conditions in the deep sea, the effects of population pressure or sudden population increase are different from what they would be in the littoral; the changed currents reduce exchange between populations. These evolutionary factors ought to change gradually with depth. The rate of evolution in the abyssal is, therefore, probably lower than in the littoral. In fact, the percentage of archaic forms in the various phyla and classes in the abyssal is considerably greater than in the littoral and the number of monotypic genera appears to increase with depth.

There is no doubt that the abyssal fauna has been originally derived from the littoral region, and it is probable that preadapted forms penetrate the abyssal more or less continuously, although this process will occasionally have been speeded up by palaeoclimatic changes. In the present state of knowledge it is not possible to assess what morphological and functional characters predispose an animal for migration into deep water. Certain facts, such as the presence of a cryptofauna in the littoral and the vertical migration of plankton, suggest the importance of light.

It would be very difficult, from a practical point of view, to assess whether any return migration from the deep sea to the littoral or epipelagial has taken place. The occurrence of several shade-loving forms in littoral caves is indeed no proof of the ascent of species from the deep littoral or bathyal.

Research on the physiology of marine organisms from an ecological

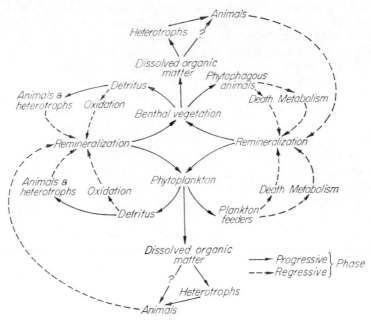

Figure 206. A food cycle (Original)

viewpoint is still only in its early stages; this is a subject which might contribute to solving the problem of how the deep sea has been colonized. Our knowledge of feeding and of food cycles (Fig. 206) is somewhat more advanced. Food cycles in the sea are essentially based upon the type and amount of plant nutrients and light, both of which are required for assimilation and primary production. The reduction of light as it passes through the water restricts assimilation to a surface layer which is very small in relation to the average depth of the sea; as the latitude increases there are periodic changes in the amount of light and these cause a seasonal periodicity in the primary production. The rate at which the combined mineral substances are again released depends upon the type and population density of the animals and heterotrophs and on the temperature and oxygen supply. Exchanges of water at all depths vary from place to place and bring about considerable local variations in the amounts of available nutrients.

The organisms not only remove dissolved substances from the water, but at the same time release their metabolic products and thus condition the medium in which they live.

Some of the released substances have biotic effects, e.g. vitamins and assimilates, while others are antibiotic, e.g. certain toxins produced during mass outbursts of phytoplankton organisms; the biological history of a water mass may therefore be important to its fauna and flora. Our present knowledge of these phenomena is, however, not sufficient to assess their effect on the general ecological situation or on the autecology and physiology of individual species.

Many holopelagic and meropelagic animals undertake extensive periodic migrations, whether vertically or in a north–south or east–west direction, and these are associated with vast spatial and temporal increases in production; such migrations may also be stimulated and their direction determined by reproduction. Many fishes, for example, travel great distances to reach spawning grounds offering the appropriate conditions, whether of depth, sediment, temperature or food supply. The development of fisheries, particularly in offshore waters, is very largely dependent upon concentrations of animals in the feeding or spawning grounds.

BIBLIOGRAPHY

ABBOTT, B. C., & D. BALLANTINE (1957): The toxin from *Gymnodinium veneficum* Ballantine. J. Mar. Biol. Ass. U. K. 36: 169–189.
ABEL, E. F. (1962): Freiwasserbeobachtungen an Fischen im Golf von Neapel als Beitrag zur Kenntnis ihrer Ökologie und ihres Verhaltens. Intern. Rev. ges. Hydrobiol., 47: 219–290.
ABEL, O. (1916): Paläobiologie der Cephalopoden aus der Gruppe der Dibranchiaten. Jena, iv+281 S.
ADAMS, J. A. (1960): A contribution to the biology and postlarval development of the Sargassum Fish, *Histrio histrio* (Linnaeus), with a discussion of the Sargassum complex. Bull. Mar. Sc. Gulf and Caribbean, 10: 55–82.
ALEXANDER, W. B. (1955) Birds of the Ocean. Putnam, London.
ALLEN, B. (1956): Excretion of organic compounds by *Chlamydomonas*. Arch. f. Mikrobiol. 24: 163–168.
ALLUAUD, C. (1926): Observations sur la faune entomologique intercotidale. Bull. Soc. Zool. France, 51: 152–154.
ALTEVOGT, R. (1957): Untersuchungen zur Biologie, Ökologie und Physiologie indischer Winkerkrabben. Z. Morph. Ökol. d. Tiere, 46: 1–110.
ANDRIASHEV, A. P. (1953): Archaic deep-sea and secondary sea-fishes and their role in zoo-geographical analysis. Essays on the general problems of ichthyology, 58–64 (Russian, cit. by. Zenkevich and Bierstein 1960).
ANKEL, W. E. (1926): Spermiozeugmenbildung durch atypische (apyrene) und typische Spermien bei *Scala* und *Janthina*. Verh. D. Zool. Ges. Leipzig, 1926: 193–202.
— (1950): Ein rankenfüßiger Krebs mit Schaumfloß (*Lepas fascicularis*). Natur u. Volk, 80: 309–320.
— (1952): *Phyllirrhoë bucephala* Per. et Les. und die Meduse *Mnestra parasites*, Krohn. Publ. Staz. Zool. Napoli, 23: 91–140.
— (1959): Beobachtungen an Pyramidelliden des Gullmar-Fjordes. Zoolog. Anz. 162: 1–21.
— (1962): Die blaue Flotte. Natur und Museum, 92: 351–366.
Anonymous (1961): Capture d'Albacore marqué. Cahiers du Pacific, No. 3: 125.
APSTEIN, K. (1905): Tierleben der Hochsee. Verlag Lipsius u. Tischer, Kiel u. Leipzig, 115 pp.
ARMSTRONG, F. A. J. (1955, 1957, 1958): Phosphorus and Silicon in seawater off Plymouth during 1954, 1955, 1956. J. Mar. Biol. Ass. U. K. 34: 223–228, 36: 317–321, 37: 371–377.
— & W. R. G. ATKINS (1950): The suspended matter of sea water. J. Mar. Biol. Ass. U. K. 29: 139–143.
ASCHOFF, J. (1955): Exogene und endogene Komponente der 24-Stunden-Periodik bei Tier und Mensch. Naturwiss., Jahrg. 42, H. 21: 569–575.
ATKINS, W. R. G. (1923a): The hydrogen ion concentration of sea-water in its relation to photosynthetic changes. J. Mar. Biol. Ass. U.K. 13: 93–118, 437–446.
— (1930): Seasonal variations in the Phosphate and Silicate content of sea-water in relation to Phytoplankton Crop. J. Mar. Biol. Ass. U.K. 16: 821–852.
AX, P. (1956): Die Einwanderung mariner Elemente der Mikrofauna in das limnische Mesopsammal der Elbe. Verh. D. Zool. Ges. Leipzig 1955: 428–435.
— (1960): Die Entdeckung neuer Organisationstypen im Tierreich. Die Neue Brehm-Bücherei, Ziemsen Verlag Wittenberg Lutherstadt, 116 pp.

BACKUS, R. H., C. S. YENTSCH & A. WING (1961): Bioluminescence in the surface waters of the sea. Nature, 192: 518–521.
BAINBRIDGE, R. (1952): Underwater observations on the swimming of marine zooplankton. J. Mar. Biol. Ass. U.K. 31: 107–112.
— (1953): Studies on the interrelationships of Zooplankton and Phytoplankton. J. Mar. Biol. Ass. U.K. 32: 385–447.
BAINBRIDGE, V. (1963): Continuous Plankton Records: Contribution towards a plankton atlas of the North Atlantic and the North Sea. Bull. Mar. Ecol., 6: 40–51
BALECH, E. (1962): Tintinnoinea y Dinoflagellata del Pacifico. Revista Museo Argentino de Ciencias Natur., Cie. Zool. 7, 1–249.
BANSE, K. (1955): Über das Verhalten von meroplanktischen Larven in geschichtetem Wasser. Kieler Meeresforsch., 11: 188–200.

Q—M.B.

BANSE, K. (1957): Über das Verhalten von Copepoden im geschichteten Wasser der Kieler Bucht. Zool. Anz. Suppl. 20: 435–444.

BARNARD, K. H. (1932): Amphipoda. Discovery Reports. 5: 1–326.

— (1963): Deep sea mollusca from West of Cape Point, South Africa. Ann. S. Africa Mus. 46: 407–452.

BARNARD, J. L., & D. J. REISH (1960): Wood-browsing habits of the harpacticoid copepod Tisbe gracilis (T. SCOTT) in Southern California. Pacific Naturalist, 1, No. 22, 4 pp.

BARNES, T. C. (1932): The physiological effect of trihydrol in water. Proceed. Nat. Acad. Sci. 18: 136–13

BARNES, H. (1959a): Oceanography and Marine Biology. A Book of Techniques. George Allen & Unwin Ltd., London.

— (1959b): Apparatus and methods of Oceanography. Vol. 1. Chemical. Interscience Publishers, New York 324 pp.

— (1962): Note on variations in the release of nauplii of Balanus balanoides with special reference to the spring diatom outburst. Crustaceana, 4: 118–122.

— & — M. (1960): Recent spread and present distribution of the barnacle Elminius modestus DARWIN in North-West Europe. Bull. Mar. Sc. Gulf and Caribbean, 135: 137–145.

BARNES, H., & H. T. POWELL, (1954): Onchidoris fusca (Müller), a predator of barnacles. J. Anim. Ecol., 23, 361–363.

BARY, B. McK. (1959): Biogeographic boundaries: The use of temperature-salinity-plankton diagrams. Intern. Oceanogr. Congress, New York, p. 132–133.

BAYER, F. M. (1963): Observations on pelagic mollusks associated with the siphonophores Velella and Physalia. Bull. Mar. Sc. Gulf and Caribbean, 13: 454–466.

Bé A. W. H. (1962): Quantitative multiple opening-and-closing plankton samplers. Deep-sea Res. 9: 144–151.

BECKER, G. W. D. KAMPF & KOHLMEYER (1957): Zur Ernährung der Holzbohrasseln der Gattung Limnoria. Naturwiss., 44: 473–474.

BECKER, G., & J. KOHLMEYER (1958): Holzzerstörung durch Meerespilze in Indien und ihre besondere Bedeutung für Fischereifahrzeuge. Arch. f. Fisch.wiss. 9: 29–40.

BEEBE, W. (1926): The Arcturus Oceanographic Expedition. Zoologica, N.Y. 8: 1–45.

BEERS, D. (1963): Relation of feeding in the sea urchin Strongylocentrotus droebachiensis to division in some of its endocommensal ciliates. Biol. Bull., 124: 1–8.

BEIN, S. J. (1954): A study of certain chromogenic bacteria isolated from 'Red Tide' water with a description of a new species. Bull. Mar. Sc. Gulf and Caribbean. 4: 110–119.

BEKLEMISHEV, C. W. (1960): Biotope and community in marine plankton. Intern. Rev. ges. Hydrobiol., 45: 269–301.

BENECKE, W. (1933): Bakteriologie des Meeres. In: ABDERHALDEN, E.: Handb. biol. Arbeitsmethoden, Abt. IX, Teil 5, 1: 717–854.

BENSON, R. H., & G. L. COLEMAN II (1963): Recent marine ostracodes from the Eastern gulf of Mexico. Univers. of Kansas Palaeont. Contrib., Arthropoda, Article 2: 1–52.

BERLAND, L. (1940): Les araignées marines. Mém. Soc. Biogéograph. 7: 347–353.

BERNARD, F. (1953): Rôle des Flagellés calcaires dans la fertilité et la sédimentation en mer profonde. Deep-Sea Research, 1: 34–46.

— (1959): Données sur l' abondance de sable corrodé dans les eaux de la Méditerranée orientale. Rev. Géograph. Phys. et Géol. Dyn., 2: 113–119.

BERNARD, F., & J. LECAL (1960): Plancton unicellulaire récolté dans l'Océan Indien par le 'Charcot' (1950) et le 'Norsel' (1955–1956). Bull. Inst. Océanogr. Monaco, No. 1166, 59 pp.

BERRILL, N. J. (1950): The Tunicata, with an Account of the British Species. The Ray Society, London, 354 pp.

BERTELSEN, E. (1951): The Ceratioid Fishes. Dana-Report No. 39: 1–276.

BERTRAM, G. L. R. (1936): Some aspects of the breakdown of coral at Ghardaga Red Sea. Proceed. Zool. Soc. London, (1936: 1011–1026.

BEVERTON, R. J. H., & S. J. HOLT (1957): On the dynamics of exploited fish populations. Fishery Investigations, Minist. Agricult., Fish., Food Ser. II, 19: 533 pp.

BIEBL, K. (1938): Trockenresistenz und osmotische Empfindlichkeit der Meeresalgen verschieden tiefer Standorte. Jahrb. wiss. Botanik 86: 350–386.

— (1939): Über die Temperaturresistenz von Meeresalgen verschiedener Klimazonen und verschieden tiefer Standorte. Jahrb. wiss. Botanik, 88: 389–120.

BLACK, W. A. P., & R. L. MITCHELL (1952): Trace elements in the common brown algae and sea water. J. Mar. Biol. Ass. U.K,. 30: 575–584.

BLACKER, R. W. (1957): Benthic animals as indicators of hydrographic conditions and climatic change in Svalbard waters. Fishery Investig. Ser. II, 20 Nr. 10, 49 pp.

BLINKS, L. R. (1955): Photosynthesis and productivity of littoral marine algae. J. Mar. Research 14 363–373.

BODEN, B. P., & E. M. KAMPA (1953): Winter cascading from an oceanic island and its biological implications. Nature, 171: 426–427.

BÖHNECKE, G. (1936): Temperatur, Salzgehalt und Dichte an der Oberfläche des Atlantischen Ozeans. Wiss. Erg. Deutsche Atlant. Exped. 'Meteor', 5.

BOGDANOV, D. V. (1963): Map of the natural zones of the ocean. Deep-Sea Research, 10: 520–523.

BOGOROV, B. G. (1946): Peculiarities of diurnal vertical migrations of zooplankton in polar seas. J. Mar. Research, 6: 25–32.

Bogorov, B. G. (1959). On the standardisation of marine plankton investigations. Intern. Rev. ges. Hydrobiol. **44**: 621–642.

Bolin, R. L. (1949): The linear distribution of intertidal organisms and its effect on their evolutionary potential. C.R. XIII. Congr. Intern. Zool. Paris 1948, 459–460.

Boolootian, R. A., & A. C. Giese (1959): The effect of latitude on the reproductive activity of *Strongylocentrotus purpuratus*. Intern. Oceanogr. Congr., New York, 216–217.

—, —, J. S. Tucker und A. Farmanfarmaian (1959): A contribution to the biology of a deep sea echinoid, *Allocentrotus fragilis* (Jackson). Biol. Bull. **116**: 362–372.

Boschma, H. (1949): Ellobiopsidae. Discovery Reports, **25**: 281–314.

Bourdillon, A. (1954): Mise en évidence d'une substance favorisant la métamorphose des arves d'*Alcyonium coralloides* (v. Koch). C. R. Sc. Acad. Sci. Paris, **239**: 1434–1436.

Bowen, V. T., & D. Sutton (1951): Comparative studies of mineral constituents of marine sponges I. The genera Dysidae, Chondrilla, Terpios. J. Mar. Research, **10**: 153–167.

Braarud, T. (1958): Counting methods for determination of standing crop of phytoplankton. Rapp. Proc. Verb. Réunions Cons. Intern. Expl. Mer, **144**: 17–19.

— (1961): Cultivation of marine organisms as a means of understanding environmental influence on populations. Oceanography; Publ. No. 67 Amer. Ass. Adv. Sci., Washington, D.C.

— & B. Foyn (1930): Beiträge zur Kenntnis des Stoffwechsels im Meere. Avhandl. Norske Vidensk. Akad. Oslo, Math. Naturw. Kl. 1930, No. **14**: 1–24.

Brandt, K. (1892): Über Anpassungserscheinungen und Art der Verbreitung von Hochseethieren. Krümmel's Reisebeschreibung d. Plankton-Exped. pp. 338–370.

Brandt, K. (1895): Biologische und faunistische Untersuchungen an Radiolarien und anderen pelagischen Tieren. Zool. Jahrb. Abt. System., Geogr. u. Biol. **9**: 25–74.

— (1899): Über den Stoffwechsel im Meere. Wissensch. Meeresunters. Kiel, **4**: 213.

— (1925): Die Produktion in den heimischen Meeren und das Wirkungsgesetz der Wachstumsfaktoren. Ber. Dtsch. wiss. Komm. Meeresforschg. N.F. **1**: 67.

Brattström, H. (1941): Studien über die Echinodermen des Gebietes zwischen Skagerrak und Ostsee, besonders des Öresundes, mit einer Übersicht über die physische Geographie. Undersökningar över Öresund, XXVII, 1–329, Lund.

Brauer, A. (1908): Die Tiefsee-Fische II. Anatomischer Teil. Wiss. Ergeb. Valdivia-Exped. **15**: 1–266

Brinkmann, A. (1913): *Bathynectes murrayi* n. gn., n. sp. Eine neue bathypelagische Nemertine mit äußeren männlichen Genitalien. Bergens Mus. Aarbok 1912, No. **9**: 1–9.

Brinton, E. (1962): The distribution of Pacific Euphausiids. Bull. Scripps Inst. Oceanogr., **8**: 51–270.

Broch, H. (1927): Methoden der marinen Biogeographie. In: Abderhalden, E.: Handb. biol. Arbeitsmeth. Abt. 9, Teil 5, Heft 1.

Brongersma-Sanders, M. (1948): The importance of upwelling water to vertebrate paleontology and oil geology. Kon. Ned. Ak. Wet. Verh. Afd. Nat. (Tweede Sectie), D I 45, No. **4**: 1–112.

— (1957): Mass mortality in the sea. Geol. Soc. Amer. Mem. 67, 1: 941–1010.

Bruce, H. E., & D. W. Hood (1959): Diurnal inorganic phosphate variations in Texas Bays. Publ. Inst. Mar. Science, **6**: 133–145.

Bruun, A. Fr. (1940): A study of a collection of the fish *Schindleria* from South Pacific waters. Dana-Rep. No. 21, 12 pp.

— (1943): The biology of *Spirula spirula* (L.). Dana-Report No. **24**: 1–46.

— (1955): The ecological zonation of the deep-sea. Proc. Unesco Symp. Phys. Oceanogr., Tokyo 1955: 160–168.

— (1956a): The abyssal fauna: its ecology, distribution and origin. Nature, London, **177**: 1105–1108.

— (1956b): Animal life of the deep-sea bottom. The Galathea Deep-Sea Expedition 1950–1952, London, 149–195.

— (1957): Deep-Sea and abyssal depths. Geol. Soc. Amer. Mem. 67, 1: 641–672.

Buddenbrock, W. v. (1934): Beobachtungen über nesselnde Tintenfische. Biol. Zentralbl. **54**: 284–287.

Bückmann, A. (1956): Marine Fischereibiologie und allgemeine Meeresbiologie. Experientia, **12**:

Bünning, E. (1963): Die physiologische Uhr. Springer Verlag, Berlin. 2. Aufl.

Bull, H. O. (1957): Behaviour: conditioned responses. In: M. E. Brown: The physiology of Fishes, **2**, Chap. III, Pt. I: 221–228.

— (1961): The role of ethology in oceanographie. Oceanography, Publ. No. 67 Amer. Ass. Adv. Sc.: 239–255.

Bullock, T. H. (1953): Predator recognition and escape responses of some intertidal gastropods in presence of starfish. Behaviour, 5:130–140.

— (1955): Compensation for temperature in the metabolism and activity of poikilotherms. Biol. Reviews, **30**: 311–342.

Burkholder, P. R. (1963): Drugs from the sea. Armed Forces Chemical Journ., **17**: 1–8.

Burkholder, P. R., & L. M. Burkholder (1958): Studies on B Vitamins in Relation to Productivity of the Bahia Fosforescente, Puerto Rico. Bull. Mar. Sci. Gulf and Caribbean, **8**, (3), 201–223.

— & — & L. R. Almodovar (1960): Antibiotic activity of some marine algae of Puerto Rico. Botanica Marina, **2**, 149–156.

Cannon, H. G. (1946): *Nebaliopsis typica*. Discovery Reports, **23**: 213–222.

Carlgren, O. (1917): Actiniaria and Zoantharia of the Danmark Exped. Medd. Grönland, **43**, 505–57.

448 BIBLIOGRAPHY

CASPERS, H. (1951): Rhythmische Erscheinungen in der Fortpflanzung von *Clunio marinus* (Dipt. Chiron.) und das Problem der lunaren Periodizität bei Organismen. Arch. f. Hydrobiol., Suppl. **18**: 415–594.

CASPERS, H. (1959): Die Einteilung der Brackwasser-Regionen in einem Ästuar. Arch. Oceanogr. Limnol., Venedig, **11**, Suppl., 153–159.

— (1959): Vorschlag einer Brackwassernomenklatur. ("The Venice System"). Int. Rev. Hydrobiol., **44**, 313–315.

CASSIE, M. (1959): An experimental study of factors inducing aggregation in marine plankton. New Zealand Jour. Sci. **2**: 339–365.

CATTELL, J. McK. (1936): The physiological effects of pressure. Biol. Reviews **11**: 441–476.

CAULLERY, M. (1922): Le parasitisme et la symbiose. Paris, Librairie Octave Doin, 400 pp.

CHABANAUD, P. (1949): Essai d'une division biogéographique du domaine océanique XIII. Congrès Intern. Zool. Paris 1948: 535–538.

CHAPMAN, G. (1953): Studies of the mesogloea of Coelenterates I. Histology and chemical properties. Quart. J. Microsc. Sc. **94**: 3. Ser., 155–176.

CHAPMAN, V. J. (1949): Seaweeds and their uses. Pitman Publish. Corp.

CHAPMAN, W. Mc. (1939): Eleven new species and three new genera of oceanic fishes, collected by the International Fisheries Commission from the northeastern Pacific. Proc. U.S. Nat. Mus. **86**: 501–542.

CHENG, C. (1941): Ecological relations between the herring and the plankton off the North-East Coast of England. Hull Bull. Mar. Ecol. **1**: 239–254.

CHESTERS, C. G., & J. A. SCOTT (1956): The production of antibiotic substances by seaweeds. Second Internat. Seaweed Symposium. Pergamon Press, New York.

CHUN, C. (1886): Über die geographische Verbreitung der pelagisch lebenden Seetiere. Zool. Anz. **9**: 55–59.

— (1890): Die pelagische Tierwelt in großen Meerestiefen. Verh. Ges. Deutsch. Naturf. u. Ärzte, **63**: 69–85.

— (1892): Die Dissogonie, eine neue Form der geschlechtlichen Zeugung. In: Festschrift 70. Geburtstag von R. LEUCKART, Leipzig, 77–108.

— (1896): Atlantis, Biologische Studien über pelagische Organismen. Bibl. Zoologica, **7**: 260 pp.

— (1903): Aus den Tiefen des Weltmeeres, 2. Aufl., Jena, 592 pp.

CLARKE, A. H. JR. (1962): Annotated list and bibliography of the abyssal marine molluscs of the world. Bull. Nation. Mus. Canada, No. 184, 114 pp.

— (1962b): On the composition, zoogeography, origin and age of the deep-sea mollusc fauna. Deep-Sea Res. **9**: 291–306.

CLARKE, G. L. (1933): Diurnal migration of plankton in the Gulf of Maine and its correlation with changes in submarine irradiation. Biol. Bull. **65**: 402–436.

CLARKE, G. L. (1938): Seasonal changes in the intensity of submarine illumination off Woods Hole, Ecol. **19**: 89–106.

— (1946): Dynamics of production in a marine area. Ecol. Monogr., **16**: 321–335.

— & D. W. BISHOP (1948): The nutritional value of marine zooplankton with a consideration of its use as an emergency food. Ecol., **29**: 54–71.

— & L. R. BRESLAU (1960): Studies on luminescent flashing in Phosphorescent Bay, Puerto Rico, and the Gulf of Naples using a portable Bathyphotometer. Bull. Inst. oceanogr. Monaco, No. 1171, 32 pp.

CLARKE, G. L., & C. J. HUBBARD (1959): Quantitative record of the luminescent flashing of oceanic animals at great depths. Limnol. Oceanogr., **4**, 163–180.

CLARKE, W. D. (1955): A new species of the genus *Heteromysis* (Crustacea, Mysidacea) from the Bahama Islands, commensal with a sea-anemone. Amer. Mus. Novit., No. 1716: 1–13.

COE, W. R. (1943): Biology of the Nemerteans of the Atlantic coast of North America. Transact. Connecticut Acad. Art. a. Sci., **35**, 129–328.

COE, W. R. (1945): Plankton of the Bermuda Oceanographic Expeditions. XI. Bathypelagic Nemerteans of the Bermuda Area and other parts of the North and South Atlantic Ocean, with evidence as to their means of dispersal. Zoologica, **30**: 145–168.

— (1953): Resurgent populations of littoral marine invertebrates and their dependence on ocean currents and tidal currents. Ecol., **34**: 225–229.

CONNELL, J. H. (1961a): Effects of competition, predation by *Thais lapillus* and other factors on natural populations of the barnacle *Balanus balanoides*. Ecological Monographs, **31**: 61–104.

— (1961b): The influence of interspecific competition and other factors on the distribution of the barnacle *Chthamalus stellatus*. Ecol., **42**:710–723.

Contributions to Plankton Symposium 1957, Measurements of primary production in the Sea. Rapp. Proc. Verb. Réun. Cons. Perm. Intern. Explor. Mer. **144**, 1958, 158 pp.

COPPER, L. H. N. (1933): Chemical constituents of biological importance in the English Channel, November, 1930 to January, 1932. Part I u. II. J. Mar. Biol. Ass. U.K., **18**: 677–753.

— (1952): Factors affecting the distribution of silicate in the North Atlantic Ocean and the formation of North Atlantic deep water. J. Mar. Biol. Ass. U.K., **30**: 511–526.

COWEY, C. B. (1956): A preliminary investigation of the variation of vitamin B_{12} in oceanic and coastal waters. J. Mar. Biol. Ass. U.K., **35**: 609–620.

CRISP, D. J., & A. J. SOUTHWARD (1959): The further spread of *Eliminius modestus* in the British Isles to 1959. J. Mar. biol. Ass. U.K., **38**, 429–437.

CUSHING, D. H. (1951): The vertical migration of planktonic Crustacea. Biol. Reviews, **26**: 158–192.

— (1959): On the nature of production in the sea. Fishery Investigations, Ser. 2, **12**: No. 6, 1–40.

DAHL, E. (1951): *Amallocystis*, ett parasitiskt flagellatsläkte från djuphavet. Fauna och Flora, 1951, 165–70.
DAHL, E. (1952): Some aspects of the ecology and zonation of the fauna on sandy beaches. Oikos, 4: 1–27.
— (1954): The distribution of deep sea Crustacea. Union intern. Sci. Biol., Ser. B, No. 16: 43–48.
— (1959): Amphipoda from depths exceeding 6000 meter. Galathea Report, 1: 211–241.
DAISLEY, K. W., & L. R. FISHER (1958): Vertical distribution of vitamin B_{12} in the sea. J. Mar. Biol. Ass. U.K. 37: 683–686.
DAKIN, W. J., & A. N. COLEFAX (1936): *Ctilopis*, a rare pelagic Nudibranch of the family *Phyllirhoidae* (BERGH). Proc. Zool. Soc. London, 1936: 455–460.
DALES, R. P. (1955): The evolution of the pelagic alciopid and phyllodocid polychaetes. Proc. Zool. Soc. London, 125: 411–420.
— (1957): Commensalism. Geol. Soc. Amer. Memoir 67, 1: 391–412.
DALES, R. P. (1966): Symbiosis in marine organisms. In *Symbiosis*, ed. Henry, S. M. Vol. 1, 299–326. New York, Academic Press.
DAMAS, D. (1905): Notes biologiques sur les copépodes de la Mer Norvegienne. Publ. de Circonstance, No. 22, Cons. perm. intern. pour l'explor. de la Mer: 1–23.
DANFORTH, C. H. (1907): A new Pteropod from New England. Proc. Boston Soc. Nat. Hist. 34: 1–21.
DAVID, P. M. (1965): Die Fauna der Meeresoberfläche. Endeavour, 24 (Nr. 92): 95–100.
DAVIS, CH. C. (1953): Concerning the flotation mechanism of *Noctiluca*. Ecology, 34: 189–192.
— (1963): On questions of production and productivity in ecology. Arch. Hydrobiol. 59: 145–161.
DEEVEY, G. B. (1952a): A survey of the zooplankton of Block Island Sound 1943–1946. Bull. Bingham Oceanogr. Collection, 13: 65–119.
— (1960a): The Zooplankton of the surface waters of the Delaware Bay Region. Bull. Bingham Oceanogr. Collection, 17: 6–53.
— (1960b): Relative effects of temperature and food on seasonal variations in length of marine copepods in some Eastern American and Western European waters. Bull. Bingham Oceanogr. Collection, 17: 54–85.
DEHNEL, P. A. (1955): Rates of growth of gastropods as a function of latitude. Physiol. Zool. 28: 115–144.
DELAMARE DEBOUTTEVILLE, CL. (1960): Biologie des eaux souterraines littorales et continentales. Paris (Hermann), 1960, 740 pp.
DELSMAN, H. C. (1939): Preliminary plankton investigations in the Java Sea. Treubia, Deel 17, Aufl. 2: 139–181.
DÉMEUSY, N. (1957): Respiratory metabolism of the fiddler crab *Uca pugilator* from two different latitudinal populations. Biol. Bull. 113: 245–253.
DENTON, E. J. (1963): Buoyancy mechanisms of sea creatures. Endeavour, 22: 3–8.
— & J. B. GILPIN-BROWN (1961a): The buoyancy of the cuttlefish, *Sepia officinalis* (L.). J. Mar. Biol. Ass U.K. 41: 319–342.
— & — (1961b): The effect of light on the buoyancy of the cuttlefish. J. Mar. Biol. Ass. U.K. 41: 343–350.
—, — & J. V. HOWARTH (1961c): The osmotic mechanism of the cuttlebone. J. Mar. Biol. Ass. U.K. 41: 351–364.
—, — (1961d): The distribution of gas and liquid within the cuttlebone. J. Mar. Biol. Ass. U.K. 41: 365–381.
DENTON, E. J., GILPIN-BROWN, J. B., & HOWARTH, J. V. (1967): On the buoyancy of *Spirula spirula*. J. Mar. Biol. Ass. 47: 181–191.
DEVEZE, M. L. (1955): Parallélisme d'évolution des populations planctoniques et bactériennes marines durant la période estival 1955. C. R. Acad. Sc. 241: 1629–1631.
DIETRICH, G. (1963): Meereskunde der Gegenwart. Naturw. Rundschau, 11: 465–473.
— W. HANSEN, W. HORN, J. JOSEPH & KALLE (1952): Ozeanographie. In: Landolt-Börnstein, Zahlenwerte u. Funktionen, 3: 426–533. Springer-Verlag, Berlin–Göttingen–Heidelberg.
— & K. KALLE (1957): Allgemeine Meereskunde. Eine Einführung in die Ozeanographie. Gebr. Borntraeger, Berlin, 492 pp.
DIETRICH, R. & W. HÖHNK, (1958): Über das Öl des submers lebenden Pilzes *Ceratostomella spec.* (Sphaeriales, Ascomycetes). Veröff. Inst. f. Meeresforschung, Bremerhaven, 5: 135–142.
— & (1960): Studien zur Chemie ozeanischer Bodenproben. I. Über die Kohlenstoff-, Stickstoff- und Phosphor-Verhältnisse in Hinsicht auf die Mykoflora. Veröff. Inst. Meeresforschg. Bremerhaven, 7, 15–35.
— & (1964): Studien etc. II. Ein Beitrag zur Kenntnis der Huminstoffe des Meeresbodens. Ebenda 9, 1–23.
DIETZ, R. S. (1948): Some oceanographic observations on operation Highjump. U.S. Navy Electronics Labor. San Diego, Cal. Usnel Report No. 55: 1–97.
DIGBY, P. S. B. (1950): The biology of the small planktonic copepods off Plymouth. J. Mar. Biol. Ass. U.K. 29: 393–416.
— The biology of the marine plankton copepods of Scoresby Sound, East Greenland. J. Animal Ecology, 23: 298–338.
DOBZHANSKY, T. (1950): Evolution in the tropics. Amer. Scient., Burlington, 38, 209–221.
DROOP, M. R. (1954): Cobalamin requirements in Chrysophyceae. Nature (London), 174: 520–521.
— (1959): Water-soluble factors in the nutrition of *Oxyrrhis marina*. J. Mar. Biol. Ass. U.K., 38: 605–620.

DUNBAR, M. S. (1951a): Resources of arctic and subarctic seas. Transact. Roy. Soc. Canada, 45: Ser. III. 61–67.
— (1951b): Eastern arctic waters. Fish. Res. Board. Canada, Bull. No. 88, 131 pp.
DURCHON, M. (1956): Role du cerveau dans la maturation génitale et le déclenchement de l'épitoquie chez les néréidiens. Ann. Sci. Nat. Zool., 11ᵉ ser., 1956.
DUURSMA, E. K. (1960): Dissolved organic carbon, nitrogen and phosphorus in the sea. Dissertation Amsterdam, Verlag J. B. Wolters, Groningen 1960, 147 pp.
— (1961): In: Netherlands Journ. Sea Res., 1: 1–147.

EALES, N. B. (1952): The littoral fauna of Great Britain, a handbook for collectors. Cambridge, University Press, XVI, 305 pp.
EDWARDS, G. A., & IRVING, L. (1943a): The influence of temperature and season upon the oxygen consumption of the sand crab, *Emerita talpoida* SAY. J. Cell. Comp. Physiol. 21: 169–182
— & — (1943b): The influences of season and temperature upon the oxygen consumption of the beach flea, *Talorchestia megalophthalma*. J. Cell. Comp. Physiol. 21: 183–189.
EIBL-EIBESFELDT, I. (1955): Über Symbiosen, Parasitismus und andere besondere zwischenartliche Beziehungen tropischer Meeresfische. Zeitschr. f. Tierpsych. 12: 203–219.
EINARSSON, H. (1945): Euphausiacea. I. Northern Atlantic Species. The Carlsberg Found. Ocean. Exped. 1928–1930. Dana-Report No. 27: 1–185.
EKMAN, S. (1946): Zur Verbreitungsgeschichte der Warmwasserechinodermen im Stillen Ozean (Asteroidea, Ophiuroidea, Echinoidea). Nova Acta Regiae Soc. Scient. Upsaliensis, Ser. IV, 14: 1–42.
— (1947): Über die Festigkeit der marinen Sedimente als Faktor der Tierverbreitung. Zool. Bidrag, Uppsala, 25: 1–20.
EKMAN, S. (1953): Zoogeography of the sea. Sidgwick & Jackson, London, XII + 417 pp.
EMILIANI, C. (1954): Temperatures of Pacific bottom waters and Polar superficial waters during the Tertiary. Science, 119: 853–855.
— & C. EDWARDS (1953): Tertiary ocean bottom temperatures. Nature (London), 171: 887–888.
—, T. MAYEDA & R. SELLI (1959): Paleotemperature analysis of the Plio-Pleistocene series of La Castella, Calabria. Intern. Oceanogr. Congr. New York. 91–92.
ENDEAN, R., R. KENNEDY & W. STEPHENSON (1956): The ecology and distribution of intertidal organisms on the rocky shores of the Queensland mainland. Austr. J. Mar. Freshw. Res. 7: 88–146.
ERCEGOVIC, A. (1957): Principes et essai d'un classement des étages benthiques. Rec. Travaux Stat. Mar. d'Endoume, 22: 17–21.
ERDMANN, J. G., MARLETT & W. E. HANSEN (1956): Survival of amino acids in marine sediments. Science, 124, 1026.
ERICSON, L. E., & L. LEWIS (1953): On the occurrence of vitamin B₁₂-factors in marine algae. Ark. Kemi, 6: 427–442.
ERICSON, D. B., & G. WOLLIN (1962): Micropalaeontology. Scientific American, 207: 97–106.
ESTERLY, C. O. (1911): Diurnal migrations of *Calanus finmarchicus* in the San Diego Region during 1909. Intern. Rev. ges. Hydrobiol. & Hydrogr. 4: 140–151.
— (1917): The occurrence of a rhythm in the geotropism of two species of plankton copepods when certain external conditions are absent. Univ. Calif. Publ. Zool. 16: 393–400.
EVANS, R. G. (1947): The intertidal ecology of Cardigan Bay. Journ. Ecology, 34, 273–309.

FAGE, L. (1941, 1942): Mysidacea, Lophogastrida I u. II. Dana-Report No. 19: 1–52 und 23: 1–67.
— (1954a): Remarques sur les conditions de vie de la faune benthique abyssale. Union intern. Sci Biol., Sér. B, No. 16: 12–19.
— (1954b): Remarques sur les pycnogonides abyssaux. Union intern. Sci. Biol., Sér. B, No. 16: 49–56.
FELDMANN, J. (1937): Recherche sur la végétation marine de la Méditerranée. La côte des Albères. Revue algologique 1937.
FISCHER, A. G. (1960): Latitudinal variations in organic diversity. Evolution, 14: 64–81.
FISH, C. J. (1936a): The biology of *Calanus finmarchicus* in the Gulf of Maine and Bay of Fundy. Biol. Bull. 70: 118–141.
— (1936b): The biology of *Pseudocalanus minutus* in the Gulf of Maine and Bay of Fundy. Biol. Bull. 70: 193–216.
— (1936c): The biology of *Oithona similis* in the Gulf of Maine and Bay of Fundy. Biol. Bull. 71: 168–187.
FISHER, J., & R. M. LOCKLEY (1954): Sea-birds. An introduction to the natural history of the sea birds of the North Atlantic. Collins, London, 320 pp.
FISHER, L. R., S. K. KON & S. Y. THOMPSON (1955): Vitamin A and carotnoids in certain invertebrates. III. Euphausiacea. J. Mar. Biol. Ass. U.K. 34: 81–100.
FLEMING, R., & T. LAEVASTU (1956): The influence of hydrographic conditions on the behaviour of fish. FAO Fish. Bull. 9: No. 4, 16 pp.
FOCKE, E. (1961): Die Rotatoriengattung *Notholca* und ihr Verhalten im Salzwasser. Keiler Meeresforschungen 17, (2), 190–205.
FOGG, G. E. (1957): The production of extracellular nitrogenous substances by a blue green alga. Proc. Roy. Soc. London, B, 139: 372–397.
FORBES, E. (1859): The Natural History of the European Seas, London.
FORD, E. (1923): Animal communities of the level sea-bottom in the water adjacent to Plymouth. J. Mar. Biol. Ass. U.K. Plymouth, 13, 164–224.

Fox, H. M. (1925): The effect of light on the vertical movement of aquatic organisms. Proceed. Cambr. Philos. Soc. Biol. Sci. 1: 219–224.
Fox, H. M. (1939): The activity and metabolism of poikilothermal animals in different latitudes. Proceed. zool. soc. London, 109A, 141–156.
— & C. A. Wingfield (1937): The activity and metabolism of poikilothermal animals in different latitudes. II. Proc. Zool. Soc. London, Ser. A, 107, 275–282.
Fraenkel, G. (1927): Biologische Beobachtungen an *Janthina*. Z. Morph. Ökol. d. Tiere, 7: 596–608.
Fraser, J. H. (1952): The Chaetognatha and other Zooplankton of the Scottish Area and their value as Biological Indicators of Hydrographical Conditions. Scottish Home Department, Marine Research No. 2.
Fraser-Brunner, A. (1951): The Ocean Sunfishes (Family Molidae). Bull. Brit. Mus. (Nat. Hist.) Zool., 1: 89–121.
Fretter, V. (1953): Experiments with radioactive strontium (Sr^{90}) on certain molluscs and polychaetes. J. Mar. Biol. Ass. U.K. 32: 367–384.
Freund, L. (1927): Anoplura pinnipediorum (Robbenläuse). Tierw. Nord-u. Ostsee, 11d, 36 pp, (Leipzig).
Friedrich, H. (1950): Versuch einer Darstellung der relativen Besiedlungsdichte in den Oberflächenschichten des Atlant. Ozeans. Kieler Meeresforsch., 7.
Friedman, I. (1957): Determination of Deuterium-Hydrogen ratios in Hawaiian waters. Tellus 4: 553–556.
Friedrich, H. (1931): Mitteilungen über vergleichende Untersuchungen über den Lichtsinn einiger mariner Copepoden. Z. vergl Physiol. 15: 121–138.
Friedrich, H. (1952a): Betrachtungen zur Synökologie des ozean ischen Pelagials. Veröff. Inst. Meeresforsch. Bremerhaven, 1: 7–36.
— (1952b): Über neuere Gesichtspunkte zur Physiologie der Biocönosen. Veröff. Inst. Meeresforsch. Bremerhaven, 1: 225–231.
— (1954): Über den Stand unserer Kenntnis der Arten im marinen Pelagial. Veröff. Inst. f. Meeresforsch. Bremerhaven, 3: 34–41.
— (1955a): Beiträge zu einer Synopsis der Gattungen der Nemertini monostilifera nebst Bestimmungsschlüssel. Z. wiss. Zool. 158: 133–192.
— (1955b): Materialien zur Frage der Artbildung in der Fauna des marinen Pelagials. Veröff. Inst. f. Meeresforsch. Bremerhaven, 3: 159–189.
— (1961): Physiological significance of light in marine ecosystems. Oceanography, Amer. Ass. Advancem. Sc., 257–270.
— & H. Lüneburg (1962): Beiträge zu einer Bibliographie „Abwasser in Meerwasser". Veröff. Inst. Meeresforsch. Bremerhaven, 8: 37–52.
Fuller, J. R. (1937): Feeding rate of *Calanus finmarchicus* in relation to environmental conditions. Biol. Bull. 72: 233–246.

Gaarder, T., & H. H. Gran (1927): Investigations of the production of plankton in the Oslo Fjord. Rapp. Procès Verb. Cons. Intern. Explor. Mer, 42: 1–48.
Gardiner, A. C. (1932–1933): Vertical distribution in *Calanus finmarchicus*. J. Mar. Biol. Ass. U.K. 18: 575–610.
— (1937): Phosphate production by planktonic animals. Journ. d. Conseil Perman. Intern. pour l'Explor. de la Mer. 12: 144–146.
Gauld, D. T., & J. E. G. Raymont (1953): The respiration of some planktonic copepods, II. The effect of temperature. J. Mar. biol. Ass. U.K., 31, 447–460.
Gehringer, J. W. (1952): High-speed plankton samplers. 2. An allmetal plankton sampler (Model Gulf III). U.S. Dept. Int., Fish Wildl. Serv., Spec. Sci. Rep.-Fish, 88.
Gerlach, S. A. (1953): Die biozönotische Gliederung der Nematodenfauna an den deutschen Küsten. Z. Morph. Ökol. Tiere, 41, 411–512.
— (1954): Das Supralitoral der sandigen Meeresküsten als Lebensraum einer Mikrofauna. Kieler Meeresforsch., 10: 121–129.
— (1958): Die Mangroveregion tropischer Küsten als Lebensraum. Z. Morph. u. Ökol. Tiere, 46: 636–730.
— (1960): Über das tropische Korallenriff als Lebensraum. Zool. Anz. Suppl. 23: 356–363.
Gessner, F. (1955): Hydrobotanik, Bd. 1, Berlin, 517 pp.
— (1959): Hydrobotanik, II. Stoffhaushalt, Berlin.
— & L. Hammer (1960): Die Primärproduktion in mediterranen Caulerpa-Cymodocea-Wiesen. Botanica Marina, 2: 157–163.
Gillbright, M. (1952): Untersuchungen zur Produktionsbiologie des Planktons in der Kieler Bucht I. Kieler Meeresforsch., 8: 175–191.
— (1954): Das Verhalten von Zooplankton—vorzugsweise von *Tintinnopsis beroidea* Entz—gegenüber thermohalinen Sprungschichten. Kurze Mitt. Inst. Fischereibiol. Hamburg, Nr. 5: 32–44.
Giltay, L. (1934): Notes ichthyologiques. VIII. Les larves de Schindler sont-elles des Hémiramphidae? Bull. Mus. Hist. Nat. Belg., 10, No. 13, 10 pp.
Glover, R. S. (1961): Biogeographical boundaries: the shapes of distributions. Oceanography, Publ. 67 Amer. Ass. Advancem. Sci. Washington, 201–228.

GISLÉN, T. (1930): Epibiosis of the Gullmar Fjord. I. Kristinebergs Zoologiska Station 1877-1927, 123 pp., 1929; II. 380 pp., 1930.
— (1949): Ecology and physiography of the littoral of the Northern Pacific. XIII. Intern. Congr. Zool., Paris, 1948, 411-414.
GOLDBERG, E. D., W. McBLAIR & K. M. TAYLOR (1951): The uptake of vanadium by tunicates. Biol. Bull., 101: 84-94.
GOLDSCHMIDT, R. (1906): Amphioxides und Amphioxus. Zool. Anz., 30: 443-448.
GORDON, M. S., & H. M. KELLY (1962): Primary productivity of an Hawaiian coral reef: a critique of flow respirometry in turbulent waters. Ecology, 43: 473-480.
GRAHAM, HERBERT W. (1941): Plankton production in relation to character of water in the open Pacific. Journ. Mar. Res. 4 (3), 189-197.
— & E. G. MOBERG (1944): Chemical results of the last cruise of the Carnegie. Carnegie Inst. of Washington, Publ. 562, V+58 pp.
GRAN, H. H. (1932): Phytoplankton. Methods and problems. J. Conseil Intern. Expl. Mer, 7: 343.
GRAVE, B. H. (1927): The natural history of Cummingia tellinoides. Biol. Bull. Woods Hole, 53.
GRAVIER, C. (1919): Pédogénèse et viviparité chez les Actiniaires. C. R. Acad. Sci. Paris, 168, 736-738.
— (1920): Larves d'Actiniaires provenant des campagnes scientifiques de S. A. S. le Prince Albert Ier de Monaco. Res. camp. sci. Monaco, 57.
GRIFFIN, D. R. (1945): Hearing and acoustic orientation in marine animals. Deep-Sea Research, Suppl. to Vol. 3: 406-417.
GRØNTVED, J. (1960): On the productivity of microbenthos and phytoplankton in some Danish fjords.— Medd. Danmarks Fisk. Havunders. N. S. 3: 55-92.
— (1962): Preliminary report on the productivity of microbenthos and phytoplankton in the Danish wadden sea.—Medd. Danmarks Fisk. Havunders. N. S. 3; 347-371.
GROSS, F. (1937): The life-history of some marine plankton diatoms.—Trans. Roy. Soc. London, B. 228, 1.
— & E. ZEUTHEN (1948): The buoyancy of plankton diatoms: a problem of cell physiology. Proceed. Roy. Soc. London, Ser. B, 135: 382-389.
GUILCHER, A. (1958): Coastal and submarine morphology. Methuen & Co. Ltd., London, 274 pp.
GULLAND, J. A., & G. R. WILLIAMSON (1962): Transatlantic journey of a tagged cod. Nature (London) 195: 921.
GUNTER, G. (1959): Pollution problems along the Gulf Coast. Transact. of the Second Seminar on Biol. Problems in Water Pollution, 1959.
— (1961): Some relations of estuarine organisms to salinity. Limnology and Oceanography, 6: 182-190.
—, W. SMITH & R. WILLIAMS (1947): Mass mortality of marine animals on the lower West coast of Florida, November 1946—January 1947. Science, 105: 256-257.
GÜNTHER, KL. (1950): Oekologische und funktionelle Anmerkungen zur Frage des Nahrungserwerbes bei Tiefseefischen mit einem Exkurs über die ökologischen Zonen und Nischen. In: Moderne Biologie, Festschr. f. Hans Nachtsheim, Berlin, 1950, 55-93.
— & K. DECKERT (1950): Wunderwelt der Tiefsee. Berlin (F. A. Herbig), 240 pp.
GURJANOWA, E. et al. (1928, 1929): Das Litoral des Kola-Fjords. Trav. Soc. Nat. Leningrad, 58: 89-143, 59: 47-152.

HAECKEL, E. (1890): Plankton-Studien. Vergleichende Untersuchungen über die Bedeutung und Zusammensetzung der pelagischen Fauna and Flora. Jena, G. Fischer.
HAECKER, V. (1908): Tiefsee-Radiolarien. Wiss. Ergebn. Deutsche Tiefsee-Exp. Valdivia, 14.
HAGMEIER, A. (1930): Eine Fluktuation von Mactra (Spisula) subtruncata Da Costa an der ostfriesischen Küste. Ber. Dtsch. wiss. Komm. Meeresforsch. N.F. 5: 126-155.
HALME, E., & T. LUKKARINEN (1960): Planktologische Untersuchungen in der Pojo-Bucht und angrenzenden Gewässern. Ann. Zool. Soc. Zool. Bot. Fennicae- „Vanamo", 22/2: 1-24.
HARDER, W. (1957): Verhalten von Organismen gegenüber Sprungschichten. Ann. biol. Ser. 3, 33: 227-232.
HARDY, A. C. (1939): Ecological investigations with the continuous plankton recorder. Hull. Bull. Mar. Ecol., 1: 1-57.
— (1953): Some problems of pelagic life. In: Essays in Marine Biology, p. 101-121, Edinburgh and London.
— (1956): The Open Sea. Its Natural History: The world of Plankton. Collins, London, XV+335 pp.
— & R. BAINBRIDGE (1951): Effect of pressure on the behaviour of decapod larvae (Crustacea). Nature, London, 167: 354-355.
— & E. R. GUNTHER (1935): The plankton of the South Georgia whaling grounds and adjacent waters, 1926-1927. Discovery Reports, 11: 1-455.
HARMS, J. W. (1948): Über ein inkretorisches Cerebralorgan bei Lumbriciden, sowie Beschreibung eines verwandten Organs bei drei neuen Lycastis-Arten. Arch. Entw. Mech. d. Org. 143: 332-346.
HARRIS, J. F. (1953): Physical factors involved in the vertical migration of plankton. Quart. J. Microsc Sc., 94, 3. Ser.: 537-550.
— (1963): The role of endogenous rhythms in vertical migrations. J. Mar. Biol. Ass. U.K., 43: 153, 166.
HARVEY, H. W. (1925): Oxidation in sea water.. J. Mar. Biol. Ass. U.K., 13: 953-969.
— (1939): Substances controlling the growth of a diatom. J. Mar. Biol. Ass. U.K. 23: 499-520.
— (1955): The Chemistry and Fertility of Sea waters. Cambridge, University Press, VIII+224 pp.

BIBLIOGRAPHY 453

HASLE, G. R. (1959): A quantitative study of phytoplankton from the equatorial Pacific. Deep-sea Research, **6**1: 38–59.
HASLER, A. D. (1954): Odour perception and orientation in fishes. J. Fish. Res. Board Canada, **11**: 107–129.
HASTINGS, J. W. (1959): Unicellular clocks. Ann. Rev. Microbiol. **13**: 297–312.
HAUENSCHILD, C. (1955): Photoperiodizität als Ursache des von der Mondphase abhängigen Metamorphose-Rhythmus bei dem Polychaeten *Platynereis Dumerilii*. Zeitschr. f. Naturforsch. **10**b: 658–662.
— (1956): Neue experimentelle Untersuchungen zum Problem der Lunarperiodizität. Naturwiss, **43**: 361–363.
HAYES, H. L., & TH. S. AUSTIN (1951): The distribution of discolored sea water. Texas J. of Science, **3**: 530–541.
HEDGPETH, J. W. (1957a): Classification of Marine Environments. Geol. Soc. America, Memoir 67, **1**: 17–28.
— (1957b): Concepts of Marine Ecology. Geol. Soc. America, Memoir **67**, 1: 29–52.
— (1957c): Marine Biogeography. Geol. Soc. America, Memoir **67**, 1 359–382.
HEEGAARD, P. (1948): Larval stages of *Meganyctiphanes* (Euphausiacea) and some phylogenetic remarks. Medd. Komm. Danmarks Fisk. og Havunters., Ser. Plankton, **5**: 1–27.
HEEZEN, B. C. & C. HOLLISTER (1964): Deep-sea current evidence from abyssal sediments. Marine Geol. **1**: 141–174.
HENDRICKS, K. (1908): Zur Kenntnis des gröberen und feineren Baues des Reusenapparates an den Kiemenböger von *Selache maxima* Cuvier. Inaug. Diss. Münster. Z. wiss. Zool. **91**: 1–87.
HENTSCHEL, E. (1921): Über den Bewuchs auf den treibenden Tangen der Sargassosee. Beiheft Jahrb. Harnb. Wiss. Anst. **38**: 1–26.
— (1939): Kinetisches and akinetisches Plankton. Naturwiss. **27**: 209–211.
— & H. WATTENBERG (1930): Plankton und Phosphat in der Oberflächenschicht des Südatlantischen Ozeans. Ann. Hydrogr, u. marit. Meteor, **53**: 273–277.
HEUSS, TH. (1940): Anton Do. Atlantis Verl., Hamburg, 319 pp.
HIATT, R. W., & STRASBURG (1960): Ecological relationships of the fish fauna on coral reefs of the Marshall Islands. Ecological Monographs, **30**: 65–127.
HIDA, T. S., & J. E. KING (1955): Vertical distribution of zooplankton in the central equatorial Pacific, July–August 1952. Spec. sci. Rep. U.S. Fish Wildlife Service (Fish.) **144**: 1–22.
HILLIS, L. W. (1959): A revision of the genus *Halimeda* (order Siphonales). Publ. Inst. Mar. Science, **6**: 321–403.
HJORT, J. (1911): Die Tiefsee-Expedition des „Michael Sars" nach dem Nordatlantik im Sommer 1910. Intern. Rev. ges. Hydrobiol. u. Hydrogr. **4**: 152–173, 335–361.
HODGKIN, E. P. (1960): Patterns of life on rocky shores. J. Roy. Soc. West Austr. **43**: 35–43.
HÖHNK, W (1952–1956): Studien zur Brack- und Seewassermykologie I-VI. Veröff. Inst. f. Meeresfschg. Bremerhaven, 1–4.
— (1955c): Niedere Pilze vom Watt und Meeresgrund (Chytridiales und Thraustochytriaceae). Naturwissenschaften, **42**: 348–349.
— (1958): Fortschritte der marinen Mykologie in jüngster Zeit. Naturw. Rundschau: 39–44.
HOFFMANN, C. (1942): Beiträge zur Vegetation des Farbstreifenwatts. Kieler Meeresforsch., **4**: 85.
— (1949): über die Lichtdurchlässigkeit dünner Sandschichten für Licht. Planta, **36**: 48–56.
HOLTHUIS, L. B. (1963): On red coloured shrimps (Decapoda, Caridea) from tropical land-locked saltwater pools. Zool. Medel. Leiden, **38**: 261–279.
HOPKINS, S. H. (1957): Parasitism. Geol. Soc. Amer. Memoir **67**, 1: 413–428.
HUBBS, C. J., (1951): Record of the Shark *Carcharhinus longimanus*, accompanied by *Naucrates* and *Remora*, from the East-Central Pacific. Pacific Science, **5**: 78–81.
HUNDT, R. (1939): Das mitteldeutsche Graptolithenmeer. Halle (Boerner), 395 pp.
HUVÉ, P. (1955): Sur les propagules planctoniques de l'hydroide *Halecium pusillum* (SARS). Bull. Inst. Océanogr. No. 1064, 12 pp.
HYMAN, L. H. (1940): The Invertebrates. Protozoa through Ctenophora. London (McGraw-Hill), 726 pp.

IHLE, J. E. W. (1937, 1939): Salpidae. Bronns Klassen u. Ordn. **3**, Suppl. Tunikaten, Abt. II. 2. Buch, Liefg. 2 u. 3: 69–240.
IMMS, A. D. (1936): On a new species of *Halobates*, a genus of pelagic Hemiptera. Sc. Rep. John Murray Exped. Brit. Mus. Nat. Hist. **4**, No. 2: 71–78.
IVANOV, A. V. (1963): Pogonophora. Acad. Press, London, XVI+479 pp.
IVLEV, V. S. (1961): Experimental ecology of the feeding of fishes. New Haven, Yale Univers. Press, VIII+302 pp.
IZHEVSKII, G. K. (1964): Oceanological principles as related to the fishery productivity of the seas. Jerusalem, 186 pp.

JACKSON, PH. (1954): Engineering and economic aspects of marine plankton harvesting. Journ. d. Conseil, **20**: 166–174.
JACOBS, W. (1935): Das Schweben der Wasserorganismen. Ergebn. d. Biologie, **11**: 131–218.
— (1937): Beobachtungen über das Schweben der Siphonophoren. Zeitschr. vergl. Physiol. **24**: 583–601.

JACOBS, W. (1938): Fliegen, Schwimmen, Schweben. Berlin, Springer-Verlag. 134 pp.
— (1944): Das Problem des spezifischen Gewichtes bei Wassertieren. Arch. f. Hydrobiol. **39**: 432 457.
JÄGERSTEN, G. (1940): Die Abhängigkeit der Metamorphose vom Substrat des Biotops bei *Protodrilus*. Ark, f. Zool., **32**, A, No. 17: 1–12.
JEPPS, M. W. (1937): On the Protozoan Parasites of *Calanus finmarchicus* in the Clyde Sea Area. Quart. J. Micr. Sc. N.S. **79**: 589–658.
JERLOV, N. G. (1951): Optical studies of ocean waters. Rep. Swedish Deep-Sea Exped. 1947–1948, **3**: 1–59.
— Particle distribution in the ocean. Rep. Swedish Deep-Sea Exped. 1947–1948, **3**: 73–97.
JESPERSEN, P. (1940): Investigations on the quantity and distribution of zooplankton in Icelandic waters. Medd. fra Komm. for Danmarks fiskeri- og havundersogelser. Serie Plankton, **3**, Nr. 5, 77 pp.
JITTS, H. R. (1959): The adsorption of phosphate by estuarine bottom deposits. Austr. Journ. Marine and Freshwater Res., **10**: 7–21.
JOHN, D. D. (1936): The Southern species of the genus *Euphausia*. Discovery Rep. **14**: 193–324.
— (1938): Crinoidea. Discovery Reports, **18**: 123–222.
JOHNSON, M. W. (1948): Sound as a tool in marine ecology, from data on biological noises and the deep scattering layer. J. Mar. Res. **7**: 443–458.
JOHNSON, R. (1955): Biologically active compounds in the sea. J. Mar. Biol. Ass. U.K. **34**: 185–195.
JOHNSTONE, J. (1908): Conditions of life in the sea, a short account of quantitative marine biological research. Cambridge University Press, XIII+332 pp.
JONES, M. L. (1961): A quantitative evaluation of the benthic fauna off Point Richmond, California. Univers. California Publ. Zool., **67**: 219–320.
JONES, N. S. (1950): Marine bottom communities. Biolog. Rev., **25**: 283–313.
JONES, W. E. (1959): Experiments on some effects of certain environmental factors on *Gracilaria verrucosa* (Hudson) Papenfuss. J. Mar. Biol. Ass. U.K., **38**.
JØRGENSEN, C. B. (1955): Quantitative aspects of filter feeding in invertebrates. Biol. Rev., **30**: 391–454.
JØRGENSEN, E. (1924): Mediterranean Tintinnidae. Rep. Danish Ocean. Expeds. 1908–1910, No. 8, vol. II, Biology 3, 1–110.
JOUBIN, L. (1928): Eléments de biologie marine. Paris (Gauthier-Villars), 355 pp.
— (1929): Notes préliminaires sur les céphalopodes des croisières du Dana (1921–1927). Ann. Inst. Océanogr. **6**: 363–394.
JUBB, R. A. (1961): The freshwater eels (*Anguilla* spp.) of Southern Africa. An introduction to their identification and biology. Ann. Cape Prov. Mus. **1**: 15–48.
JULOU, DR. (1952): Variations de l'acuité visuelle en basse luminance sous l'influence de l'hormone méla-nophore (intermédine). Revue Chibret, **9**: 1–48 (cf. R. Motais 1961).

KAMPTNER, E. (1941): Die Coccolithineen der Südwestküste von Istrien. Ann. Natur Hist. Mus. Wien **51**: 54–149.
KÄNDLER, R. (1953): Hydrographische Untersuchungen zum Abwasserproblem in den Buchten und Förden der Ostseeküste Schleswig-Holsteins. Kieler Meeresforsch., **9**: 176–200.
KAIN, J. M., & G. E. FOGG (1958–1960): Studies on the growth of marine phytoplankton. I, II, III. J. Mar. Biol. Ass. U.K., **37**: 397–413, 781–788, **39**: 33–50.
KALLE, K. (1937): Meereskundliche chemische Untersuchungen mit Hilfe des Zeißschen Pulfrich-Photometers VI. Die Bestimmung des Nitrits und des „Gelbstoffs". Ann. Hydrogr. u. Marit. Meteorol. **65**: 276–282.
— (1943): Der Stoffhaushalt des Meeres. Leipzig. Akadem. Verlagsges. VII+263 pp.
— (1960): Die rätselhafte und „unheimliche" Naturerscheinung des „explodierenden" und des „rotie-renden" Meeresleuchtens—eine Folge lokaler Seebeben? Dtsch. Hydrogr. Zeitschr., **13**: 49–77.
— (1953): Der Einfluß des englischen Küstenwassers auf den Chemismus der Wasserkörper in der südlichen Nordsee. Ber. Dtsch. Wiss Komm. Meeresforschg. **13**: 130–135.
— (1956): Chemisch-hydrographische Untersuchungen in der inneren Deutschen Bucht. Dtsch. Hy-drogr. Zeitschr. **9**: 55–65.
KANWISHER, J. (1957): Freezing and drying in intertidal algae. Biol. Bull. **113**: 275–285.
KELLOGG, W. N. (1958): Echo ranging in the porpoise. Perception of objects by reflected sound is de-monstrated for the first time in marine animals. Science, **128**: 982–988.
— (1959): Auditory perception of submerged objects by porpoises. J. Accoustical Soc. Am. **31**: 1–6.
KEMP, S. (1917): Notes on the Fauna of the Matlah River in Gangetic Delta. Rec. Indian Mus. Calcutta, **13**: 233–241.
KERR, GR. (1931). Notes upon the Dana specimens of *Spirula* and upon certain problems of Cephalopod morphology. The Danish „Dana" Exped. 1920–1922, No. 8: 1–34.
KETCHUM, B. H. (1951): Plankton Algae and their biological significance. In: Manual of Phycology—an Introduction to the Algae and their Biology, G. M. Smith, Waltham, Mass. USA, 335–346.
KINNE, O. (1954a): Experimentelle Untersuchungen über den Einfluß des Salzgehaltes auf die Hitzeresi-stenz von Brackwassertieren. Zool. Anz., **152**: 10–16.
— (1954b): Interspezifische Sterilpaarung als konkurrenzökologischer Faktor bei Gammariden (Crus-tacea, Peracarida). Naturwissensch., **41**: 434.
— (1956): Uber den Wert kombinierter Untersuchungen (im Biotop und im Zuchtversuch) für die ökologische Analyse. Naturwiss,. **43**: 8–9.

KINNE, O. & E. M. (1962a): Rates of development in embryos of a Cyprinodont fish exposed to different temperature-salinity-oxygen-combinations. Canad. J. Zool. **40**: 231-253.
— (1962b): Irreversible nongenetic adaptation. Comp. biochem. Physiol. **5**: 265-282.
— (1963a): Über den Einfluß des Salzgehaltes auf verschiedene Lebensprozesse des Knochenfisches *Cyprinoden macularius*. Veröff. Inst. f. Meeresforsch. Bremerhaven, Sonderbd., 49-66.
— (1963b): The effects of temperature and salinity on marine and brackish water animals. Oceanogr. Mar. Biol. Ann. Rev. **1**: 301-340.
— (1959): Ecological data on the amphipod *Gammarus duebeni*. A Monograph. Veröff. Inst. f. Meeresforsch. Bremerhaven, **6**: 177-202.
KINZELBACH, R. (1965): Die Blaue Schwimmkrabbe (*Callinectes sapidus*), ein Neubürger im Mittelmeer. Natur und Museum, **95**: 293-296.
KINZER, J. (1962): Ein einfacher Schließmechanismus für die Planktonröhre „Hai". Kurze Mitt. Inst. f. Fischereibiol. Univers. Hamburg, Nr. **12**: 13-17.
KIRKEGAARD, J. B. (1954): The zoogeography of the abyssal Polychaetes. Union intern. Sci. Biol., Sér. B, No. **16**: 40-42.
KIRSTEUER, E. (1963): Zur Ökologie systematischer Einheiten bei Nemertinen. Zool. Anz. **170**: 343-354.
KLAUSEWITZ, W. (1961): Das Farbkleid der Korallenfische. Natur u. Volk, **91**: 204-215.
— (1962): Wie schwimmen Haifische? Natur u. Museum, **92**: 219—226.
— (1962): Röhrenaale im Roten Meer. Natur u. Museum, **92**: 96-98.
KNIGHT-JONES, E. W., & S. Z. QASIM (1955): Response of some marine plankton animals to changes in hydrostatic pressure. Nature, **175**: 941-943.
— & J. MOYSE (1961): Intraspecific competition in sedentary marine animals. Symposia Soc. Exper. Bio., Nr. **15**: 72-95.
KNOX, G. A. (1960): Littoral ecology and biogeography of the southern oceans. Proceed. Roy. Soc. London, Ser. B, **152**: 577-624.
KNUDSEN, J. (1950): Egg capsules and development of some marine prosobranchs from tropical West-Africa. Atlantide Rep., **1**: 85-130.
KNOX, G. A. (1961): The study of marine bottom communities. Proc. Roy. Soc. N.Z. **89**: 167-182.
KNUDSEN, V. O., R. S. ALFORD & J. W. EMLING (1948): Underwater ambient noise. J. Mar. Res., **7**: 410.
KOE, B. K., D. L. FOX & L. ZECHMEISTER (1950): The nature of some fluorescing substances contained in a deep-sea mud. Arch. of Biochem. **27**: 449-452.
KOFOID, C. A., & A. S. CAMPBELL (1929): A conspectus of the marine and fresh-water Ciliata belonging to the suborder Tintinnoinea, with descriptions of new species principally from the Agassiz Expedition to the eastern tropical Pacific 1904-1905. Californ. Univers. Public. Zool. **34**, 404 pp.
KOHLMEYER, I. (1963): Parasitisch und epiphytische Pilze auf Meeresalgen. Nova Hedwigia, **6**: 127-146.
KORRINGA, P. (1957): Lunar periodicity. Geol. Soc. America, Memoir **67**, **1**: 917-934.
KOSSWIG, C. (1950): Erythräische Fische im Mittelmeer und an der Grenze der Ägäis, Syllegomena biologica, Festschr. Kleinschmidt 203-212.
— (1955): Beitrag zur Faunengeschichte des Mittelmeeres. Publ. Staz. Zool. Napoli, **28**: 78-88.
— (1959): Genetische Analyse stammesgeschichtlicher Einheiten. Verh. Deutsche Zool. Ges. Münster, 1959: 42-73.
KOTTHAUS, A. (1952): *Hoplostethus islandicus*, nov. spec. (Acanthopterygia, Abt. Beryciformes, Fam. Trachichthyidae) aus den isländischen Gewässern. Helgoländer Wiss. Meeresunters. **4**: 62-87.
KRAMP, P. L. (1942): Ctenophora. The Godthaab Exped. 1928. Meddelel. om Gronland, **80**, No. 9, 19 pp.
— (1959): The hydromedusae of the Atlantic Ocean and adjacent waters. Dana Rep., **46**: 1-283.
KRAUSS, W. (1961): Wellen in der Tiefe des Meeres I. Umschau in Wissensch. u. Technik, **61**: 276-278, 328-331.
KREPS, E. (1934): Organic catalysts or enzymes in sea water. Lancashire Sea-Fisheries Laboratory, James Johnstone Memorial Volume: 193-202, Univ. Press Liverpool.
— & N. VERJBINSKAYA (1930): Seasonal changes in the phosphate and nitrate content and in hydrogen ion concentration in the Barents Sea. Journ. Cons. Intern. Explor. Mer, **5**: 329-346.
KREY, J. (1939): Die Bestimmung des Chlorophylls in Meerwasser-Schöpfproben. J. Cons. Int. Expl. Mer., **14**: 201.
KREY, J. (1952a): Die Untersuchung des Eiweißgehaltes in kleinen Planktonproben. Kieler Meeresforschungen **8**: 164-172.
— (1952b): Die Charakterisierung von Wasserkörpern durch optische Messungen. Arch. f. Hydrobiol. **46**: 1-14.
— (1953)a): Über die Fruchtbarkeit des Meeres. Veröff. Inst. f. Meeresforsch. Bremerhaven, **2**: 1-13.
— (1953b): Plankton- und Sestonuntersuchungen in der südwestlichen Nordsee auf der Fahrt der „Gauss" Februar/März 1952. Ber. Dt. Wiss. Komm, Meeresforsch. **13**: 136-153.
— (1954): Beziehungen zwischen Phytoplankton, Temperarursprungschicht und Trübungsschirm in der Nordsee im August 1952. Kieler Meeresforsch., **10**: 3-18.
— (1956): Die Trophie küstennaher Meeresgebiete. Kieler Meeresforsch. **12**: 46-64.
— (1961): Der Detritus im Meere. J. Cons. Intern. Explor. Mer., **26**: 263-280.
KRISS, A. E. (1961): Meeres-Mikrobiologie. Fischer Verlag, Jena, IX+570 pp.
— & M. I. NOWOSHILOFF (1954): Sind Hefeorganismen Meer-und Ozeanbewohner? Akad. d. Wissensch. d. UdSSR, Mikrobiol., **23**: Ausgabe 5.

KRISS, A. E. M. N. LEBEDEVA & I. N. MITZKEVICH (1960): Micro-organisms as indicators of hydrological phenomena in seas and oceans II. Deep Sea Research, 6: 173–183.

KROGH, A. (1933a): Conditions of Life in the Ocean. Ecological Monographs. 4: 421–429.

KÜHL, H. (1952): Über die Hydrographie von Wattenpfützen. Helgol. Wiss. Meeresunters., 4: 101–106.

— & H. MANN (1958): Das Verhalten anorganischer Stickstoffverbindungen im Mündungsgebiet eines Flusses. Arch. f. Fischereiwiss., 9: 9–16.

— & H. MANN (1961): Vergleichende hydrochemische Untersuchungen an den Mündungen deutscher Flüsse. Verh. Internat. Verein. Limnologie, 14: 451–458.

— & H. Mann (1962): Über das Zooplankton der Unterelbe. Veröff. Inst. f. Meeresforsch. Bremerhaven, 8: 53–69.

KUHL, W. (1938): Chaetognatha. Bronns Klassen u. Ornungen, Bd, IV, Abt. 4, 2. Buch. Teil 1.

KYLIN, H. (1917): Über die Kälteresistenz der Meeresalgen. Ber. Deutsch. Bot. Ges., 35.

LABAN, A., J. M. PÉRÈS & J. PICARD (1963): La photographie sousmarine profonde et son exploitation scientifique. Bull. Inst. Océan. Monaco, 60 (1258), 32 pp.

LABOREL, J. (1960): Contribution à l'étude directe des peuplements benthiques sciaphiles sur substrat rocheux en Méditerranée. Rec. Trav. Stat. Mar. Endourne, 33: 117–173.

LAGARDE, E. (1963): Métabolisme de l'azote minéral en milieu marin. Vie et Milieu, 14: 37–51.

LAUGHTON, A. S. (1957): Photography of the seafloor. Endeavour, 18: 178–185.

LAURSEN, D. (1953): The genus Janthina. Carlsberg Found. Oceanogr. Exped. 1928–1930. Dana-Report No. 38, 40 pp.

LEAVITT, B. B. (1938): The quantitative vertical distribution of macrozooplankton in the Atlantic Ocean basin. Biol. Bull. 74: 376–394.

LEDANOIS, Y. (1959): Adaptations morphologiques et biologiques des poissons des massifs coralliens. Bull. de l'I. Fr. A. N., 21: 1304–1325.

LEFÈVRE, M. H., & M. NISBET (1948): Action des substances excrétées en culture par certaines espèces d'algues sur le métabolisme d'autres espèces d'algue. Proc. Intern. Assoc. Limnol. 10: 259–264.

LEMOINE, P. (1940): Les algues calcaires de la zone néritique. Mem. Soc. Biogéogr. 7: 75–138.

LENGERKEN, H.v. (1929): Halophile und halobionte Coleoptera. Tierw. Nord-u. Ostsee, 11e, 32 pp. (Leipzig).

LEVRING, T. (1960): Submarines Licht und die Algenvegetation. Botanica Marina, 1: 67–73.

LEWIN, R. A. (1954): A marine Stichococcus sp. which requires vitamin B_{12} (Cobalamin). J. Gen. Microbiol. 10: 93 96.

LEWIS, A. G. (1959): The vertical distribution of some inshore copepods in relation to experimentally produced conditions of light and temperature. Bull. Mar. Sci. Gulf a. Carribbean, 9: 69–78.

LIGHTWARDT, R. W. (1961): A stomach fungus in Callianassa spp. (Decapoda) from Chile. Lunds Univers. Arsskr. N. F. Ard. 2, 57, Nr. 6.

LILLICK, L. C. (1940): Phytoplankton and planktonic protozoa of the offshore waters of the Gulf of Maine. 11. Qualitative composition of the planktonic flora. Transact. Am. Phil. Soc. N.S. 31.

LIMBAUGH, C. (1961): Cleaning symbiosis. Scient. Amer., 205: 32–49.

LIMBAUGH, C., H. PEDERSON & F. A. CHACE JR. (1961): Shrimps that clean fishes. Bull. Mar. Sc. Gulf and Caribbean, 11: 237–257.

LINKE, O. (1939): Die Biota des Jadebusens. Helgol. Wiss. Meeresunters. 1: 201–348.

LOCHER, F. W. (1953): Ein Beitrag zum Problem der Tiefseesande im westlichen Teil des äquatorialen Atlantiks. Rep. Swed. Deep Sea Exped. 1947–1948, 7, (5): 209–225.

LOEB, J. (1894): On the influence of light on the periodical depth-migrations of pelagic animals. Bull. U.S. Fish. Comm., 23: 65–68.

LOHMANN, H. (1909): Über die Quellen der Nahrung der Meerestiere und PÜTTERS Untersuchungen hierüber. Intern. Rev. Hydrobiol. Hydrogr. 2: 10–30.

LOHMANN, H. (1912): Untersuchungen über das Pflanzen- und Tierleben der Hochsee. Zugleich ein Bericht über die biologischen Arbeiten auf der Fahrt der "Deutschland" von Bremerhaven nach Buenos Aires in der Zeit vom 7. Mai bis 7. September 1911. Veröff. Inst. Meereskde. Berlin, 1: 1–92.

LOHMANN, H. (1914): Die Appendicularien der Valdivia-Expedition. Verhandl. Deutsche Zool. Ges. Freiburg, 157–192.

— (1920): Die Bevölkerung des Ozeans mit Plankton. Arch. f. Biontologie, 4.

— (1922): Oesia disjuncta WALCOTT, eine Appendicularie aus dem Kambrium. Mitt. Zool. Staatsinst. u. Zool. Mus. Hamburg 38: 69–74.

LOOSANOFF, V. L., & J. B. ENGLE (1947): Effect of different concentrations of microorganisms on the feeding of oysters (O. virginica). Fishery Bull. Fish- and Wildlife Service, 51: 31–57.

LORENZ, J. R. (1863): Physikalische Verhältnisse und Verteilung der Organismen im Quarnerischen Golfe. Wien 1863, XII + 379 pp.

LOWNDES, A. G. (1935): The swimming and feeding of certain Calanoid Copepods. Proceed. Zool. Soc. London. 687–715.

LUCAS, C. E. (1936): On certain interrelations between phytoplankton and zooplankton under experimental conditions. J. Cons. Intern. Explor. Mer. 11: 348–362.

— (1938): Some aspects of integration in plankton communities. Cons. Intern. Explor. Mer. 13: 309–322.

— (1947): The ecological effects of external metabolites. Biol. Rev. 22: 270–295.

Lucas, C. E. (1949): External metabolites and ecological adaptations. Symposia Soc. exper. Biol. No. III. Growth, 336–356.
— (1955): External metabolites in the sea. Deep Sea Research Suppl. to Vol. **3**: 139–148.
— (1958): External metabolites and productivity. Rapp. et Procès Verb. des Réunions, **144**: 154–158.
Lühmann, M. (1954): Über intermediäre Formen zwischen *Anarrhichas minor* Olafs. und *A. lupus* L. (Teleostei). Ber. Dtsch. Wiss. Komm. Meeresforschg., **13**: 310–326.
Lundbeck, J. (1958): Der Zusammenbruch der kalifornischen Sardinenfischerei im Lichte moderner populationsdynamischer Untersuchungsmethoden. Archiv für Fischereiwissenschaft **9**: 40–45.
Lüneburg, H. (1953): Beiträge zur Hydrographie der Wesermündung. II. Die Probleme der Sinkstoffverteilung in der Wesermündung. Veröff. Inst. f. Meeresforsch. Bremerhaven, **2**: 15–51.

Maas, O. (1911): Abgüsse rezenter Tiefseemedusen zum Vergleich mit Fossilien aus der Kreide. Verhdl. Deutsche Zool. Ges. **21**: 186–192.
MacGinitie, G. E., u. N. (1949): Natural history of marine animals. New York, MacGraw-Hill, 473 pp.
Mackie, G. O. (1962): Factors affecting the distribution of *Velella* (Chondrophora). Int. Rev. ges. Hydrobiol., **47**: 26–32.
McFarland, W. N. (1959): Standing crop, chlorophyll content and *in situ* metabolism of a giant kelp community in Southern California. Publ. Inst. Mar. Sci., **6**: 109–132.
McIntyre, A. D. (1956): The use of trawl, grab and camera in estimating marine benthos. J. Mar. Biol. Ass. U.K. **35**: 419–429.
MacIntyre, R. J. (1963): The supra-littoral fringe of New Zealand sand beaches. Transact. Roy. Soc. New Zealand, General, **1**: 89–103.
Mackintosh, N. A. (1937): The seasonal circulation of the antarctic macroplankton. Discovery Report, **16**: 365–412.
McNulty, J. K., R. C. Work & H. B. Moore (1962a): Level sea bottom communities in Biscayne Bay and neighboring areas. Bull. Mar. Sc. Gulf and Caribbean, **12**: 204–233.
— (1962b): Some relationships between the infauna of the level bottom and the sediment in South Florida. Bull. Mar. Sc. Gulf and Caribbean, **12**: 322–332.
Madsen, F. J. (1954): Some general remarks on the distribution of the echinoderm fauna of the deep-sea. Union intern. Sci. Biol., Sér. B. No. **16**: 30–37.
— (1961a): The Porcellanasteridae. A monographic revision of an abyssal group of sea-stars. Galathea Report, **4**, 33–176.
— (1961b): On the zoogeography and origin of the abyssal fauna, in view of the knowledge of the Porcellanasteridae. Galathea Report, **4**: 177–218.
Man, J. G. de (1920): The Decapoda of the Siboga-Exped., Part IV, Res. Explor. du "Siboga"., Livr. 87, Mon. **39**a3, 318 pp.
Marr, J. W. S. (1938): On the operation of large plankton nets. Discovery Reports, **18**: 105–120.
Marshall, N. B. (1951): Bathypelagic fishes as sound scatterers in the ocean. J. Mar. Res. **10**: 1–17.
— (1953):The balanoid biome-type of intertidal rocky shores. Ecology, **34**: 434–436.
— (1954): Aspects of deep sea biology. Hutchinson, London, 380 pp.
Marshall, S. M., A. G. Nicholls & A. P. Orr (1934): On the biology of *Calanus finmarchicus*. V. Seasonal distribution, size, weight and chemical composition in Loch Striven in 1933 and their relation to the phytoplankton. J. Mar. Biol. Ass. U.K. **19**: 793–828.
Marshall, S. M., Nicholls, A. G., & A. P. Orr, (1935): On the biology of *Calanus finmarchicus*. VI. Oxygen consumption in relation to environmental conditions. J. Mar. Biol. Ass. U.K., **20**: 1–28.
— & A. P. Orr (1952): On the biology of *Calanus finmarchicus*. VII. Factors affecting egg production. J. Mar. Biol. Ass. U.K. **30**: 527–547.
— & — (1955a): On the biology of *Calanus finmarchicus*. VIII. Food uptake, assimilation and excretion in adult and stage V Calanus. J. Mar. Biol. Ass. U.K. **34**: 495–529.
— & — (1955b): The biology of a marine copepod. London u. Edinburgh, VI+188 pp.
Marshall, S. M., & A. P. Orr, (1962): Carbohydrate as a measure of phytoplankton. J. Mar. biol. Ass. U.K., **42**: 511–19.
Masacchia, X. J., & M. R. Clark (1957): Effects of elevated temperatures on tissue chemistry of the arctic sculpin *Myoxocephalus quadricornis*. Physiol. Zool., **30**: 12–17.
Mather, F.-J. (1960): Recaptures of tuna, marlin and sailfish tagged in the Western North Atlantic. Copeia. 1960: 149–151.
Mayer, A. G. (1910): Medusae of the world. Carnegie Inst. of Wash. Publ. No. **109**: 1–3.
Meisenheimer, J. (1905): Pteropoda. Wiss. Ergebn. Dt. Tiefsee-Exped., „Valdivia“. **9**: 1–314.
Menard, H. W. (1960): Consolidated slabs on the floor of the Eastern Pacific. Deep-Sea Research, **7**: 35–41.
Menzies, R. J., & J. Imbrie (1958): On the antiquity of the deep sea bottom fauna. Oikos **9**: 192–201.
—, — & B. C. Heezen (1961): Further considerations regarding the antiquity of the abyssal fauna with evidence for a changing abyssal environment. Deep-Sea-Research, **8**: 79–94.
Mertens, R. (1927): Meeresechsen. Ber. Senckenb. Ges. Frankfurt, **57**: 78–91.
Mertens, R. (1962): Ist die Färbung der „blauen Flotte“ eine Tarntracht? Natur u. Museum, **92**: 413–414.
Meyer, P. F., & K. Kalle (1950): Die biologische Umstimmung in der Ostsee in den letzten Jahrzehnten, eine Folge hydrographischer Wasserumschichtungen? Arch. f. Fischereiwiss., **2**: 2.

Meyers, S. P., & E. S. Reynolds (1960): Occurrence of lignicolous fungi in Northern Atlantic and Pacific marine localities. Canad. J. Bot. **38**: 217–226.

Michael, E. L. (1911): Classification and vertical distribution of the Chaetognatha of the San Diego region. Univ. Calif. Publ. Zool., **8**: 21–186.

Michael, E. L. (1916): Dependence of marine biology upon hydrography and necessity of quantitative biological research. Univ. Calif. Publ. Zool. **15**: 1–23.

Millar, R. H. (1959): Ascidiacea. Galathea Report, **1**: 189–210.

Miyazaki, J. (1938): On a substance which is contained in green algae and induces spawning action of the male oyster (Preliminary note). Bull. Jap. Soc. Sci. Fish. **7**: 137–138.

Möbius, K. (1871): Die wirbellosen Tiere der Ostsee. Jahresber. Comm. wiss. Unters. Deutsch. Meere in Kiel.

— (1877): Die Auster und die Austernwirtschaft, Berlin.

Mohr, E. (1954): Fliegende Fische. Die Neue Brehm-Bücherei, A. Ziemsen Verlag, Wittenberg, 55 pp.

— (1957): Sirenen oder Seekühe. Die Neue Brehm-Bücherei, A. Ziemsen Verlag, Wittenberg, 61 pp.

Molander, A. R. (1928): Investigations into the vertical distribution of the fauna of the bottom deposits in the Gullmar Fjord. Svenska Hydrogr.-biol. Komm. Skrifter, Ny Ser.: Hydrogr., **6**.

— (1962): Studies of the fauna in the fjords of Bohuslän with reference to the distribution of different associations. Arkiv f. Zool., Ser. 2, **15**: 1–64.

Molinier, R., & J. Picard (1952): Recherche sur les herbiers de Phanérogames marines du littoral méditerranéen francais. Ann. Inst. Océanogr., **27**, H. 3.

Moore, H. B. (1950): The relation between the scattering layer and the Euphausiacea. Biol. Bull. **99**: 181–212.

— (1952): Physical factors affecting the distribution of Euphausids in the North Atlantic. Bull. Mar. Sc. Gulf a. Caribbean, **1**: 278–305.

— (1958): Marine Ecology. Wiley & Sons, New York, XI+493 pp.

Moore, D. R., & H. R. Bullis (1960): A deep-water coral reef in the Gulf of Mexico. Bull. Mar. Sc. Gulf a. Caribbean, **10**: 125–128.

Mortensen, T. (1933): Ophiuroidea. Danish Ingolf Exped., **4** (8) 121.

Morton, J. E. (1959): The habits and feeding organs of *Dentalium entalis*. J. Mar. Biol. Ass. U.K. **38**: 225–238.

Motais, R. (1961): Sur la pigmentation mélanique des poissons profonds et la teneur de leur hypophyse en principe mélanophorique. Bull. Inst. Océanogr. No. 1198, 28 pp.

Müller, C. D. (1960): Fauna und Sediment der Leybucht; biologisch-bodenkundliche Wattuntersuchungen mit Stellungnahme zur Landgewinnung. Jahresber. 1959 d. Forschungsstelle Norderney, **11**: 39–178.

Müller, D. (1962): Überjahres-und lunarperiodische Erscheinungen bei einigen Braunalgen. Botan. Marina, **4**: 140.

Müller Melchers, F. C. (1952): *Biddulphia chinensis* Grev. as indicator of ocean currents. Comm. botan. Museo Hist. Nat. Montevideo, **2**: p. 1–14.

Murray, J. (1895): A summary of the scientific results obtained at the sounding, dredging and trawling stations of HMS Challenger. Challenger Rep., Summary of Res., **2**: 797–1608.

Murray, J., & Hjort, J. (1912): The depths of the Ocean. London (Macmillan Co), XX: 821.

Nathanson, A. (1906): Über die Bedeutung vertikaler Wasserbewegungen für die Produktion des Planktons im Meere. Abh. Math. Phys. Klasse Kgl. Sächs. Ges. Wiss., **29**: 358–441.

Neill, W. T. (1958): The occurrence of amphibians and reptiles in saltwater areas, and a bibliography. Bull. Mar. Sci. Gulf and Caribbean, **8**: 1–97.

Neumann, D. (1960): Osmotische Resistenz und Osmoregulation der Flußdeckelschnecke *Theodoxus fluviatilis* L. Biol. Zentralbl. **79**: 585–605.

— (1963): Über die Steuerung der lunaren Schwärmperiodik der Mücke *Clunio marinus*. Verh. Dt. Zool. Ges. in Wien 1962: 275–285.

Neumann, G. (1944): Das schwarze Meer. Ein ozeanographischer Überblick. Zeitschr. Ges. Erdkde Bln. 1944, 92–114.

Nicholls, A. G. (1935): Copepods from the interstitial fauna of a sandy beach. J. Mar. Biol. Ass. U.K., **20**: 379–406.

Nightingale, H. W. (1937): Red water organisms. Argus Press, Seattle, Washington.

Nigrelli, R. F. (1958): Dutchman's baccy juice or growth-promoting and growth-inhibiting substances of marine origin. Trans. N.Y. Acad. Sci., Ser. 11, **20**: 248–262.

Nikitine, B. (1929): Les migrations verticales saisonnières des organismes planktoniques dans la mer Noire. Bull. Inst. Océan. Monaco, No. 540, 24 pp.

Noddack, J. & W. (1940): Die Häufigkeiten der Schwermetalle in Meerestieren. Arkiv f. Zool., **32 A**, No. 4: 1–35.

Norman, J. R., & F. C. Fraser (1948): Giant fishes, whales and dolphins. London (Putnam), XXII+375 pp.

Nybelin, O. (1954): Sur la distribution géographique et bathymétrique des Brotulidés trouvés audessous de 2000 mètres de profondeur. Union Intern. Sc. Biol., Sér. B, No. 16, 65–71.

Odhner, N. H. (1936): Nudibranchia Dendronotacea. Mém. Mus. Roy. Hist. Nat. Belgique, II. Ser. **3**: 1055–1128.

Odum, H. T., & E. P. (1955): Trophic structure and productivity of a windward coral reef community on Eniwetok Atoll. Ecological Monographs, 25: 291–320.

Ohshima, H. (1930): Preliminary note on *Entovalva semperi* sp. nov., a commensal bivalve living attached to the body of a synaptid. Annot. zool. Jap. 13: 25–27.

Olesen, P. E., A. Maretzki & L. A. Almodovar (1964): An investigation of antimicrobial substances from marine algae. Botan. Marina, 6: 224–232.

Osterhout, W. J. V. (1930): The accumulation of electrolytes. II. Suggestions as to the nature of accumulation in *Valonia*. J. Gen. Physiol. 14: 285.

Ostwald, W. (1902): Zur Theorie des Planktons. Biolog. Zentralbl. 22, 596–605, 609–638.

Ostwald, W. (1903a): Theoretische Planktonstudien. Zool. Jahrb. System., Georgr., Biol. 18: 1–62.

— (1903b): Zur Theorie der Schwebevorgänge sowie der spezifischen Gewichtsbestimmungen schwebender Organismen. Pflügers Arch. 94: 251–272.

Owre, H. B. (1960): Plankton of the Florida current. Part VI. The Chaetognatha. Bull. Mar. Sc. Gulf a. Caribbean, 10: 255–322.

Paasche, E. (1960): On the relationship between primary production and standing stock of phytoplankton. Journ. Conseil Perm. Intern. Expl. Mer., 26: 33–48.

Panning, A., & N. Peters (1933): Die chinesische Wollhandkrabbe (*Eriocheir sinensis* H. Milne Edwards) in Deutschland. Zool. Anz., 101: 265–271.

Pardi, L. (1957): Modificazione sperimentale della direzione di fuga negli antipodi e orientamento solare. Z. f. Tierpsych., 14: 261–275.

— & F. (1952): Die Sonne also Kompaß bei *Talitrus saltator* Montagu (Amphipoda, Talitridae). Naturwissenschaften, 39: 262–263.

Parker, J. (1960): Seasonal changes in cold-hardiness of *Fucus vesiculosus*. Biol. Bull. 119: 474–478.

Parr, A. E. (1934): Problems of the Sargasso Sea. Bull. New York Zool. Soc. 37: 39–46.

— (1937): Concluding report on fishes. Bull. Bingham oceanogr. Coll. 3: 1–79.

— (1939): Quantitative observations on the pelagic Sargassum vegetation of the Western North Atlantic. Bull. Bingham Oceanogr. Coll., 6, Art. 7, 94 pp.

Parsons, T. R., & J. D. H. Strickland (1960): Proximate analysis of marine standing crops. Nature, 184, 2038.

Pax, F. (1962): Meeresprodukte. Ein Handwörterbuch der marinen Rohstoffe. Gebr. Borntraeger, Berlin, XII + 460 pp.

Pearson, E. A. (1960): Proceedings of the first international conference on Waste Disposal in The Marine Environment, Berkeley 1959. Pergamon Press, London, New York, Paris, 569 pp.

Peiss, C. N., & J. Field (1950): The respiratory metabolism of excised tissues of warm- and cold-adapted fishes. Biol. Bull. 99: 213–224.

Pelseneer, P. (1888): Report on the Pteropoda collected by H.M.S. *Challenger* during the years 1873/76. Rep. Scient. Res. Challenger, Zool., 23.

Pennak, R. W. (1942): Ecology of some copepods inhabiting intertidal beaches near Woods Hole, Massachusetts. Ecology 23: 446–456.

Pereira, J. R. (1925): On the combined toxic action of light and eosin. Jour Exper. Zool., 42: 257.

Pérès, J. M. (1957): Essai de classement des communautés benthiques marines du globe. Rec. Travaux Stat. Mar. d'Endoume, 22: 23–54.

— (1958): Trois plongées dans le canyon du Cap Sicié, effectués avec le bathyscaphe F.N.R.S. III de la Marine Nationale. Bull. Inst. Océanogr. No. 1115, 20 pp.

Pérès, J. M. (1960): La „soucoupe plongeante", engin de prospection biologique sous marine. Deep-Sea Research, 7: 208–214.

— (1961): Océanographie biologique et biologie marine I. Paris, Presses universitaires de France, VI + 541 pp.

Pérès, J. M., & L. Deveze (1963): Océanographie biologique et biologie marine. II. La vie pélagique. Presses Universitaries de France, Paris, 514 pp.

— & R. Molinier (1957): Compte-rendu des scéances du colloque tenu par le comité du benthos de la commission internationale pour l'exploration scientifique de la mer méditerranée. Rec. Travaux Stat. Mar. d'Endoume, 22: 5–13.

— & J. Picard (1958): Manuel de bionomie benthique de la mer méditerranée. Rec. Travaux Stat. Mar. d'Endoume, 23: 7–122.

Perkins, E. J. (1958): The food relationships of the microbenthos, with particular reference to that found at Whitstable. Ann. Mag. Nat. Hist. Ser. 13, 1: 6–77.

Petersen, C. G. J. (1911, 1913): Valuation of the sea. I. Animal life of the sea bottom, its food and quantity. II. The animal communities of the sea bottom and their importance for marine zoogeography. Rep. Danish Biol. Station, 20, 81 pp., 21, 44 pp. Kopenhagen.

Pettersson, H., H. Höglund & S. Landberg (1934): Submarine daylight and the photosynthesis of phytoplankton. Göteborgs Kungl. vetenskaps-och vittevhets- samhälles handlingar. Femte följden. 4,5: 1–17.

Pflugfelder, O. (1933): Landpolychäten aus Niederländisch-Indien. Zool. Anz. 105: 65–76.

Pickford, G. E. (1946, 1949): *Vampyroteuthis infernalis* Chun, an archaic dibranchiate Cephalopod. I u. II. Dana-Report No. 29 u 32, Copenhagen.

Plaine, H. L. (1952): A variation in the distribution of a spionid polychaete in the Woods Hole Region. Ecology, 33: 121–123.

PLUNKETT, M. A. (1957): The qualitative determination of some organic compounds in marine sediments. Deep-Sea Research 5: 259–262.
— & N. W. RAKESTRAW (1955): Dissolved organic matter in the sea. Deep-Sea Research, Suppl. to Vol. 3: 12–14.
POMEROY, L. R. (1960): Primary productivity of Boca Ciega Bay, Florida. Bull. Mar. Sci. Gulf and Caribbean, 10: 1–10.
POSTMA, H., & K. KALLE (1955): Die Entstehung von Trübungszonen im Unterlauf der Flüsse, speziell im Hinblick auf die Verhältnisse in der Unterelbe. Dt. Hydrogr. Zeitschr. 8: 137–144.
PRASAD, R. R. (1958): A note on the occurrence and feeding habits of Noctiluca and their effects on the plankton community and fisheries. Proceed. Indian Acad. Sci. 47: 331–337.
PRAT, H. (1940): Observations bionomiques sur les rivages atlantiques de l'Amerique du Nord et les iles voisines. Mém. Soc. Biogéogr., 7: 253–277.
PRATT, R., H. MAUTNER, G. M. GARDNER, HSIEN SHA & J. DUFRENOY (1951): Antibiotic activity of seaweed extracts. J. Amer. Pharm. Assoc. (Sci. Ed.) 40: 575.
PRECHT, H., J. CHRISTOPHERSEN & H. HENSEL (1955): Temperatur und Leben. Springer-Verlag, Berlin, 514 pp.
PRESCOTT, G. W. (1948): Objectionable Algae with reference to the killing of fish and other animals. Hydrobiol. 1: 1–13.
PROVASOLI, L., & I. J. PINTNER (1953): Ecological implications of in vitro nutritional requirements of algal flagellates. Ann. New York Acad. Sci. 56: 839–851.
PRUDHOE, ST. (1950): On the taxonomy of two species of pelagic Polyclad Turbellarians. Ann. Mag. Nat. Hist. Ser. 12, 3: 710–716.
PRUVOT-FOL, M. A. (1925): Contribution à l'étude du genre Janthina. CR. Séanc. Acad. Scs. 181: 56–57.
PURASJOKI, K. J. (1953): Zwei verbesserte Apparate zum Auffangen von Kleintieren des See und Meeresbodens. Arch. Soc. Zool. Botan. Fenn. „Vanamo" 8: 110–114.

RADOVICH, J. (1961): Relationships of some marine organisms of the Northeast Pacific to water temperatures particularly during 1957 through 1959. Fish. Bull. No. 112, State Calif. Dptm. Fish a. Game Mar. Resources Oper., 62 pp.
RANADE, M. R. (1957): Observations on the resistance of Tigriopus fulvus (FISCHER) to changes in temperature and salinity. J. Mar. Biol. Ass. U.K., 36: 115–119.
RANDALL, J. E. (1965): Grazing effect on sea grasses by herbivorous reef fishes in the West Indies. Ecology, 46: 255–260.
RAY, D. L. (1959): Marine boring and fouling organisms. Seattle, University of Washington Press, XII+536 pp.
RAY, S. M. & W. B. WILSON (1957): Effects of unialgal and bacteria-free cultures of Gymnodinium brevis on fish. Fish and Wildlife Serv. 57, Fish. Bull. 123, 469–496.
RAYMONT, J. E. G. (1963): Plankton and productivity in the oceans. Pergamon Press, Oxford. VII+660 pp.
REDFIELD, A. C. (1958): The inadequacy of experiment in marine biology. Perspectives in marine biology, Symposium at Scripps Institution 1956: 17–26.
REES, C. B. (1940): A preliminary study of the ecology of a mud flat. J. mar. biol. Ass. U.K. 24: 185–199.
REES, W. J. (1954): The Macrotritopus Problem. Bull. Brit. Mus. (Nat. Hist.) Zool. 2: 69–99.
— (1957): Evolutionary trends in the classification of capitate hydroids and medusae. Brit. Mus. (Nat. Hist.), Zool., 4: 455–534.
REGAN, T. C., & E. TREWAVAS (1932): Deep-sea Angler-fishes (Ceratioidea). Rep. Carlsberg Ocean. Exped. 1928–30, 2: 113 pp.
REICHENOW (1949): Lehrbuch der Protozoenkunde, 1. Teil. 6. Aufl. Leipzig, II+408 pp.
REISH, D. J. (1959a): A discussion of the importance of the screen size in washing quantitative marine bottom samples. Ecology, 40: 307–309.
— (1959b): An ecological study of pollution in Los Angeles-Long Beach Harbors, California. Allan Hancock Foundation Publ., Occas. Pap. No. 22: 1–19.
— & J. L. BARNARD (1960): Field toxicity tests in marine waters utilizing the polychaetous annelid Capitella capitata (FABRICIUS). Pacific Naturalist, 1, No. 21, 8 pp.
REISINGER, E. (1960): Die Lösung des Parergodrilus-Problems. Z. Morph. Ökol. Tiere 48: 517–544.
REMANE, A. (1933): Verteilung und Organisation der benthonischen Mikrofauna der Kieler Bucht. Wiss. Meeresunters., Kiel 21: 161–221.
— (1934): Die Brackwasserfauna. Verh. Dt. Zool. Ges., 36: 34–74.
— (1940): Einführung in die zoologische Ökologie der Nord- under Ostsee. Tierw. Nord- u. Ostsee, Ia, 238 pp.
— (1950): Das Vordringen limnischer Tierarten in das Meeresgebiet der Nord- und Ostsee. Kieler Meeresforsch., 7: 5–23.
— (1955): Die Brackwasser-Submergenz und die Umkomposition der Coenosen in Belt- und Ostsee. Kieler Meeresforsch., 11: 59–73.
— (1963): Verschiedenheiten der biologischen Entwicklung im Meer und Süßwasser. Veröff. Inst. Meeresforsch. Bremerhaven, Sonderband: 122–141.
— & E. SCHULZ (1935): Das Küstengrundwasser als Lebensraum. Schriften Naturw. Ver. f. Schleswig-Holstein, 20: 399–408.

REMANE, A. & C. SCHLIEPER (1958): Die Biologie des Brackwassers. Die Binnengewässer, **22**: Schweizerbart, Struttgart, 348 pp.
RENSCH, B. (1954): Neuere Probleme der Abstammungslehre. Die transspezifische Evolution. Enke-Verlag, Stuttgart, VII+436 pp.
RICHARDS, F. A. (1957): Oxygen in the ocean. Geol. Soc. America, Mem. 67, **1**: 185–238.
RICHTER, G. (1961): Die Radula der Atlantiden (Heteropoda, Prosobranchia) und ihre Bedeutung für die Systematik und Evolution der Familie. Z. Morph. Ökol. d. Tiere, **50**: 163–238.
RICKETTS, E. F., & J. CALVIN (1962): Between Pacific Tides. 3. Edition, Revisions by J. W. HEDGPETH, Stanford, Calif., XIII+516 pp.
RIEDL, R. (1955): Über die Isolation der lebenden Mikrofauna aus marinen Schlammböden. Zool. Anz., **155**: 263–275.
— (1956): Zur Kenntnis der Turbellarien adriatischer Schlammböder sowie ihrer geographischen und faunistischen Beziehungen. Thalassia Jugoslavica, **1**: 69–180.
— (1956): Über Tierleben in Höhlen unter dem Meeresspiegel. Verh. D. Zool. Ges. in Erlangen 1955: 429–440.
— (1960): Ein Kinemeter zur Beobachtung von Dredgen in beliebigen Tiefen. Inter. Revue ges. Hydrobiol., **45**: 155–167.
— (1960): Neue nordatlantische Formen von adriatischen Schlammböden. Zool. Anz., **165**: 297–311.
— (1963): Probleme und Methoden der Erforschung des litoralen Benthos. Verh. Dt. Zool. Ges. in Wien 1962: 505–567.
RILEY, G. A. (1938, 1939): Plankton studies, I, II. J. Mar. Res., **1**: 335 **2**: 145–62.
ROBERG, M. (1930): Ein Beitrag zur Stoffwechselphysiologie der Grünalgen. Jb. Wiss. Bot., **72**: 369–384.
ROBERTS, J. L. (1957): Thermal acclimation of metabolism in the crab *Pachygrapsus crassipes* RANDALL. II. Mechanisms and the influence of season and latitude. Physiol. Zool., **30**: 242–255.
ROCH, F. (1924): Experimentelle Untersuchungen an *Cordylophora caspia* (Pallas) (= *lacustris* Allman). Zeitschr. f. Morphol. u. Ökol. d. Tiere, **2**: 350–670.
ROCHFORD, D. J. (1951): Studies in Australian estuarine hydrology. I. Introductory and comparative features. Austral. J. Mar. and Freshwater Res., **2**: 1–116.
RODEWALD, M. (1952–1959): Beiträge zur Klimaschwankung im Meere. 1. bis 11. Beitrag. Dt. Hydrogr. Zeitschr. 9–12.
RODRIGUEZ, G. (1959): The marine communities of Margarita Island, Venezuela. Bull. Mar. Sci. Gulf a. Caribbean, **9**: 237–280.
ROMER, A. SH. (1955): Fish origins—fresh or salt water? Deep-Sea Research, Suppl. to Vol. **3**: 261–280.
ROSE, M. (1925): Contribution à l'étude de la biologie du plankton; le problème des migrations verticales journalières. Arch. Zool. Exper. Gén. **64**: 387–542.
ROSE, M. (1926): Le plankton et ses relations avec la température, la salinité et la profondeur. Ann. Inst. Océanogr. Paris, N. S. **3**: 161–242.
ROSS, H. (1957): Untersuchungen über das Vorkommen antimikrobielle Substanzen in Meeresalgen. Kieler Meeresforschg. **13**: 41–58.
ROWAN, WM. (1938): Light and seasonal reproduction in animals. Biol. Rev. **13**: 374–402.
RUNNSTRÖM, S. (1936): Die Anpassung der Fortpflanzung und Entwicklung mariner Tiere an die Temperaturverhältnisse verschiedener Verbreitungsgebiete. Bergens Mus. Aarb. Nr. 3, 36 pp.
RUSSELL, E. S. (1962): The diversity of animals. An evolutionary study. Acta Biotheor. **13**, Suppl. 1 (Bibi. Biotheor. **12**), 151 pp.
RUSSELL, F. S. (1925/1926): The vertical distribution of marine macroplankton. An observation on diurnal changes. J. Mar. Biol. Ass. U.K. **13**: 769–809, **14**, 101–159, 387–440.
RUSSELL, F. S. (1927): The vertical distribution of plankton in the sea. Biol. Rev. **2**: 213–262.
— (1935): A review of some aspects of zooplankton research. Cons. Intern. Explor. Mer., Rapp. et Procès Verb. **95**: 5–30.
— & W. J. REES (1960): The viviparous Scyphomedusa *Stygiomedusa fabulosa* RUSSELL, J. Mar. Brit. Ass. U.K. **39**: 303–317.
— & C. M. YONGE (1963): The seas. Our knowledge of life in the sea and how it is gained. Warne & Co., London u. New York, XII+376 pp.
RYTHER, J. H. (1954): Inhibitory effects of phytoplankton upon the feeding of *Daphnia magna* with reference to growth, reproduction, and survival. Ecology, **35**: 522–533.
— & R. R. L. GUILLARD (1959): Enrichment experiments as a means of studying nutrients limiting to phytoplankton production. Deep-Sea Research, **6**: 65–69.
RZEPISHEVSKY, I. K. (1962): Conditions of the mass liberation of the nauplii of the common barnacle, *Balanus balanoides* (L.), in the Eastern Murman. Int. Rev. ges. Hydrobiol., **47**: 471–479.

SCAGEL, R. F. (1959): Culture studies of benthonic algae in the northeast pacific. Oceanogr., Amer. Assoc. Advancement of Sc.; 203–204.
SCHÄFER, W. (1951): Der „kritische Raum", Maßeinheit und Maß für die mögliche Bevölkerungsdichte innerhalb einer Art. Verh. Dt. Zool. Ges., 1951: 391–395.
— (1954): Form und Funktion der Brachyuren-Schere. Abh. Senckenb. Naturf. Ges., **489**: 1–65.
SCHÄFER, W. (1962): Aktuo-Paläontologie nach Studien in der Nordsee. Frankfurt/M., Verlag Kramer, VIII+666 pp.
SCHEFFER, V. B. (1958): Seals, sea-lions and walrusses, a review of the Pinnipedia. Stanford Univers. Press., Stanford Calif., X+179 pp.

SCHELLENBERG, A. (1929): Die abyssalen und pelagischen Gammariden. Bull. Mus. Compar. Zool. Harvard. 69: 193-201.

SCHILLER, J. (1916): Über neue Arten Membranverkieselung bei *Meringosphaera*. Arch. f. Protistenkde. 36: 198-208.

— (1931): Über autochthone pflanzliche Organismen in der Tiefsee. Biol. Centralbl. 51: 329-334.

SCHLIEPER, C. (1958): see REMANE, A., u C. SCHLIEPER 1958.

— (1963): Neuere Aspekte der biologischen Tiefseeforschung. Die Umschau in Naturw. u. Technik, 63: 457-461.

— (1963): Biologische Wirkungen hoher Wasserdrucke. Veröff. Inst. f. Meeresforsch. Bremerhaven, Sonderband. 1963: 31-48.

SCHMIDT, J. (1922): Live specimens of *Spirula*. Nature, London, 110: 788-790.

SCHMIDT, J. (1925): The breeding places of the eel. Smithson. Rep. for 1924: 279-316.

SCHMIDT, U. (1958): Die deutschen Köhleranlandungen 1946/47 bis 1956/57 aus norwegischen un isländischen Gewässern und ihre Abhängigkeit vom Fischbestand. Ber. Dt. Wiss. Komm. Meeresforschg. 15: 145-158.

SCHMIDT, W. J. (1962a): Mycelitesbefall an Stacheln von lebenden Seeigeln. Zool. Anz. 169: 245-252.

— (1962b): Über *Mycelites*-Befall an Zähnen fossiler Haie. Int. Rev. ges. Hydrobiol., 47: 587-601.

SCHMIDT-NIELSEN, K. (1959): Salt glands. Scientific American 200: 109-116.

SCHNAKENBECK, W. (1955): Der Kiemenreusenapparat vom Riesenhai (*Cetorhinus maximus*). Zool. Anz. 154: 99-108.

SCHULZ, E. (1937): Das Farbstreifenwatt und seine Fauna, eine ökologische biozönotische Untersuchung an der Nordsee. Kieler Meeresforsch., 2: 359.

SCHUSTER, R. (1962): Das marine Litoral als Lebensraum terrestrischer Kleinarthropoden. Int. Rev. ges. Hydrobiol. 47: 359-412.

SCOTT, G. T., & H. R. HAYWOOD (1955): Sodium and potassium regulation in *Ulva lactuca* and *Valonia macrphysa*. Electrolytes in Biolog. Systems: 35-64.

SEGERSTRÅLE, S. E. (1951): The recent increase in salinity off the coasts of Finland and its influence upon the fauna. J. Conseil Intern. Explor. Mer, 17: 103-110.

SEGERSTRÅLE, S. (1957): On the immigration of the glacial relicts of Northern Europe, with remarks on their prehistory. Comment. biol. Helsingf., 16: 117 pp.

— (1960): Fluctuations in the abundance of benthic animals in the Baltic area. Soc. Sci. Fennica, Comment. Biol. 23: 1-19.

SEILACHER, A. (1953): Der Brandungssand als Lebensraum in Gegenwart und Vorzeit. Natur u. Volk, 83: 263-272.

SEWELL, R. B. SEYMOUR (1948): The free-swimming planktonic Copepoda; Systematic account. Sc. Rep. John Murray Exp. (Brit. Mus. Nat. Hist.) 8: 1-303.

SHEPARD, F. P. (1959): The earth beneath the sea. London, Oxford, University Press, 275 pp.

SIEBECK, O. (1960): Untersuchungen über die Vertikalwanderungen planktischer Crustaceen unter Berücksichtigung der Strahlungsverhältnisse. Int. Revue ges. Hydrobiol., 45: 381-454.

SIEPMANN, R. (1959): Ein Beitrag zur saprophytischen Pilzflora des Wattes der Wesermündung. Veröff. Inst. f. Meeresforschg., Bremerhaven, 6: 213-301.

SIEPMANN, R., & W. HÖHNK (1962): Über Hefen und einige Pilze (Fungi imp., Hyphales) aus dem Nordatalantik. Veröff. Inst. Meeresforsch. Bremerhaven, 8: 79-97.

SIEWING, R. (1951): Besteht eine engere Verwandschaft zwischen Isopoden und Amphipoden? Zool. Anz. 147: 166-180.

SIMROTH, H. (1914): Pelagische Gastropoden-Larven der Deutschen Südpolar-Expedition 1901-1903. Dt. Südpolar Exped. 1901-1903, 15 (Zool. 7): 145-160.

SKOGSBERG, T. (1920): Studies on marine ostracoda, Pt. I. Zool. Beitr. Uppsala, Suppl. Bd. 1, 787 pp.

SLASTENENKO, E. P. (1959): Zoogeographical review of the Black Sea fish fauna. Hydrobiol., 14: 177-188.

SLIJPER, E. J. (1962): Whales. Hutchinson, London.

SLOBODKIN, J. B. (1953): A possible initial condition for Red Tides on the coast of Florida. Journ. Mar. Res. 12: 148-155.

SØMME, J. D. (1933): A possible relation between the production of animal plankton and the current-system of the sea. American Naturalist, 67: 30-52.

SØMME, J. D. (1934): Animal plankton of the Norwegian coast waters and the open sea. I. Production of *Calanus finmarchicus* (Gunner.) and *Calanus hyperboreus* (Kröyer) in the Lofoten area. Fiskeridiv. Skr. Havundersog. 4, (9) 1-163.

SOROKIN, J. I. (1960): Vertical distribution of phytoplankton and the primary production in the sea. J. du Conseil. 26: 49-56.

SOUTHWARD, A. J. (1963): The distribution of some plankton animals in the English Channel and approaches. III. Theories about long-term biological changes, including fish. J. mar. biol. Ass. U.K., 43: 1-29.

SPÄRCK, R. (1935): On the importance of quantitative investigation of the bottom fauna in marine biology. Journ. Conseil, 10: 3-19.

SPÄRCK, R. (1936): On the relation between metabolism and temperature in some marine lamellibranchs, and its zoogeographical significance. Biol. Medell., 13: No. 5: 22 pp.

SPOONER, G. M. & H. B. MOORE (1940): The ecology of the Tamar estuary. VI. An account of the macrofauna of the intertidal muds. J. Mar. Biol. Ass. U.K., 24: 283-330.

STANBURY, F. A. (1931): The effect of light of different intensities, reduced selectively and non-selectively on the rate of growth of *Nitzschia closterium*. J. Mar. Biol. Ass. U.K. **17**: 633–653.

STARMÜHLNER, F. (1956): Zur Molluskenfauna des Felslitorals und submariner Höhlen am Capo di Sorrento (2. Teil). Ergebnisse der Österreichischen Tyrrhenia-Expedition 1952. Österr. Zool. Zeitschr. **6**: 631–713.

STARR, T. J. (1958): Notes on a toxin from *Gymnodinium breve*. Texas Rep. Biol. Medic. **16**: 500–507.

— M. E. JONES & D. MARTINEZ (1957): The production of vitamin B_{12}-active substances by marine bacteria. Limnology and Oceanogr., **2**: 114–119.

STCHECTRINA, Z. G. (1958): Foraminifera of the Kuril-Kamchatka depression. Trud. Inst. Okean., **27**: 161–179.

STEELE, J. H. (1958): Production studies in the northern North Sea. Rapp. Proc. Verbaux Cons. Intern. Explor. Mer, **144**: 79–84.

STEEMANN NIELSEN, E. (1952): The use of radio-active Carbon (C^{14}) for measuring organic production in the sea. Journ. du Conseil Perman. Intern. Expl. Mer, **18**: 117–140.

— (1955): The production of antibiotics by plankton algae and its effects upon bacterial activities in the sea. Papers in Marine Biol. and Oceanogr., Suppl. to vol. 3 of Deep Sea Research, 281–286.

— (1955): Production of organic matter in the oceans. J. Mar. Res. Sears Found. **14**: 374–386.

& E. A.—JENSEN (1957): Primary oceanic production. The autotrophic production of organic matter in the oceans. Galathea Report, **1**: 49–136.

— (1958): Experimental methods for measuring organic production in the sea. Rapp. Proc. Verbaux Cons. Intern. Explor. Mer, **144**: 38–46.

— (1960): Productivity of the oceans. Ann. Rev. Plant Physiol. **11**: 341–362.

— & V. KR. HANSEN (1959): Measurements with the carbon-14 technique of the respiration rates in natural populations of phytoplankton. Deep-Sea Res., **5**: 222.

STEPHEN, A. C. (1938a): Temperature and the incidence of certain species in western European waters in 1932–1934. J. Animal Ecology, **7**: 125–129.

STEPHEN, A. C. (1938b): Production of large broods in certain marine lamellibranchs with a possible relation to weather conditions. J. Animal Ecology, **7**: 130–143.

STEPHENSON, T. A. (1958): Coral reefs regarded as seashores. Proc. XV. Intern. Congr. Zool., London, 244–246.

STEPHENSON, T. A. & A. (1949): The universal features of zonation between tide-marks on rocky coasts. J. Ecol. **37**: 289–305.

STEPHENSON, W., & R. B. SEARLES (1960): Experimental studies on the ecology of intertidal environments at Heron Island. Austral. J. Mar. Freshwater Res. **2**: 241–267.

STEUER, A. (1910): Planktonkunde. Leipzig und Berlin (Teubner) XV + 723.

STEUER, A. (1933a): Zur planmäßigen Erforschung der geographischen Verbreitung des Haliplanktons, besonders der Copepoden. Zoogeographica, **1**: 269–299.

STIASNY, G. (1913): Das Plankton des Meeres. Göschen, Leipzig.

STICKNEY, A. P., & L. D. STRINGER (1957): A study of the invertebrate bottom fauna of Greenwich Bay, Rhode Island. Ecology, **38**: 111–122.

STOCK, J. H. (1963): South African deep-sea Pycnogonida, with descriptions of five new species. Ann. S. Afr. Mus., **46**: 321–340.

STOHLER, R. (1959): The red tide of 1958 at Ensenada, Baja California, Mexico. The Veliger, **2**: 32–35.

STÖRMER, L. (1929): Copepods from the Michael Sars Expedition, 1924. Rapp. et Proc. Verbaux, Conseil Perm. Intern. pour l'Expl. de la Mer. **54**, 57 pp.

STRENZKE, K. (1956): Ökologie der Wassertiere. Handb. d. Biol., **3**: 115–192.

STRICKLAND, J. D. H. (1958): Solar radiation penetrating the ocean. J. Fish. Res. Board Canada, **15**: 453.

STUBBINGS, H. G. (1937): Phyllirhoidae. Sc. Rep. John Murray Exped. (Brit. Mus. Nat. Hist.) **5**: 1–14.

SUYEHIRO, Y. & al. (1957): Movements of the fish in response to sound stimuli with reference to sound-intensity. Japan. J. Ichthyol. **6**: 136–140.

SVERDRUP, H. U., M. W. JOHNSON & R. H. FLEMING (1942): The oceans. Prentice-Hall, New York, 1087 pp.

SZIDAT, L. (1961): Versuch einer Zoogeographie des Süd-Atlantik mit Hilfe von Leitparasiten der Meeresfische. In: Parasitologische Schriftenreihe. G. Fischer Jena, 98 pp.

TAKATSUKI, S. (1928): The heart pulsation of oysters in tropical seas compared with that of those living in seas of the temperate zone. Rec. Oceanogr. works Tokyo, **1**: 102–112.

TANAKA, T., & Y. NOZAWA (1960): One red algal parasite from Japan. Mem. Fac. Fish. Kagoshima Univ. **9**: 107–112.

TÅNING, A. V. (1949): On the breeding places and abundance of the Red Fish (*Sebastes*) in the North Atlantic. Journ. du Conseil, **16**: 85–95.

TASHIAN, R. E. (1956): Geographic variation in the respiratory metabolism and temperature coefficient in tropical and temperature forms of the fiddler crab, *Uca pugnax*. Zoologica, **41**: 39–47.

TAVOLGA, W. N. (1956): Visual, chemical and sound stimuli as cues in the sex discriminatory behavior of the Gobiid Fish *Bathygobius soporator*. Zoologica, **41**: 49–64.

TESCH, J. J. (1946 u. 1948): The thecosomatous Pteropods. The Carlsberg Found. Ocean. Exped. 1928–1930 Dana Report Nr. **28 u. 30.**

TESCH, J. J. (1950): The Gymnosomata II. Carlsberg Found. Oceanogr. Exped. 1928–1930, Dana-Report Nr. 36, 55 pp.

THIENEMANN, A. (1950): Verbreitungsgeschichte der Süßwassertierwelt Europas. In: Die Binnengewässer 18, XVI+809 S. Stuttgart, Schweizerbart.

THIRUPAD, P. U. V., & C. V. G. REDDY (1959): Seasonal variations of the hydrological factors of the Madras coastal waters. Indian J. Fish., 6: 298–305.

THOMPSON, H. (1948): Pelagic Tunicates of Australia. Commonwealth Council for Scientific and Industrial Research, Australia, Melbourne, 196 pp.

THORSON, G. (1933): Investigations on shallow water animal communities in the Franz-Joseph Fjord (East Greenland) and adjacent waters. Meddel. om Gronland, 100, 70 pp.

— (1936): The larval development, growth, and metabolism of arctic marine bottom invertebrates compared with those of other seas. Medel. Gronland, 100, 155 pp.

— (1946): Reproduction and larval development of Danish marine bottom invertebrates, with special reference to the planktonic larvae in the sound (Öresund). Medd. Komm. Danmarks Fisk Havunders, Ser. Plankton, 4, Nr. 1, 523 pp.

— (1950): Reproductive and larval ecology of marine bottom invertebrates. Biol. Rev. 25: 1–45.

— (1952): Zur jetzigen Lage der marinen Bodentier-Ökologie. Verh. Dtsch. Zool. Ges. in Wilhelmshaven 1951: 276–327.

THORSON, G. (1955): Modern aspects of marine level-bottom animal communities. J. Mar. Res. Sears Found. 14: 387–397.

— (1957): Bottom communities (sublittoral or shallow shelf). Mem. Geol. Soc. Am. 67, 1: 461–534.

— (1958): Parallel level-bottom communities, their temperature adaptation, and their „balance" between predators and food animals. 67–86 in : Perspectives in marine biology, ed. A. A. BUZZATI-TRAVERSO. Univ. Calif. Press: 1–621.

TIMMERMANN, G. (1932): Biogeographische Untersuchungen über die Lebensgemeinschaft des treibenden Golfkrautes. Zeitschr. Morph. u Ökol. d. Tiere, 25: 288–335.

TOLBERT, N. E., & L. P. ZILL (1956): Excretion of glycolic acid by algae during photosynthesis. J. Biol. Chem., 222: 895–906.

TREGOUBOFF, G., & M. ROSE (1957): Manuel de planctonologie méditerranéenne. I+II. Centre Nat. Rech. Sci., 587 pp. Paris.

TREWAVAS, E. (1933): On the structure of two oceanic fishes, Cyema atrum GUNTHER and Opisthoproctus soleatus VAILL. Proc. Zool. Soc. London, 1933, pt. 3: 601–614.

TSUJITA, T. (1955): Comparative studies on the red tide appeared in the waters adjacent to Western Japan. Records Oceanogr. Works Japan, 2, No. 3: 19–27.

TUCKER, D. W. (1959): A new solution to the Atlantic eel problem. Nature, London, 183: 495–501.

TULLY, J. P., A. J. DODIMEAD & S. TABATA (1960): An anomalous increase of temperature in the ocean off the Pacific coast of Canada through 1957 and 1958. J. Fish. Res. Board Canada, 17: 61–80.

ULRICH, J. (1964): Tiefseekuppen in den Weltmeeren. Umschau Naturw. u. Technik, 64. Jahrg., 334–338.

USCHAKOV, P. V. (1963): Quelques particularités de la bionomie benthique de l'Antarctique de l'Est. Cah. Biol. Marine, 4: 81–89.

USSING, H. H. (1938): The biology of some important plankton animals in the fjords of East Greenland. Medd. om Gronland, 100: 1–108.

UTERMÖHL, H. (1931): Neue Wege in der quantitativen Erfassung des Planktons mit besonderer Berücksichtigung des Ultraplanktons. Verh. intern. Verein. theor. angew. Limnol., 5: 567–96.

VACCA, D. D., & R. A. WALSH (1954): The antibacterial extract obtained from Ascophyllum nodosum. J. Amer. Pharm. Assoc. (Sci. Ed.) 43: 24–26.

VACCARO, R. F., M. BRIGGS, C. CAVEY, B. KETCHUM (1950): Viability of Escherichia coli in sea water. Amer. Journ. Publ. Health, 40: 1257–1266.

VACELET, J., & CL. LÉVI (1958): Un cas de survivance en Méditerranée du groupe d'éponges fossiles des Pharétronides. C. R. Acad. Sci. Paris, 246: 318–320.

VALLENTYNE, J. R. (1957): The molecular nature of organic matter in lakes and oceans, with lesser reference to sewage and terrestrial soils. J. Fish. Research Board Canada, 14: 33–82.

VATOVA, A. (1949): La fauna benthonica dell'Alto e medio Adriatico. Nova Thalassia, 1: 1–110.

VERNBERG, F. J. (1959): Studies on the physiological variation between tropical and temperate zone fiddler crabs of the genus Uca. II. Oxygen consumption of whole organisms. Biol. Bull., 117: 163–184.

— & R. E. TASHIAN (1959): Studies on the physiological variation between tropical and temperate zone fiddler crabs of the genus Uca. I. Thermal death limits. Ecology, 40: 589–593.

VINOGRADOV, M. E., & N. M. VORONINA (1962): Influence of the oxygen deficit on the distribution of plankton in the Arabian Sea. Deep-Sea Res., 9: 523–30.

VIRVILLE, A. D., DE (1940): Les zones de végétation sur le littoral atlantique. Mém. Soc. Biogéograph., 7: 205–251.

VOLBEHR, K. (1963): Arznei aus dem Meere vom 17. Jahrhundert bis heute. Bremerhaven, Ditzen u. Co., 62 pp.

VOLZ, P. (1938): Studien über das „Knallen" der Alpheiden. Nach Untersuchungen an Alpheus dentipes Guérin und Synalpheus laevimanus (Heller). Z. Morph. Oekol. Tiere, 34: 272–316.

Voss, G. L., & N. A. (1955): An ecological survey of Soldier Key, Biscayne Bay, Florida. Bull. Mar. Sci. Gulf a. Caribbean, **5**: 203–229.
WAGNER, H. (1939): Biologische Beobachtungen an *Antennarius marmoratus* Gth. Zool. Anz. **126**: 285–297.
WAHLERT, G., & H. VON (1961): Le comportement de nettoyage de *Crenilabrus melanocercus* (Labridae, Pisces) en Méditerranée. Vie et Milieu, **12**: 1–10.
— & — (1962): Beobachtungen und Bemerkungen zum Putzverhalten von Mittelmeerfischen. Veröff. Inst. Meeresforsch. Bremerhaven, **8**: 71–77.
WALFORD, L. A. (1958): Living resources of the sea. New York, Ronald Press Comp. XV+321 pp.
WARBURTON, F. E. (1953): Antagonism between different species of hydroids on the same shell. Ecology, **34**: 193–194.
WASMUND, E. (1938): Entwicklung der Naturforschung unter Wasser im Tauchgerät. Geol. d. Meere u. Binnengewässer, **2**: 87–151.
WATERMAN, T. H. (1954): Polarization patterns in submarine illumination. Science, **120**: 927–932.
WATERMAN, T. H., R. F. NUNNEMACHER, F. A. CHACE & G. L. CLARKE (1939): Diurnal vertical migrations of deep-water plankton. Biol. Bull. **76**: 256–279.
WATTENBERG, H. (1936): Kohlensäure und Kalziumkarbonat in Meere. Fortschr. Mineral., Kristallogr. u. Petrogr. **20**: 168–195.
— (1957): Die Verteilung des Phosphats im Atlantischen Ozean. Wiss. Ergeb. Dt. Atlant. Exped. „Meteor" 1925–1927, **9**, 2. Lfg.: 133–180.
WEBB, J. L. (1958): The ecology of Lagos lagoon III. The life history of *Branchiostoma nigeriense* Webb. Phil. Trans. Roy. Soc. London, B **241**: 335–353.
WEGNER, R. N. (1959): Der Schädelbau der Lederschildkröte *Dermochelys coriacea* LINNÉ (1766). Abh. Dt. Akad. Wiss. Berlin, Kl. f. Chemie, Geol., Biol., Jahrg. 1959, Nr. **4**, 80 pp.
WELLS, H. W., & M. J. (1961): Observations on *Pinnaxodes floridensis*, a new species of pinnotherid crustacean commensal in holothurians. Bull. Mar. Sci. Gulf a. Caribbean, **11**: 267–279.
WELSH, J. H. (1935): Further evidence of a diurnal rhythm in the movement of the pigment cells in the eyes of crustaceans. Biol. Bull. **68**: 247–252.
—, F. A. CHACE JR & R. T. NUNNEMACHER (1937): The diurnal migration of deep-water animals. Biol. Bull. **73**: 185–196.
WERNER, B. (1948): Die amerikanische Pantoffelschnecke *Crepidula fornicata* L. im Nordfriesischen Wattenmeer. Zool. Jahrb. Systematik etc., **77**: 449–488.
— (1953a): Über den Nahrungserwerb der Calyptraeidae (Gastropoda Prosobranchia). Morphologie, Histologie und Funktion der am Nahrungserwerb beteiligten Organe. Helgol. Wiss. Meeresunters., **4**: 260–315.
— (1953b): Ausbildungsstufen der Filtrationsmechanismen bei filtrierenden Prosobranchiern. Verh. Dt. Zool. Ges. in Freiburg: 529–546.
— (1954a): Über die Fortpflanzung der Anthomeduse *Margelopsis haeckeli* HARTLAUB durch Subitanund Dauereier und die Abhängigkeit ihrer Bildung von äußeren Faktoren. Verh. Dt. Zool. Ges. 1954: 124–133.
— (1954b): Eine Beobachtung über die Wanderung von *Arenicola marina* L. (Polychaeta sedentaria). Helgoländer Wiss. Meeresunters. **5**: 93–102.
— (1956): Über die Winterwanderung von *Arenicola marina* L. (Polychaeta sedentaria). Helgoländer Wiss. Meeresunters., **5**: 353–378.
WESTLAKE, D. F. (1963): Comparisons of plant productivity. Biol. Rev. **38**: 385–425.
WIBORG, K. F. (1940): The production of zooplankton in the Oslo Fjord in 1933–34 with special reference to the copepods. Hvalrad. Skr., **21**: 1–87.
WIESER, W. (1958): Occurrence of *Protohydra leuckarti* in Puget Sound. Pacific Science, **12**: 106–108.
— (1960): Meeresökologie, Fortschr. d. Zool. **12**: 336–378.
WILSON, D. P. (1951): A biological difference between natural sea waters. J. Mar. biol. Ass. U.K. **30**: 1–19.
— (1954): The attractive factor in the settlement of *Ophelia bicornis* SAVIGNY. J. Mar. Biol. Ass. U.K. **33**: 361–380.
— & F. A. J. ARMSTRONG (1958): Biological differences between sea-waters: experiments in 1954 and 1955. J. Mar. Biol. Ass. U.K., **37**: 331–348.
WILSON, I. M. (1960): Marine fungi: a review of the present position. Proc. Linnean Soc. **171**: 53–70.
WIMPENNY, R. S. (1938): Diurnal variation in the feeding and breeding of Zooplankton related to the numerical balance of the Zoo-Phytoplankton Community. Journ. du Conseil, **13**: 323–337.
WIMPENNY, R. S. (1966): The plankton of the sea. Faber, London.
WINGE, O. (1928): The Sargasso Sea, its boundaries and vegetation. Rep. Dan. Ocean. Exped. 1908–1910, **3**, 34 pp.
WINGFIELD, C. A. (1939): The activity and metabolism of poikilothermal animals in different latitudes. IV. Proc. Zool. Soc. London, Ser. A, **109**: 103–108.
WOHLENBERG, E. (1939): Die Nutzanwendung biologischer Erkenntnisse im Wattenmeer zugunsten der praktischen Landgewinnung an der deutschen Nordseeküste. Rapp. et Procès-Verbaux, **104**.
— (1954): Sinkstoff, Sediment und Anwachs am Hindenburgdamm. Die Küste, **2**: 33–94.
WOHLSCHLAG, D. E. (1960): Metabolism of an antarctic fish and the phenomenon of cold adaptation. Ecology **41**: 287–292.
WOLFF, T. (1960): The hadal community, an introduction. Deep-Sea Research, **6**: 95–124.

WOLTERECK, R. (1905): Bemerkungen zur Entwicklung von Narcomedusen und Siphonophoren. Verh. Dt. Zool. Ges. **15**: 106–122.

WOOD, E. J. F. (1965): Marine microbial ecology. Chapman & Hall Ltd., London, 320 pp.

WOODCOCK, A. H. (1944): A theory of surface water motion deduced from the wind-induced motion of *Physalia*. J. Mar. Res., **5**: 196–206.

WOODCOCK, A. (1950): Subsurface pelagic Sargassum. J. Mar. Res., **9**: 77–92.

WÜST, G. (1964): The major deep-sea expeditions and research vessels 1873–1960. Progress in Oceanogr., **2**: 3–52.

YENTSCH, C. S. & J. H. RYTHER (1959): Absorption curves of acetone extracts of deep water particulate matter. Deep-Sea Res., **6**: 72–73.

YONGE, C. M. (1926): Ciliary feeding mechanisms in the thecosomatous Pteropods. Journ. Linn. Soc. of London, **36**: 417–429.

— (1944): Experimental analysis of the association between invertebrates and unicellular algae. Biol. Rev., **19**: 68–80.

— (1957): Symbiosis. Geol. Soc. America, Memoir **67**, 1: 429–442.

ZANDER, E. (1907): Das Kiemenfilter bei Tiefseefischen. Zeitschr, wiss. Zool. **85**: 157–182.

ZENKEVITCH, L. A. (1949): La structure biologique de l'océan. C. R. XIII. Congres Intern. Zoologie. Paris, 522–529.

— (1954): Erforschungen der Tiefseefauna im nordwestlichen Teil des Stillen Ozeans. Union Intern. c. Biol., Ser. B, No. **16**: 72–84.

— (1955): Oceanographic research conducted by the USSR in the North-West Pacific. Proceed. Unesco Symposium Phys. Oceanogr., Tokyo 1955: 251–252.

— (1958): Certain zoological problems connected with the study of the abyssal and ultra-abyssal zones of the ocean. Proc. XV. Int. Congr. Zool., London, 215–218.

— & J. A. BIRSTEIN (1956): Studies on the deep water fauna and related problems. Deep-Sea Res., **4**: 54–64.

— & — (1960): On the problem of the antiquity of the deep-sea fauna. Deep-Sea Res., **7**: 10–23.

ZOBELL, CL. E. (1941): Studies on marine bacteria I. The cultural requirements of heterotrophic aerobes. J. Mar. Res., **4**: 42–75.

— (1946): Marine Microbiology. Waltham, Mass., USA. XV+240 pp.

— (1954): The occurrence of bacteria in the deep sea and their significance for animal life. Intern. Union Biol. Ser. B, **16**: 20–26.

ZOBELL, C. E., & C. B. FELTHAM (1938): bacteria as food for certain marine invertebrates. J. Mar. Res. **1**: 312.

— & C. B. FELTHAM (1942): The bacterial flora of a marine mud flat as an ecological factor. Ecology, **23**: 69–77.

— & R. Y. MORITA (1959): Deep-Sea bacteria. Galathea Report, **1**: 139–154.

ZYL, R. P. VAN (1959): A preliminary study of the salps and doliolids of the West and South coast of South Africa. Commerce and Industry, Investig. Rep. No. 40, 31 pp. South Africa.

INDEX